AF608509

Graduate Texts in Physics

Graduate Texts in Physics publishes core learning/teaching material for graduate- and advanced-level undergraduate courses on topics of current and emerging fields within physics, both pure and applied. These textbooks serve students at the MS- or PhD-level and their instructors as comprehensive sources of principles, definitions, derivations, experiments and applications (as relevant) for their mastery and teaching, respectively. International in scope and relevance, the textbooks correspond to course syllabi sufficiently to serve as required reading. Their didactic style, comprehensiveness and coverage of fundamental material also make them suitable as introductions or references for scientists entering, or requiring timely knowledge of, a research field.

Timo Felser • Simone Montangero
Editors

Introduction to Tensor Network Methods

From Many-Body Quantum Systems to Machine Learning

Second Edition

Editors

Timo Felser
Tensor AI Solutions GmbH
Pfaffenhofen an der Roth, Germany

Simone Montangero
Department of Physics and Astronomy
University of Padova
Padua, Italy

ISSN 1868-4513 ISSN 1868-4521 (electronic)
Graduate Texts in Physics
ISBN 978-3-032-17634-9 ISBN 978-3-032-17635-6 (eBook)
https://doi.org/10.1007/978-3-032-17635-6

This Springer imprint is published by the registered company Springer Nature Switzerland AG
The registered company address is: Gewerbestrasse 11, 6330 Cham, Switzerland

To my family,
a bundle of brave and amazing persons,
spread around but always united together.

S.M.

Preface to the Second Edition

In the last few years, the field of tensor network method has undergone a very fast development. Since the publication of the first edition of this book, there have been several exciting developments in different directions. At the same time, I moved from Ulm University back to Italy, where I am now heading the Quantum Information and Matter group at the Department of Physics and Astronomy of Padova University. The research group has grown substantially which allowed me to explore different research directions, most of them devoted to the development and exploitation of new features of tensor network methods. Meanwhile, the Italian part of the EU Recovery Plan—NextGenerationEU has impacted my research as well, as it funded the Italian National Center for High-Performance Computing, Big Data, and Quantum Computing (Foundation ICSC), which has boosted the Italian activities on quantum computation, has increased the interaction and cohesion between universities, national research centers and industries in the field of quantum computing, and, as far as concern us, has greatly tightened our interactions with the CINECA, the Italian HPC center. At the same time, in 2022, Padova University funded via an open call a Quantum Computing and Simulation Center (QCSC) which I have the honor to direct. The QCSC was co-funded by several Quantum Centers and universities in Italy, together with the Italian National Institute for Nuclear Physics (INFN), and develops coordinated activities on quantum information among those institutions.

This thriving environment put us in the conditions to explore some of the exciting developments of tensor network methods, especially concerning future quantum computations and simulations. Indeed, tensor network methods are becoming more and more relevant as tools to benchmark, verify, and guide the development of future quantum computers and simulators. On the one hand, tensor network quantum computing emulations are becoming the ultimate quantum advantage certification tools. The ability of tensor network emulators to simulate quantum computations with hundreds of moderately entangled qubits makes them the perfect tool to discern where the quantum computing advantage threshold lays. In the last years, starting from the first claim of Google's quantum supremacy experiment in 2018—right after the publication of the first edition of this book—we have seen an increasing number of experiments claiming to have reached quantum advantage that have been seriously challenged by tensor network emulations. This friendly competition is currently ongoing and has surely contributed to improving our profound understanding of the limitations and the potential of current quantum chips and quantum computations.

Alongside, tensor network methods have kept attacking the most serious challenge they have been facing for a couple of decades, that is, their effectiveness in describing many-body quantum systems with dimensions higher than one. Indeed, since the introduction of the Density Matrix Renormalization Group, tensor network methods have been recognized as the method of choice for describing one-dimensional systems in regimes where Monte Carlo methods are hindered by the sign problem, i.e., described by a negative or imaginary action. However, for higher dimensional systems, the situation is different. On the one side, there are many rigorous and important theoretical results related to tensor network ansatzes for high-dimensional systems, like as PEPS and MERA. On the other side, their high algorithmic complexity prevented their application in many practical uses, and their adoption is not yet as ubiquitous as methods for one-dimensional systems, i.e., based on Matrix Product States.

For many years, the difficulties of tensor network methods in studying high-dimensional systems have prevented their application in a large number of fundamental problems, such as the study of lattice gauge theories at finite chemical potential or out of equilibrium. One of the exciting developments of the last years has been a combination of new developments that are starting to unlock the possibility of studying them. Indeed, a combination of algorithmic developments, new tensor network ansatzes (e.g., the augmented Tree Tensor Networks), and computational improvements (the use of modern programming languages, of GPUs, and HPC environments, etc.) are such that simulations of 2D and even 3D dimensional systems are becoming available and start to produce interesting results in different fields, from condensed matter theory to high-energy physics. Once again, the results that will be obtained in the years to come will guide future research in different fields and will serve as future highly nontrivial benchmarks for future quantum computers and simulators that will eventually realize Feynman's vision.

On a parallel direction, tensor network methods are finding increasing direct applications in computer science problems. Indeed, since the seminal work of Stoudenmire and Schwab in 2016, it is becoming increasingly evident that the interplay of tensor network methods and machine learning can lead to very interesting developments, possibly opening up new routes toward the solution of some long-standing limitations of neural networks, from their explainability to their energy and computational extreme needs. Moreover, tensor network machine learning is a sub-class of quantum machine learning, that is, it is a perfect bridge between quantum and classical machine learning, and as such, it is the perfect playground to study their interplay and properties such as scalability, resilience to errors, dependence on the entanglement of the trial function, different architectures, and so on. Similarly, thanks to the new developments and the new ability to efficiently simulate high-dimensional systems, tensor network methods can be used to attack hard combinatorial classical optimizations, one of the main attractive real-world applications of quantum computers for industries. Once again, they can be used to develop new classical optimizers that might become competitive and relevant for industrial applications, and naturally bridge the quantum and classical optimizers, providing a perfect playground to study hard instances, their scalability to thousands of variables, the role of entanglement, and various strategies and architectures.

The aforementioned developments urged us to update the previous version of the book, which now includes some of the latest developments in the field that we consider promising for the future. We have added new chapters to introduce the new concepts and one part of the book with more advanced technical concepts. The new Part III of the book contains all the technical details to prepare the interested reader to understand and code any tensor network algorithm in a loopless structure, that is, in a Tree Tensor Network (TTN) ansatz. This ansatz is shown to be powerful enough to describe highly relevant physics in dimensions higher than one, with very high efficiency. Moreover, whenever necessary, it is possible to extend this ansatz via the augmented TTN (aTTN) that encodes the area law in higher dimensions keeping the algorithmic complexity of the overall computation to the minimum, the bond dimension to the fourth power. This improvement paves the way for many exciting applications, some of which are described in the new chapters in Part IV of the book: Hamiltonian Lattice Gauge theories, both Abelian and Non-Abelian, described in detail in the new Chap. 10. Finally, in the new Chap. 12, the reader is introduced to Tensor Network Machine learning a completely new field of application of Tensor Network Methods that has been recently unveiled.

Most of the book's new content comes from the Ph.D. theses of two brilliant Ph.D. students who, under my supervision, have developed new tensor network methods and applied them to highly challenging problems: Timo Felser and Giovanni Cataldi have just entered this fascinating world and have already demonstrated their value.

Most of the ideas and applications presented in this book have been explored, coded, and applied by members of my group in the last 20 years. The result of this accumulated knowledge is now available as the open-source software suite QuantumTEA[1] for anybody who wants to run or further develop it. It is also preinstalled on Leonardo, the CINECA supercomputer for ease of use. This could not be possible without the hard daily work and support of Daniel Jaschke who is leading this effort.

The research group I have the pleasure of coordinating is made up of amazing people, from which I get the stimuli to keep going every day, learning something new, and getting excited by new developments. They are too many to be thanked here, but the group could not smoothly work without the constant efforts of Marco Di Liberto, Carmelo Mordini, Ilaria Siloi, and Pietro Silvi.

[1] Ref. [1]

Finally, I would like to thank the Physics and Astronomy Department of Padova University, which supported me from the first day I arrived: the past and present directors Francesca Soramel and Flavio Seno, all my colleagues, and the administrative and technical staff for the very stimulating, cooperative, and open environment they contributed to creating. It is impressive how solutions to practical problems can be always found if working together toward a common goal.

Padova, Italy
December 2024

Simone Montangero

After Simone's overview of the progress in the field of tensor networks following the first edition, let me give some background to my journey with tensor network methods, which ultimately concluded several contributions in this new edition.

The implementation and development of tensor networks for solving cutting-edge problems in research have become an interdisciplinary field combining elements from mathematics, such as linear algebra and graph theory, with high-performance computing and Quantum Information. I started my journey of tensor network developments back in 2015 in Simone's group—at the time at the University of Ulm. After attending his lecture on quantum computational physics myself as a physics student, I was hooked by the passion for simulating quantum systems on a classical computer—especially since my interest in numerical simulations existed a long time before: Starting in the motorsports team of the *Einstein Automotive e.V.* which builds a competitive race car each year, I had the pleasure to dive deep into numerical simulations ranging from solid-state simulations of car components and their structural optimization up to computational fluid dynamics (CFD)-simulations for optimizing the aerodynamic parts—all areas where tensor computation form the mathematical foundation.

My first encounter with tensor networks was the implementation of non-Abelian symmetries and a time evolution (concretely a TEBD) algorithm for Matrix-Product States—the arguably easiest, non-trivial geometry of a tensor network. The exploitation of symmetries—as illustrated in Chap. 6—has become a crucial, state-of-the-art add-on for many tensor network algorithms nowadays deployed to investigate quantum many-body systems. It reduces the numerical resources needed and thereby helps to push forward the limitations of simulating quantum systems on classical computers via Tensor Networks.

During my PhD, I had the great pleasure of getting a set of challenging problems to solve under the supervision of Simone. The target was to develop a tensor network algorithm for higher systems' dimensions, i.e., 2D and 3D, that is efficiently contractable and encodes the area law by construction. Several tensor network geometries have been established over the last decades, the most prominent of which being the *Matrix Product states* (MPS), the *Tree Tensor Networks* (TTN), the *Multiscale Entanglement Renormalization Ansatz* (MERA), and the *Projected Entangled Pair States* (PEPS). However, when going to higher system dimensions, the algorithms either do not encode the area law—and thus eventually fail to accurately capture the underlying entanglement with increasing system size failing to faithfully describe the quantum states' properties—or are numerically inefficient—up to the point where the computation of the quantum state or certain properties becomes practically infeasible. This conflict limited the scalability of tensor network algorithms in simulating higher dimensional quantum systems for a long time. After several years of development, we introduced the augmented Tree Tensor Network (aTTN) as a novel approach to simulating high-dimensional systems which is introduced in the new Chap. 9. Along the way, we applied TTN methods to problems from various fields ranging from low-energy to high-energy to medical physics. These applications include the simulation of Lattice Gauge Theories in higher dimensions and the applications of tensor networks to optimizing radiotherapy plans in cancer treatments as an alternative optimization approach to simulated or quantum annealing. Based on these research endeavors, the new Chap. 8 provides an in-depth overview of the implementation TTN algorithms for any dimensionality of the underlying systems and includes an HPC perspective to boost their performance.

During that time, the tensor network community grew quite large with different novel application areas arising. For instance, in quantum gravity, MERA has been linked to the geometry of space via the Anti-de Sitter (AdS)/Conformal Field Theory (CFT) correspondence which in return has been brought in context to research of black holes by, to some extent, resolving the black hole information paradox. Furthermore, tensor networks have become increasingly of interest to the applied math community. In this community, the idea of Matrix Product States was rediscovered as *Tensor Trains* and Tree Tensor Networks have been proposed as *Hirarchical Tucker Trees*, thus highlighting the potential of tensor networks as a general tool for information processing. This potential of tensor networks was further proven by its application to Machine Learning tasks as a novel competitor to state-of-the-art neural networks. The first approaches of machine learning with tensor networks already yielded comparable results when performing supervised learning on standardized datasets. During my PhD time, I had the pleasure of getting into contact with Miles Stoudenmire, one of the first to publish together with D. Schwab the concept of tensor network machine learning using MPS. In this context, I transferred the concept to general loopless-TN geometries, which later on provided the basis for one of the first applications of tensor network machine learning to real data from the LHCb hardron collider at CERN. In the new Chap. 12, we introduce the concept of tensor network machine learning, giving technical insights into the great potential of the method for information processing.

After my PhD, I headed a research transfer project to spin off the method of tensor network machine learning into the company Tensor AI Solutions GmbH, which applies it to industry applications across various sectors. It turns out that tensor networks are a great tool for addressing current challenges in the lack of transparency of AI systems, providing an interpretable AI alternative to neural networks. We applied them as an explainable AI to projects ranging from credit scoring, where the learned patterns and the decisions made have to be accessible, over autonomous driving, where tensor networks are more robust against adversarial attacks or data poisoning compared to neural networks, up to real-time process optimization, where the ability to compress the network post-learning is a great benefit to adapt the performance at low cost to the requirements of the hardware. Furthermore, it has been shown that tensor networks as a numerical tool can address combinatorial optimization problems with the potential to improve areas like medical treatment, logistics optimization or energy management. Nevertheless, since in the last years, quantum technologies have emerged at the brink of transferring from research to industry, a large amount of investments have been made all over the world to foster this technology. In this light, tensor networks have emerged as the main working horse for simulating quantum circuits, i.e., as quantum simulators, provide challenging benchmarks for the development of quantum computers and are about to empower the field of quantum machine learning. All of these emerging application areas and the perspective impact of tensor networks in research and industry highlight the significance of this method in the future.

During my tensor network journey, I had the great pleasure of working with outstanding people who had an important impact on me and with whom I share valuable memories. Therefore, I would like to thank all of them for putting and keeping me on this journey with the common passion we shared. These include among others F. Tschirsich, M. Gerster, P.Silvi, F.Schrodi, G. Bishop, L. Kohn with whom I spent most of my working time at the University of Ulm; A. Daley as early inspiration with discussions on TEBD algorithms giving me the confidence to proceed with research in this direction; S. Notarnicola, M. Collura, G. Magnifico, with whom I had the pleasure to work at the University of Padova and enjoyed countless of inspiring dinner discussions; G. Morigi who among others made my PhD research as co-tutelle between University of Saarland and the University of Padova possible; M. Stoundenmire who inspired me to go in the direction of tensor network machine learning and supported me with a lot of inspiring discussions; F. Schinnerling, N. Rach, M. Trenti with whom I worked extensively together in our spin-off project and J. Ankerhold who supported the project at the University of Ulm. Last but not least, I have to thank my co-author S. Montangero who put his trust in me when I started my tensor network journey and who accompanied me throughout my tensor network path as a mentor, a stimulating research partner, and a trusted friend. Without his impact, my journey would certainly not have been the same.

Ulm, Germany
December 2024

Timo Felser

Preface to the First Edition

In the last years, a number of theoretical and numerical tools have been developed by a thriving community formed by people coming from different backgrounds—condensed matter, quantum optics, quantum information, high-energy, and high-performance computing—which are opening new paths towards the exploration of correlated quantum matter. The field I have the pleasure to work in is mostly based and supported by the numerical simulation performed via Tensor Network Methods, which sprang from the Density Matrix Renormalization Group, introduced by S. White more than twenty years ago. White's contribution, as the name itself suggests, was based on the powerful theoretical construction introduced in the seventies on the Renormalization Group and critical phenomena by K. Wilson. After a first decade where Density Matrix Renormalization Group has been applied mostly in condensed matter physics, starting from the new millennium, Tensor Network Methods have been developed and adapted to a constantly increasing number of research fields, ranging from quantum information and quantum chemistry to lattice gauge theories.

This book contains the notes of the course I delivered at Ulm University from 2010 to 2016 on computational quantum physics. I planned the course trying to fulfil two very demanding requirements. From the one hand, it is structured in such a way that a student would be able to follow it, even without previous knowledge on programming, scientific calculations and only knowing the basics of the underlying theory, that is, a basic course in quantum mechanics. On the other hand, I aimed not only to introduce the students to the fascinating field of computational physics, but also to achieve that, at the end of the course, they could being able to attack one of the main open problem in modern physics—the quantum many-body problem. Thus, I designed a course that would bring the interested students from the very first steps into the world of computational physics to its cutting edge, at the point where they could—with a little additional effort—dive in into the world of research. Indeed, a student that would take such a challenge with a proper knowhow, will surely have plenty occasions to enjoy this fast expanding field and will belong to an exciting and growing community. In the last ten years, I have already witnessed many students that are succeeding in such process, which resulted in many fruitful scientific collaborations and interesting publications.

The book is structured in parts, each of them divided in chapters, according to the original plan of the course. The first part and the appendices introduce the basic concepts needed in any computational physics courses: software and hardware, programming guidelines, linear algebra and differential calculus. They are presented in a self consistent way and accompanied by exercises that will make it easer to climb the steepest learning curve of the second part of the course. The central part presents the core of the course, focus on Tensor Network methods. Although in my course I briefly introduced also other successful numerical approaches (Monte Carlo, Hartree-Fock, Density functional theory, etc.) they are not included here as many excellent books present them, much better than what I could do here in a limited space. Next, I introduce elements of group theory and the consequences of symmetries in the correlated quantum world and on tensor networks. From the one hand, the appearance of different phase of matter, of spontaneous symmetry breaking and quantum phase transitions where correlations diverge, present the highest difficulties for their numerical description. On the other hand, symmetries can be exploited to simplify the system description and to speed up the computation: Symmetric tensor networks are fundamental to be able to perform state-of-the-art simulations. This is indeed true for global symmetries like the conservation of particles number in fermionic and bosonic systems or the magnetization in spin systems. However, recently it has been shown that also gauge symmetries can be embedded in the tensor network description, paving the way to very efficient simulation of lattice gauge theories, which are going to impact the modern research in condensed matter theory and high energy physics. Finally, the last part reviews the applications of the tools introduced here to the study of phases of quantum matter and their characterization by means of correlation functions and entanglement measures. Moreover, I present some results on out-of-equilibrium phenomena, such as adiabatic quantum computation, the Kibble-Zurek mechanism, and the application of quantum optimal control theory to many-body quantum dynamics.

In conclusion, this book can be used as a textbook for graduate computational quantum physics courses. Every chapter ends with exercises, most of which has been designed as weekly exercise for the students, to be evaluated in terms of programming, analysis and presentation of the results. At the end of the course, the students should be able to write a tensor network program to begin to explore the physics of many-body quantum systems. The course closed with each student choosing a final project to be performed in one month: the subject could also be related to other exams or laboratory activities, preparation for their Master Thesis or as a part of their Ph.D. studies. Indeed, the scope of the course was to prepare them to these challenges, and in most cases they demonstrated the success of such program.

The contents presented here are unavoidably based on my particular view of the field, and in particular the focus is on the recent development of Tensor Network method I had the pleasure to participate in. Most of the contents, as acknowledge hereafter, are based on excellent books, review and research papers of collaborators and colleagues in the field. However, to the best of my knowledge, at the time I am writing these lines, there is nowhere a self-contained book presenting all the elements needed to train students to work with Tensor Network Method, despite the growing request of Ph.D. students with such skills. Finally, I think that this book could serve as a useful reference and guide to relevant literature for researchers working in the field; or as a starting point for colleagues that are entering in it.

I would like to thank the many people that made this possible and that accompanied my life and scientific career until now. The colleagues at Ulm, J. Ankerhold, M. Freyberger, J. Hecker-Denschlag, S. Huelga, F. Jelezko, B. Nayedov, M. Plenio, F. Schmidt-Kaler, J. Stockburger, and W. Schleich, made Ulm University a special and unique place to develop new scientific research. The support of the Center for Integrated Quantum Science & Technologies and of its director T. Calarco have been very precious in these last years. The scientific discussions with the other IQST fellows in Ulm and Stuttgart have been a very important stimulus.

The Institute for Quantum Optics and Quantum Information of the Austrian Academy of Science in Innsbruck played a unique role in my scientific life in the last years: especially since I became an IQOQI visiting fellow, I could fully profit from their hospitality and inspiration full environment. I am greatly thankful to Peter Zoller for his friendly guidance, his availability, and the many enjoyable discussions.

A special thank to Giovanna Morigi that always supported me and to Saarland University that hosted my Heisenberg Fellowship. It has been a very fruitful and exciting time in a very pleasant environment: I am sure it will bring to a long-standing successful scientific collaboration.

The colleagues involved in the QUANTERA project I have the pleasure to coordinate (QTFLAG) are working hard to develop the next generation of tensor network methods and experiments to investigate lattice gauge theories. I am sure that in the years to come they will inspire many of the readers of this book and myself with new exciting ideas and tools: many thanks to M. Dalmonte, I. Cirac, E. Rico, M. Lewenstein, F. Verstraete, U.-J. Wiese, L. Tagliacozzo, M.-C. Bañuls, K. Jansen, A. Celi, B. Reznik, R. Blatt, L. Fallani, J. Catani, C. Muschik, J. Zakrzewski, K. Van Acoleyen, and M. Wingate.

This book could not have been written without the many persons that spent some years working with me, making my everyday life full of stimuli. I would like to thank P. Silvi, F. Tschirsich, D. Jaschke, M. Gerster, T. Felser, L. Kohn, Y. Sauer, H. Wegener, F. Schrodi, J. Zoller, M. Keck, T. Pichler, N. Rach, R. Said, J. Cui, V. Mukherjee, M. Müller, W. Weiss, A. Negretti, I. Brouzos, T. Caneva, D. Bleh, M. Murphy, and P. Doria, for the uncountable hours of good physics.

In the last twenty years, I had the pleasure to collaborate with many other persons that, in different ways, have been present during my career development and showed me different interesting points of view: I am greatly thankful to G. Benenti, G. Casati, M. Rizzi, G. De Chiara, G. Santoro, D. Rossini, M. Palma, P. Falci, E. Paladino, F. Cataliotti, S. Pascazio, J. Prior, F. Illuminati, L. Carr, P. Zanardi, G. Pupillo, S. Lloyd, R. Onofrio, M. Lukin, L. Viola, J. Schmiedmayer, M. Tusk and to many others not included here. In addition to being always there, Saro Fazio also has the merit to be the first to point to me the original papers on the density matrix renormalization group. From that discussion many years ago at Scuola Normale Superiore, a lot of exciting developments followed, and many others are still in sight.

Finally, it would not be possible for me to concentrate on my research without the full support of my family that every day gives me the strength of pursuing new challenges. Thank you, I owe you all. A huge thank to C. Montangero for his help in improving the computer science part of the text, once more renovating the never-ending scientific confrontation between physicist and computer scientists, here enriched by a father-son debate. We did pretty well. With the advent of quantum technologies, I am sure there will be plenty of other occasions to keep on this enriching dispute between these two branches of science.

Despite the feedback I received on the draft from many colleagues, the remaining errors and typos in the text are my sole responsibility. I will appreciate any comment or feedback to improve the book in the future.

Ulm, Germany
July 2018

Simone Montangero

Contents

List of Symbols and Abbreviations

DMRG	Density Matrix Renormalization Group
LGT	Lattice Gauge Theories
MF	Mean Field
MPO	Matrix Product Operator
MPS	Matrix Product State
RG	Renormalization Group
SVD	Singular Value Decomposition
TN	Tensor Networks
TTN	Tree Tensor Network
α_j	Index for local basis
β	Spontaneous magnetization
β_T	Inverse temperature $1/k_B T$
D	System dimensionality
d	Local Hilbert space dimension
η	Anomalous dimension
C	Correlations
H	Hamiltonian
k_B	Boltzmann constant
λ	Matrix eigenvalue
m	Auxiliary bond dimension
N	Problem dimension or number of lattice sites
ν	correlation length divergence exponent
$O(.)$	Order of approximation
ψ	Wave function
ρ	Density matrix
S	Entropy
$\hat{S}$	Spin operator
T	System temperature
z	Dynamical-scaling exponent

1 Introduction

Timo Felser and Simone Montangero

The many-body quantum problem is, in its different formulations, at the core of our understanding of nature and modern technologies [2]. On the one hand, it encompasses the modeling of the fundamental constituents of the universe [3] and quantum matter [4,5]. On the other hand, the development of most future technology relies on our capability of describing and understanding many-body quantum systems: among others, the electronic structure problem with impact on chemistry and drugs research [6], the development of quantum technologies [7, 9, 625], and the engineering of new materials, e.g. more efficient light harvesting materials or displaying topological order [10–12].

The solution of the many-body quantum problem is thus an overarching goal of modern research in physics. Finding an approach to efficiently solve it would have an inestimable value and would impact directly or indirectly all fields in natural sciences—from medicine and chemistry to engineering and high-energy physics–, and it will pave the way to an Eldorado of technological developments that will enormously impact our everyday's life. However, despite the enormous efforts spent and the impressive developments of analytical and numerical methods experienced in the last decades, the many-body quantum problem remains one of the most significant challenges we face.

The complexity of the many-body quantum problem immediately appears as the dimension of the space of possible configurations of N primary constituents of the system (number of quantum particles, lattice sites, etc.), increases exponentially with N. Indeed, if each elementary constituents is characterized by d possible dif-

T. Felser (✉)
Tensor AI Solutions GmbH, Pfaffenhofen, Germany
e-mail: timo.felser@tensor-solutions.com

S. Montangero (✉)
Università di Padova, Padova, Italy
e-mail: simone.montangero@unipd.it

T. Felser, S. Montangero (eds.), *Introduction to Tensor Network Methods*,
Graduate Texts in Physics, https://doi.org/10.1007/978-3-032-17635-6_1

ferent states $\{\alpha_i\}_{i=1}^d$ (spin configurations, electronic or molecular orbitals, harmonic oscillator or angular momentum levels, etc.), the number of possible configurations for the whole system is d^N. In quantum mechanics the fundamental object of interest is the wave function of the system, the amplitude of probability distribution $\psi(\alpha_1 \ldots \alpha_N)$, whose modulus square gives the probability of the system to be in N-body configuration $\alpha_1 \ldots \alpha_N$: that is, the object of interest is the vector containing the amplitude of probability for each possible system configuration. Typically, one has to update the vector of probabilities to follow the system time evolution or to find the eigenvector corresponding to the ground state of the system. To date, only a small number of exactly solvable models have been identified [13] and exact diagonalization techniques [14] are restricted to few-body problems: thus the thermodynamical or the large N limit, which in many cases represents the interesting scenario, is often out of reach.

In conclusion, the exact treatment of the many-body quantum problem is almost always not available. However, many approximate analytical and numerical methods have been developed to attack it: a powerful class of methods is based on the semiclassical treatment of the quantum degrees of freedom [15], such as mean-field techniques that disregard or take into accounts only a small part of the quantum correlations of the system [16],e.g., Hartree-Fock methods and its generalizations [134]. It is well known, however, that these approaches typically perform optimally in high spatial dimensions, while they suffer in low dimensionality [18], especially in 1D where entanglement and quantum fluctuations play a dominant role. Another class of hugely successful techniques—mostly referred to as Monte Carlo methods—have been extensively used in the past decades o solve the many-body problem and in particular to study lattice gauge theories [19]. Monte Carlo methods capture the quantum and statistical content of the system by reconstructing the model partition functions by means of statistical samplings. However, Monte Carlo methods are severely limited whenever a negative or a complex action appears (sign problem) [20], e.g., while studying systems at finite chemical potential or out of equilibrium [21]. In many cases, careful, sophisticated choices of the statistical sampling algorithm can mitigate the sign problem. However, it has been proven that the sign problem is NP-hard, and thus a generic polynomial algorithm to study via Monte Carlo systems with complex action almost certainly does not exist [22].

An alternative approach to solve the many-body quantum problem is a class of numerical methods—at the center of this book—based on the renormalization group (RG) paradigm, that is, the idea of truncating the exponential increasing Hilbert space based on energy considerations. The underlying hypothesis is that the low-energy physics of a system is mainly determined by the low-energy sectors of the system components [23]. In standard RG algorithms, the system size is then increased iteratively, and at each iteration, the high-energy sectors of the different parts of the system are discarded. This powerful idea has been further developed in the context of condensed matter systems where the density matrix renormalization group (DMRG) has been introduced [24]. Finally, it has been extensively further developed within the quantum information community with the development of tensor network (TN) methods [25–32]. As reported in details in this book, TN methods provide a recipe to describe efficiently many-body quantum systems with a controlled level of approximation as a function of the used resources, to optimally describe the relevant information in the system wave function [25]. Indeed, the TN approach is based on the fact that the tensor structure of the Hilbert space on which the wave function $\psi(\alpha_1\alpha_2\ldots\alpha_N)$ is defined, introduces some structures. That is, typically ψ is not a fully general rank-N tensor and thus it might allow for an efficient representation. Moreover, in special but significant cases as the ground states of short-range interacting Hamiltonians, one can show that the entanglement present in the system obeys some scaling laws—area laws of entanglement [33–38]. These scaling laws imply the possibility to introduce classes of states—TN ansätze, defined via their tensor structure—which can efficiently representing the system ground state [25, 26, 29, 31, 39–44, 46, 167]. Indeed, it is possible to decompose the most natural tensor structure defined by the rank-N tensor $\psi(\alpha_1\alpha_2\ldots\alpha_N)$, in a network of lower rank tensors, which are equivalent unless approximations are introduced. Notice that while introducing a TN ansatz, one fixes the tensor structure, but not the dimension of the tensor indices nor the elements of each tensor: they are the variables to be optimized to build the most efficient representation of the system under study still capturing the system physical properties.

Starting from the reformulation of the DMRG algorithm in terms of the Matrix Product State (MPS) [48, 49, 182]—a one-dimensional tensor network ansatz—several classes of tensor networks, displaying various geometries and topologies, have been introduced over the years for different purposes, to include in the tensor structure the available physical information on the system or to allow a more efficient algorithmic optimization [25, 26, 50–52, 54, 187, 202]. It is important to highlight that TN methods complements Monte Carlo methods as they do not suffer from the sign problem, allowing the simulation of out of equilibrium phenomena or at a finite chemical potential [55–58, 58–77].

Indeed, a fascinating very recent field of application of tensor networks is lattice gauge theories [78, 79]. Since their introduction by K. Wilson [80], lattice gauge theories (LGTs) have found widespread application in the investigation of non-perturbative effects in gauge theories, with applications in different fields ranging from high-energy physics to condensed matter theory and quantum science [81–85]. LGTs are characterized by an extensive number of local symmetries and locally conserved quantities, as a consequence of the invariance of the theory under specific sets of transformations. The most common formulation of LGTs is the Lagrangian one, which is an ideal setup for Monte Carlo simulations [81–84]. Hereafter, we discuss instead the Hamiltonian formulation of LGT introduced by Kogut and Susskind [86, 354] as it is better suited for applications of tensor network methods and of quantum simulators [78, 79]. In particular, we focus on a compact formulation of the gauge fields, the so-called quantum link models [78, 88–90]. Quantum link models share many features with the Wilson formulation and converge to it in a particular limit [91], keep gauge invariance exact on the lattice, and are highly related to problems in frustrated quantum magnetism [4, 92].

Before concluding this introduction, it is also due mentioning that it exists another independent approach to solve the many-body problem, put forward by R. Feynman more than thirty years ago [93]. In modern words, he proposed to develop a dedicated quantum computer—a quantum simulator—to solve the quantum many-body problem, exploiting the fact that the exponential growth of the Hilbert space of the system under study is automatically matched by another many-body quantum system. Although hereafter we will not concentrate on this approach, we highlight that in the last decades quantum information science and the experimental setups have reached a level the of control that allows different platforms (cold atoms, trapped ions, superconducting qubits, etc.) to perform quantum simulations of several many-body phenomena [94, 95]. Paradigmatic classes of Hamiltonians of short-range interacting models such as the Ising and Heisenberg models for spin systems and the Hubbard or Bose-Hubbard model describing particles on a lattice have been quantum simulated [96–98], and also systems with long-range interactions have been engineered, exploiting Coulomb interaction of trapped ions and Rydberg atom simulators [99–104]. Several proposals for quantum simulations of lattice gauge theories have been put forward [77, 79, 99, 101, 105–118] and a seminal experimental has been performed in a ion trap quantum simulator [119].

In conclusion, although tensor network methods and quantum simulators are relatively young and shall still reach full maturity, most probably future investigations on many-body quantum systems will be performed by means of three independent ways: either performing an experiment to directly interrogate the system of interest, or performing a simulation on a classical computer (either with Monte Carlo or tensor network methods), and finally performing a quantum simulation in a controlled environment. Given the complementarity of the approaches—not only in range of validity and efficiency but also in terms of complexity and resources needed for their realization, reproducibility, access to different direct or indirect information—most probably future research will be based on a combination of them, each exploiting its advantages and strengths and complementing and certifying the results of the others.

This book aims to introduce the reader to one of these pillars of future research—tensor network methods—from their basics to their application to lattice gauge theories, to bridge the gap between basics simulation methods for quantum physics and some of the most promising numerical methods that will most probably guide and support the research in fundamental quantum science and the development of quantum technologies.

The first part of the book reviews the most important tools available to solve numerically the single-body quantum problem, and that will form the basis for the most advanced techniques presented in the second part of the book: Chap. 2 presents the linear algebra tools needed to diagonalize a Hamiltonian and some more advanced methods (Lancsoz) to diagonalize large matrices. In Chap. 3, we review the standard tools of numerical calculus, the numerical integration and the solution of partial differential equations—the Schrödinger equation—which can be used as standalone tools and will be part of tensor network algorithms.

The second part presents the core of the book, the tensor network methods: We start in Chap. 4 reviewing the numerical real-space renormalization group and its generalization, the DMRG. In Chap. 5 we introduce the tools and methods to perform numerical studies (both at and out-of-equilibrium) of one-dimensional systems of hundreds of components, mainly based on Matrix Product States. Chapter 6 introduces more advanced concepts, in particular, the exploitation of symmetries to enhance the performances of the numerical simulations and to describe lattice gauge theories in the quantum link formulation.

In the second edition of the book we have included an additional part—Part III—that presents the technical details of the concepts introduced previously. The reader aiming at mastering tensor network methods and performing state of the art simulations will find here the necessary toolkit: Chap. 7 presents the standard algorithms routinely applied to Matrix Product States ansatz for one dimensional systems, while Chap. 8 presents its generalization to Tree Tensor Networks (TTN), that unlocks the ability of investigating higher dimensional systems. The last chapter of this part presents recent developments aiming at going beyond the limitations of TTNs, generalizing the ansatz to make it compatible with area law scaling of entanglement.

Finally, the last part of the book presents some applications of tensor network methods: in Chap. 10, we report TN methods applications to the study of quantum phases of matter and the different quantities that can be used and computed efficiently to characterize them: correlation functions and entanglement measures. Chapter 11 specializes the aforementioned application to lattice gauge theories, connecting tensor network methods to high-energy physics. We first present some important preliminary concepts related to the discretization of quantum fields, the continuum limit and Lorentz invariance in this context, and how to introduce Gauge invariant local bases. Finally, the details of such construction together with some results are presented for two paradigmatic theories. The chapter ends with some consideration on the future of quantum and quantum-inspired simulations of such theories. Chapter 12 presents some successful applications of TN methods to out-of-equilibrium phenomena: adiabatic quantum computation, the Kibble-Zurek mechanism, and the optimal control of many-body quantum processes. The last chapter introduces the basis of machine learning in general and specializes it for the development of a Matrix Product State classificator, introducing the interested reader to the fast developing field of Tensor Networks Machine learning.

The appendices present some basics concepts that any physicists approaching the computational methods should know: the basics of classical computer hardware and architectures and good fundamental practices to produce reliable and useful software.

Part I
Prologue

2 Linear Algebra

Simone Montangero

In this chapter, we attack one of the simplest but ubiquitous problems that a computational physicist typically faces, the solution of an eigenvalues problem. In quantum physics, the eigenvalues problem usually arise to solve a given Hamiltonian describing a single- or many-body quantum systems. Indeed, apart from a few classes of problems for which analytical solutions are known (from the particle in the box to the Bethe Ansatz solutions), the search for the ground and excited states of a quantum system shall be performed numerically. The numerical solution of the eigenvalues problem equals to the diagonalization of a matrix representing the Hamiltonian in a suitable basis, independently from the fact that the original system is discrete, e.g. a collection of qubits or atomic levels, or continuous, e.g. atoms or ions in free space or traps, or Bose-Einstein condensates. The latter scenarios shall be recast into discrete system either via space discretization (see Appendix B) or via second quantization. Finally, a plethora of problems in math, physics, and computer science can be recast in eigenvalues problems, thus hereafter we first introduce the problem and then present, in order of complexity, the state-of-art approaches for global and partial eigensolvers. As we will see, the eigensolver strategies are based on linear algebra operations that, for the sake of completeness, we briefly recall hereafter, also to set the notation: we will use throughout the book standard quantum mechanics notation and assume that the variables are complex. We base the quick overview presented here and in the next chapter (and appendixes) on many different excellent books and reviews, such as, e.g. [120–129].

S. Montangero (✉)
Università di Padova, Padova, Italy
e-mail: simone.montangero@unipd.it

T. Felser, S. Montangero (eds.), *Introduction to Tensor Network Methods*,
Graduate Texts in Physics, https://doi.org/10.1007/978-3-032-17635-6_2

2.1 System of Linear Equations

A system of linear equations can be represented in operator form

$$\hat{O}|x\rangle = |y\rangle; \tag{2.1}$$

where $\hat{O}$ is a linear operator, and the vectors $|x\rangle$, $|y\rangle$ live in a Hilbert space $\mathcal{H} \equiv \mathbb{C}^N$. Given $\hat{O}$ and $|y\rangle$, one can use the properties of equivalency of systems of linear equations to solve for the vector $|x\rangle$ which lead to the *Gaussian elimination*: (1) the multiplication of lines of the system by a factor $f_{i,j}$, and (2) the replacement of one line with the line itself added term by term with another one does not change the system result. Thus, any system of linear equations of the form of Eq. (2.1) can be recast in an equivalent (with the same solution) system where the operator has an upper triangular form

$$\hat{U}|x\rangle = |y'\rangle; \qquad \text{where} \qquad \hat{U} = \begin{pmatrix} U_{11} & U_{12} & U_{13} & \dots & U_{1N} \\ 0 & U_{22} & U_{23} & \dots & U_{2N} \\ 0 & 0 & U_{33} & \dots & U_{3N} \\ & & & \ddots & \vdots \\ 0 & 0 & 0 & 0 & U_{NN} \end{pmatrix}. \tag{2.2}$$

Given the operator $\hat{U}$, the solution of the system of linear equations is easily obtained via back substitution as follows:

$$x_k = \left(y_k - \sum_{j=k+1}^{N} U_{kj} x_j \right) / U_{kk}. \tag{2.3}$$

The previous lines not only describe the Gaussian elimination method, but provide the description of the algorithm to solve any system of N linear equations that can be straightforwardly coded. A fundamental property of algorithms is their algorithmic complexity, defined as the number of operations needed to solve the problem as a function of the input size, here the number of equations N (see Appendix B for more on software complexity and an introduction to good software development). A simple estimate results in a total number of operations that scales as $O(N^3)$: the reduction of each of the $N(N-1)/2$ elements of the lower part of the matrix in Eq. (2.2) to zero requires $O(N)$ operations, that is, a total of $O(N^3)$. The back-substitution described in Eq. (2.3) is of lower order, as it requires $O(N^2)$ operations.

Notice that the algorithm for solving a system of linear equation provides also a method to compute the inverse of a matrix: indeed by definition the inverse of the matrix A is such that $\hat{O}^{-1}\hat{O} = \hat{O}\hat{O}^{-1} = \mathbb{1}$, then, we can identify

$$\hat{O}^{-1} \equiv \begin{pmatrix} | & | & \dots & | & | \\ x^1 & x^2 & \dots & x^{N-1} & x^N \\ | & | & \dots & | & | \end{pmatrix}, \qquad \mathbb{1} \equiv \begin{pmatrix} | & | & \dots & | & | \\ y^1 & y^2 & \dots & y^{N-1} & y^N; \\ | & | & \dots & | & | \end{pmatrix} \tag{2.4}$$

where $|x^j\rangle$ and $|y^j\rangle$ forms the column vectors of the two matrices respectively. In conclusion, the condition to find the inverse of the matrix becomes simply a set of linear systems of equations:

$$\hat{O}|x^j\rangle = |y^j\rangle; \tag{2.5}$$

which can be solved with the elementary construction of the augmented rectangular matrix, formed by the operator $\hat{O}$ and the identity on its side. Gaussian eliminating the off-diagonal terms of the operator $\hat{O}$ results in the computation of the operator $\hat{O}^{-1}$.

2.1.1 LU Reduction

The Gaussian elimination reviewed in the previous section is only one of a family of reduction methods to triangular matrices. Another reduction that plays a pivotal role in the numerical computation of ground state properties of quantum systems is the LU decomposition [122]. Indeed, one can show straightforwardly that any operator can be written as a left and right triangular operators

$$\hat{O} = \hat{L}\,\hat{U} \qquad \text{such that} \qquad \hat{L} = \begin{pmatrix} 1 & 0 & \dots & 0 & 0 \\ f_{2,1} & 1 & \dots & 0 & 0 \\ f_{3,1} & f_{3,2} & \dots & 0 & 0 \\ & & \vdots & & \\ f_{N,1} & f_{N,2} & \dots & f_{N,N-1} & 1 \end{pmatrix}, \tag{2.6}$$

where the coefficients $f_{i,j}$ are those used perform the gaussian elimination of the operator $\hat{O}$ into the matrix $\hat{U}$ given in Eq. (2.2).

As we will see later in the chapter, the usefulness of such reduction becomes clear whenever one needs to solve many different systems of linear equations which differs only by the vector $|y\rangle$. Indeed, in such a case, the original problem $\hat{O}|x\rangle = |y\rangle$ can be recast in two systems of linear equations

$$\begin{cases} \hat{L}|z\rangle = |y\rangle \\ \hat{U}|x\rangle = |z\rangle \end{cases}; \tag{2.7}$$

whose solution only requires $O(N^2)$ operations as only the back substitution is needed, being both operators already triangular ones. Thus, the algorithm to solve different systems of linear equations with the same $\hat{O}$ and different $|y\rangle$, is as follows:

1. LU-reduce the operator $\hat{O}$ (scales as $O(N^3)$);
2. Perform two back substitution (scales as $O(N^2)$).

This approach has no drawback, as the first step scales as the full Gaussian elimination. Thus, it does not introduce any overhead in computational time nor memory as the two triangular matrices can be stored with the same amount of memory as the original one. Moreover, if the number of systems of equations to be solved M (with the same $\hat{O}$ operator and different $|y\rangle$) is such that $M \gg N$, it is much more convenient as $M \cdot O(N^3) \gg 1 \cdot O(N^3) + M \cdot O(N^2)$.

2.2 Eigenvalue Problem

In this section, we introduce the most common methods to find the eigenvalues of an operator $\hat{O}$ acting on the Hilbert space $\mathcal{H} \equiv \mathbb{C}^N$. Hereafter, we specialize for the case of Hermitian operators for the obvious prominent role they play in quantum mechanical problems. However, what follows can be generalized straightforwardly for more general cases. For example, the manipulation of non-Hermitian matrices naturally appears when studying open quantum systems.

We aim to solve the eigenvalue problem defined as

$$\hat{O}|x_j\rangle = \lambda_j|x_j\rangle \qquad j = 1, \dots N, \tag{2.8}$$

where hereafter we consider orthonormal eigenstates $\langle x_j|x_k\rangle = \delta_{j,k}$ and ordered (real) eigenvalues $|\lambda_N| > |\lambda_{N-1}| > \dots |\lambda_1|$.

2.2.1 Power Methods

The set of eingenstates $|x_j\rangle$ form a complete basis for the Hilbert space $\mathcal{H}$, thus any given (random) state $|y\rangle \in \mathcal{H}$ can be written as $|y\rangle = \sum y_j|y_j\rangle$, where $y_j = \langle y_j|y\rangle$. Thus, the application of the operator $\hat{O}$ to an initial random state $|y\rangle$ results into

$$\hat{O}|y\rangle = \sum y_j\lambda_j|y_j\rangle. \tag{2.9}$$

That is, after k applications of the operator $\hat{O}$ one obtains the vector state

$$\hat{O}^k|y\rangle = \sum y_j\lambda_j^k|y_j\rangle. \tag{2.10}$$

The net result of the application of the operator to the initial state is then to change the relative projections onto the eigenvectors, according to the ratio between the eigenvalues, that is, to rotate the state towards the eigenstate of maximal eigenvalue: the sequence of states for $k \to \infty$ tends to:

$$|y^k\rangle = \hat{O}^k|y\rangle \to |x_N\rangle, \tag{2.11}$$

and correspondingly the ratio $||y^k\rangle|/||y^{k-1}\rangle| \to \lambda_N^k/\lambda_N^{k-1} = \lambda_N$. In conclusion, one can find the most significant eigenstate and eigenvector with multiple applications of the operator to an initial random vector, and computing the ratio between two subsequent vector norms. We will show later how to generalize this procedure to find all the rest of the spectrum and the remaining eigenvalues.

This procedure is working properly, can be implemented easily, and works in most scenarios. However, its efficiency is drastically reduced whenever the system has almost degenerate eigenvalues (which turns out to be a class of fundamental physical processes, e.g., as we will see in Chap. 10 systems that display critical behavior at the thermodynamical limit). Indeed, the speed of convergence of the sequence in Eq. (2.11) depends on the ratio λ_N/λ_{N-1}, which, if small, can practically prevent the algorithm to converge. To cure for this problem, and also speed up the algorithm convergence, the *inverse power method* has been introduced. The main idea stems from the fact that Eq. (2.8) is equivalent to

$$\hat{O}^{-1}|x_j\rangle = \lambda_j^{-1}|x_j\rangle, \tag{2.12}$$

thus, applying $\hat{O}^{-1}$ one can find the smallest eigenvalue and eigenvector of $\hat{O}$. As before, the speed of convergence to the final result is given by the ratio λ_2/λ_1. Apparently, there is no big difference between the two methods apart from the fact that they target the opposite extreme of the spectrum. However, the important difference is that in the latter the extreme eigenvalue is at the denominator of the speed of convergence ratio: this in many scenarios allows us to enhance the speed of convergence drastically. Indeed, in many cases, it is possible to estimate the lowest eigenvector of an operator with some perturbative argument. Another typical scenario is, when looking for the ground state energy of a Hamiltonian as a function of a parameter g, allows exploiting the previously computed result for $g - \Delta g$. In conclusion, one can redefine the offset (e.g., setting a different zero value for the energy) of the eigenvalue problem as

$$(\hat{O} - q\mathbb{1})^{-1}|x_j\rangle = (\lambda_j - q)^{-1}|x_j\rangle, \tag{2.13}$$

where q is the estimate of the smallest eigenvalue λ_1. The speed of convergence of the sequence

$$|y^k\rangle = (\hat{O} - q\mathbb{1})^{-1}|y^{k-1}\rangle \tag{2.14}$$

is now given by $(\lambda_2 - q)/(\lambda_1 - q)$ which can be highly enhanced if $\lambda_1 \sim q$. The parameter q is often referred as the *guess* value in the diagonalization subroutine and providing a good guess can make the difference between standard and professional results.

The careful reader has surely spotted that in the presentation above a crucial point has been not carefully addressed: indeed to compute Eq. (2.14), the operator $(\hat{O} - q\mathbb{1})^{-1}$ is needed. As we have seen in the previous section, this step can be done straightforwardly, however, once again there are scenarios where it is preferable to avoid it. In particular, writing the matrix representation of the operator $\hat{O}$ shall be avoided if possible as it is highly inefficient. In particular, whenever the operator has a structure or representation by a sparse matrix, the implementation of a code computing the action of the operator on the state is more efficient than calculating the matrix representation and then performing the matrix-vector multiplication. In standard subroutines, such as LAPACK, these subroutines are often called *amul* and are typically the only part of code left to be written [130]. Indeed, storing a matrix requires the square of the memory than storing a vector, reaching the memory limitation much faster: this is particularly true for many-body quantum systems where the Hamiltonians contain typically only few-body interactions, and they are sparse matrices with well-defined tensor structure. Therefore, if we want to avoid to compute the inverse of the operator explicitly, we can proceed to rewrite Eq.(2.14) as

$$(\hat{O} - q\mathbb{1})|y^{k+1}\rangle = |y^k\rangle \tag{2.15}$$

and identify it as a system of linear equations: given the state at the k-th interaction, one shall find the state at the next one. To compute the whole series, we shall solve the same system of linear equation for different state vectors $|y_k\rangle$, which is exactly the scenario studied in the previous section, where we introduced the LU decomposition. Thus, one first shall compute the LU decomposition and then perform a Gaussian elimination for each iteration of the sequence in Eq. (2.15). Notice that despite the fact that both LU decomposition and the inversion scales as $O(N^3)$, the former has a prefactor of one third with respect to the second one, thus it is still more convenient. Moreover, LU decomposition preserves sparseness of the matrix while the inversion does not allow exploiting the problem structure for efficiency.

Finally, once the first eigenvector and its correspondent eigenvalue has been computed, the procedure can be repeated to compute the next ones, provided that we eliminate the components periodically along $|x_0\rangle$: that is, after the initial random state is created, one shall redefine it as

$$|y_0\rangle' = |y_0\rangle - \langle x_0|y_0\rangle\,|x_0\rangle; \tag{2.16}$$

such that $|y_0\rangle' \perp |x_0\rangle$. The sequence in Eq. (2.15) will converge to the next smallest eigenvalue contained in the expansion $|y\rangle = \sum y_j|y_j\rangle$. Notice that this orthogonalization shall be repeated during the sequence as numerical errors might reintroduce components along $|x_0\rangle$ which would grow and prevent the algorithm from working properly.

2.3 Tridiagonal Matrices

Tridiagonal matrices, defined as the class of matrices that have elements different from zero only on the diagonal and the upper- and lower-diagonals, play a prominent role in physics as they naturally appear in wide class of problems such as the single body quantum problem (see Chap. 3) and in tight binding models. Moreover, as we will see later on in Sect. 2.4, most matrices can be written in such a form via a change of basis that can be efficiently computed. Finally, as we recall briefly hereafter, it is highly efficient to solve the eigenvalue problem for tridiagonal matrices. Indeed, given a tridiagonal matrix of the form

$$T_N = \begin{pmatrix} d_1 & b_1 & 0 & 0 & \dots & 0 & 0 \\ d_1 & d_2 & b_2 & 0 & \dots & 0 & 0 \\ 0 & b_2 & d_3 & 0 & \dots & 0 & 0 \\ & & & & \ddots & & \\ 0 & 0 & 0 & 0 & \dots & b_{N-1} & b_{N-1} \\ 0 & 0 & 0 & 0 & \dots & b_{N-1} & d_N \end{pmatrix} \tag{2.17}$$

the characteristic polynomial obeys the recurrence equation

$$\det(T_N - \lambda\mathbb{1}) = (d_N - \lambda)\det(T_{N-1} - \lambda\mathbb{1}) - d_{N-1}^2\det(T_{N-2} - \lambda\mathbb{1}). \tag{2.18}$$

Thus, we can define the one-dimensional function $f_N(\lambda) = (d_N - \lambda) f_{N-1}(\lambda) - d_{N-1}^2 f_{N-2}(\lambda)$ with $f_0(\lambda) = 1$ and use very efficient standard methods to find its roots, that is, the eigenvalues of the original matrix in Eq. (2.2) [122].

2.4 Lanczos Methods

The Lanczos diagonalization algorithm aims to find the transformation from a general matrix to a tridiagonal form, so that the eigenproblem can be solved efficiently [131]. Moreover, it allows one to operate without writing explicitly the whole matrix, thus enormously enhancing its field of applicability. As for the power methods, it starts with a random vector in the relevant Hilbert space $|y_1\rangle$ and begins applying the operator to be diagonalized to it $|y_2\rangle = \hat{O}|y_1\rangle$. It then decomposes the new state $|y_2\rangle$ in its normalized perpendicular and parallel components with respect to the original state $|y_1\rangle$, i.e.

$$\hat{O}|y_1\rangle = d_1|y_1\rangle + b_1|y_2\rangle, \tag{2.19}$$

where by construction $d_1 = \left\langle y_1 \middle| \hat{O} y_1 \right\rangle$, $b_1|y_2\rangle \equiv \hat{O}|y_1\rangle - d_1|y_1\rangle$, and the coefficient b_1 can be found imposing $|\langle b_1 y_2 | b_1 y_2\rangle|^2 = 1$. Obtained the first two vectors, the operation is iterated another time, with the additional care of decomposing the new state in the components orthogonal and parallel with respect to both previous states. We then obtain

$$b_2|y_3\rangle = \hat{O}|y_2\rangle - b_1|y_1\rangle - d_2|y_2\rangle \tag{2.20}$$

where, as before, $b_1 = \left\langle y_1 \middle| \hat{O} y_2 \right\rangle$, $d_2 = \left\langle y_2 \middle| \hat{O} y_2 \right\rangle$ and the coefficient b_2 can be calculated imposing the normalization of $b_2|y_3\rangle$.

After the first two iterations of the algorithm, we obtained the values of the coefficients d_1, d_2, b_1, and b_2 as well as the states $|y_1\rangle$, $|y_2\rangle$, and $|y_3\rangle$. The main point of the algorithm appears in its beauty at the third iteration: applying once more the operator $\hat{O}$ to the state $|y_3\rangle$ and decomposing it to the parallel and perpendicular components with respect to all previous computed states, we obtain

$$\hat{O}|y_3\rangle = r|y_1\rangle + b_2|y_2\rangle + a_3|y_3\rangle + b_3|y_4\rangle. \tag{2.21}$$

The first coefficient is defined as before as $r = \left\langle y_1 \middle| \hat{O} y_3 \right\rangle$, which is equal to zero as $\left\langle y_1 \middle| \hat{O} y_3 \right\rangle = \left\langle \hat{O} y_1 \middle| y_3 \right\rangle = (d_1 \langle y_1| + b_1^* \langle y_2|)|y_3\rangle = 0$, as the operator $\hat{O}$ is Hermitian, and the orthogonality relation $|y_3\rangle \perp |y_2\rangle, |y_1\rangle$ holds by construction. In general, one can show that for any successive iteration, it holds

$$\hat{O}|y_k\rangle = b_{k-1}|y_{k-1}\rangle + a_k|y_k\rangle + b_{k-1}|y_{k+1}\rangle. \tag{2.22}$$

In conclusion, we define the matrix composed by the first N vectors $|y_k\rangle$ as

$$Y = \begin{pmatrix} | & | & & | \\ |y_1\rangle & |y_2\rangle & \dots & |y_N\rangle \\ | & | & & | \end{pmatrix}, \tag{2.23}$$

and the relation of Eq. (2.22) implies that $T_N = Y^\dagger \hat{O} Y$ where T_N is a tridiagonal matrix as defined in Eq. (2.17).

Notice that to compute the coefficients of the tridiagonal matrix T_N, one does not need to write explicitly the matrix Y, nor an explicit full matrix representation of the original operator $\hat{O}$ if—as it is often the case—the matrix is sparse or has some structure to be exploited. Indeed, the coefficients in T_N can be obtained by means of Eq. (2.22): at every steps one need to store the vectors $|y_{k-1}\rangle$, $|y_k\rangle$, $|y_{k+1}\rangle$ and eventually two vectors of coefficients a_k and b_k. Once these coefficients are calculated, the approaches reviewed in Sect. 2.3 can be applied to find the eigenvalues and eigenvectors of the operator $\hat{O}$ in its tridiagonal form T_N. Thus, with the Lancosz approach, huge matrices can be diagonalized, even those that cannot be stored in memory. We refer the interested reader to the extensive literature on Lancsoz methods and the Krylov subspaces they are based on, see, e.g., [132].

2.5 Problems

1. Consider a random Hermitian matrix A of size N.
   (a) Write a program that initializes A and performs an LU reduction. How does it scale with N?
   (b) Diagonalize A and store the N eigenvalues λ_i in increasing order.
   (c) Compute the normalized spacings between eigenvalues $s_i = \Delta\lambda_i / \bar{\Delta\lambda}$ where

$$\Delta\lambda_i = \lambda_{i+1} - \lambda_i,$$

   and $\bar{\Delta\lambda}$ is the average $\Delta\lambda_i$.
   (d) Compute the average spacing $\bar{\Delta\lambda}$ over a different number of levels around λ_i (i.e. $N/100$, $N/50$, $N/10 \ldots N$) and compare the results of the next exercise for the different choices.
2. Study $P(s)$, the distribution of the s_i defined in the previous exercise, accumulating values of s_i from different random matrices of size.
   (a) Compute $P(s)$ for a random hermitian matrix.
   (b) Compute $P(s)$ for a diagonal matrix with random real entries.
   (c) Fit the corresponding distributions with the function:

$$P(s) = a s^{\alpha} \exp\left(-b s^{\beta}\right)$$

   comparing them with the predictions of Random Matrix Theory [133].
3. Consider the Ising Hamiltonian defined in Eq. (4.4).
   (a) Write a program that computes the $2^N \times 2^N$ matrix H for different N.
   (b) Diagonalize H for different $N = 1, \ldots, N$ and $\lambda \in [0 : 3]$.
   (c) Plot the first k levels as a function of λ for different N. Comment on the spectrum.

3 Numerical Calculus

Simone Montangero

In this chapter, we review the standard numerical methods to perform differential and integral calculus using finite precision mathematics. Indeed, the price to pay to invoke the powerful help of numerical methods is to introduce a finite precision cutoff which shall be taken carefully into account as it enters as a numerical error. Nevertheless, whenever this error is kept under control and correctly bounded, e.g., to machine precision (see Appendix A), we say that the final result is numerically exact. Hereafter, we will review the most successful numerical approaches to solve integrals of functions and partial differential equations, and specify them to solve the Schrödinger equation for few-body systems, that is, whenever it is possible to write explicitly the system wave function and the operators acting on them. These methods form the basis for the numerical approaches based on tensor network methods that we will introduce in the next parts of the book.

3.1 Classical Quadrature

The most intuitive approach to compute numerically the integral of a one dimensional real function $f(x)$ follows straightforwardly the Riemann integration, that is, to define a grid $\{x_i\}_{i=1}^{N}$ in the support space of the function f and to approximate the function f with $N-1$ trapezoids as

$$\int_a^b f(x)dx \simeq \frac{1}{2}\sum_{i=1}^{N-1}[f(x_i)+f(x_{i+1})](x_i - x_{i+1}). \tag{3.1}$$

S. Montangero (✉)
Università di Padova, Padova, Italy
e-mail: simone.montangero@unipd.it

T. Felser, S. Montangero (eds.), *Introduction to Tensor Network Methods*,
Graduate Texts in Physics, https://doi.org/10.1007/978-3-032-17635-6_3

If we assume the lattice spacing to be equidistant, $h = x_i - x_{i+1}$, we obtain the *trapezoidal rule*

$$\int_a^b f(x)dx \simeq h(\frac{1}{2}f_1 + f_2 + f_3 + \ldots f_{N-2} + f_{N-1} + \frac{1}{2}f_N) \equiv h \sum_{i=1}^{N-1} w_i f_i \quad (3.2)$$

where $f(x_i) \equiv f_i$ and $w_i = 1, 1/2$. Thus, the integral following the trapezoidal rule is given as the sum of the function values f_i evaluated at points x_i, weighted with coefficients w_i; before multiplying the result by h. By definition the result is exact in the limit of the lattice spacing h going to zero. However, it is fundamental to know the error introduced for every finite h. The order of the error can be evaluated by means of a Taylor expansion: the error is introduced by the approximation of each area of bases h under the function f with a trapezoid, while there is no additional error in the sum of each of them. It is then sufficient to compute the error for a single trapezoid, that without loss of generality we assume to be centered in zero. We then evaluate the area of the trapezoid with two functions evaluations, at $-h/2$ and $h/2$ and compare the result with the analytical integral obtained via Taylor expansion around zero, that is

$$\begin{aligned}\int_{-h/2}^{h/2} f(x)dx &\simeq \int_{-h/2}^{h/2} f(0) + hf'(0)x + f''(0)\frac{x^2}{2} + \ldots dx \\ &= hf(0) + \frac{h^3}{3}f''(0)\ldots \quad (3.3)\end{aligned}$$

The value of $f(0)$ can be expressed as a function of the two extremal points $f(-h/2)$ and $f(h/2)$ again Taylor expanding them and summing the two expression: $f(0) = \frac{1}{2}(f(-h/2) + f(h/2)) - h^2/8f''(0) + \ldots$. Finally, one obtains the dependence of the value of the integral as a function of the extremal points

$$\int_{-h/2}^{h/2} f(x)dx \simeq \frac{h}{2}[(f(-h/2) + f(h/2)) + \frac{5h^2}{12}f''(0) + \ldots], \quad (3.4)$$

which compared with Eq. (3.2) shows that the trapezoidal rule is correct up to second order in h, as one disregard the term $\frac{5h^2}{12}f''(0)$ and higher orders ones. This is a not surprising result at all, as approximating the integral as a sum of trapezoids is equivalent to approximate linearly the function $f(x)$ within each interval.

It is clearly possible to improve the accuracy of the integration algorithm simply keeping higher orders in the Taylor expansion of $f(x)$. Repeating straightforwardly the previous steps and considering three function evaluation points, which are necessary to approximate at second order a function, one obtains

$$\int_{-h}^{h} f(x)dx \simeq \frac{1}{3}(f(-h) + 4f(0) + f(h)), \quad (3.5)$$

that is the *Simpson's rule* which prescribes $w_1 = w_N = 1/3$, $w_i = 4/3$ and $w_i = 2/3$ for even and odd terms respectively. The precision of such prescription can be double checked as before: the higher order expansion reads

$$\int_{-h}^{h} f(x)dx \simeq 2hf(0) + \frac{h^3}{3}f'(0) + \frac{h^3}{3}f''(0) + \ldots \tag{3.6}$$

and inserting the relation $f''(0) = (f(h) - 2f(0) + f(-h))/h^2 + \mathcal{O}(h^2)$ one obtains the Simpson's rules neglecting $\mathcal{O}(h^4)$ terms. Notice that the approximation should be in principle at second order, however, the symmetry of the problem grants an additional order of precision.

This way of proceeding can be extended to higher orders. However, it quickly becomes quite lengthy and intricate. It is more convenient to reformulate the problem in terms of a system of linear equations, which can be solved with standard tools. Indeed, one can impose directly the condition introduced in Eq. (3.2), that is, the equivalence between the integral of the function and the sum of N weighted functions evaluations $w_i f_i$ up to some order $M + 1 \leq N$. In conclusion, one can impose the following conditions for functions of up to order $f(x) = x^M$

$$\int_a^b x^0 dx = b - a \equiv \sum w_i \tag{3.7}$$

$$\int_a^b x^1 dx = (b^2 - a^2)/2 \equiv \sum w_i x_i \tag{3.8}$$

$$\int_a^b x^2 dx = (b^3 - a^3)/3 \equiv \sum w_i x_i^2 \tag{3.9}$$

$$\vdots$$

$$\int_a^b x^M dx = (b^{M+1} - a^{M+1})/(M+1) \equiv \sum w_i x_i^M . \tag{3.10}$$

It can be easily checked that the first two equations give rise to a system of linear equations for the coefficients w_i whose solution (under the assumption $a = -h/2$ and $b = h/2$) is the trapezoidal rule. Similarly, the first three equations return Simpson's rule. This approach also highlights that the full potential of this numerical approximation is not fully exploited. Indeed, setting a priori the lattice spacing h is not convenient: to increase the numerical integration precision, we could include the positions x_i as variables. Thus, Eq. (3.9) can be solved not only for the weights w_i but also for the estimation points x_i: having doubled the number of variables, it is possible to solve more equations with the same number of function evaluations. For example, imposing $x_1 = -x_2$ (but not $x_1 = -h/2$) and $|a| = |b|$, the system of four equations

$$\begin{cases} b - a = w_1 + w_2 \\ 0 = w_1 x_1 + w_2 x_2 \\ (b^3 - a^3)/3 = w_1 x_1^2 + w_2 x_2^2 \\ 0 = w_1 x_1^3 + w_2 x_2^3 \end{cases} \tag{3.11}$$

can be solved obtaining $w_1 = w_2 = (b - a)/2$ and $x_1 = -x_2 = \sqrt{(b^3 - a^3)/(3(b-a))}$. Thus, just changing the positions of the points sampling the function, the approximation can be improved from $\mathcal{O}(h^2)$ to $\mathcal{O}(h^4)$. The generalization of such procedure to any order M goes under the name of Gaussian integration as we will review in the next section.

3.2 Gaussian Integration

In the previous section, we have seen that it is possible to write and solve a system of N linear equations for the weights w_i of the N function sampling points f_i. In the case of $N = 2$, we also showed that by properly choosing the sampling points—thus abandon a standard equidistant lattice—it is possible to double the precision of the integral without increasing the computational cost. Hereafter, we will show that this is indeed always possible thanks to the properties of polynomials basis functions such as Legendre polynomials. Indeed, by sampling the function in N points and properly choosing the weights w_i, we have shown that it is possible to solve the first N Eqs. (3.7)–(3.10). In general, the following equations define a set of polynomials Z_j of the variables x_i

$$Z_0(x_1, x_2, \dots x_N) \equiv \sum w_i - \int_a^b x^0 dx = 0 \tag{3.12}$$

$$Z_1(x_1, x_2, \dots x_N) \equiv \sum w_i x_i - \int_a^b x^0 x^1 dx = 0 \tag{3.13}$$

$$\vdots$$

$$Z_{N-1}(x_1, x_2, \dots x_N) \equiv \sum w_i\, x_i^{N-1} - \int_a^b x^{N-1} dx = 0 \tag{3.14}$$

$$Z_N(x_1, x_2, \dots x_N) \equiv \sum w_i\, x_i^N - \int_a^b x^N dx \neq 0 \tag{3.15}$$

$$Z_{N+1}(x_1, x_2, \dots x_N) \equiv \sum w_i\, x_i^{N+1} - \int_a^b x^{N+1} dx \neq 0 \tag{3.16}$$

$$\vdots$$

$$Z_{N+K}(x_1, x_2, \dots x_N) \equiv \sum w_i\, x_i^{N+K} - \int_a^b x^{N+K} dx \neq 0 \tag{3.17}$$

for which, in general, $Z_j(x_1, x_2, \dots x_N) \neq 0\ \forall j > N-1$ and $Z_j(x_1, x_2, \dots x_N) = 0\ \forall j \leq N-1$ for a proper choice of the weights w_i. Our task is then to find the set of positions $\{x_1, x_2, \dots x_N\}$ that set to zero the maximal number of polynomials Z_j, effectively increasing the order of the approximation of the numerical integration. From now on, we rescale the integration bounds to $a = -1$ and $b = 1$ and we will exploit the Legendre polynomials $L_n(x) \equiv \sum_{m=0}^{n} p_{n,m} x^m$ and their properties, in particular,

1. L_n is a polynomial of order n with n zeros x_i such that $L_n(x_i) = 0$, with $L_0 \equiv 1$.
2. L_n forms an orthogonal basis for the space of functions in $[-1 : 1]$: any other polynomial can be expanded in this basis and in particular $x^m = \sum_{n=0}^{m} q_{m,n} L_n(x)$, where $q_{m,n}$ is the matrix inverse of $p_{n,m}$.
3. L_n is orthogonal to x^m with $m < n$ as $\langle x^m | L_n(x) \rangle = \sum_{j=0}^{m} q_{m,j} \langle L_j(x) | L_n(x) \rangle = 0\ \forall j < n$.

We are now ready to look for a solution of the problem we have introduced at the beginning of this section: to find a set of $\{\bar{x}_i\}$ such that, for example, $Z_N(\bar{x}_i) = 0$. To do so, we multiply each of the first $N+1$ Eqs. (3.12)-(3.15) for the coefficient $p_{n,j}$ and sum them up, obtaining the relation

$$p_{n,n} Z_N(x_1, x_2, \dots x_N) = \sum_j \left[p_{n,j} \sum w_i x_i^j - \int_{-1}^{1} p_{n,j} x^j dx \right] \tag{3.18}$$

$$= \sum_i w_i L_n(x_i) - \int_{-1}^{1} L_n(x) dx. \tag{3.19}$$

The second term of Eq. (3.19) is identically zero as $\int_{-1}^{1} L_n(x)dx = \langle L_0(x) | L_n(x) \rangle = 0$. Finally, the polynomial $L_n(x)$ has exactly N zeros, thus if one choose them as $\{\bar{x}_i\}$, one obtains $Z_N(x_1, x_2, \dots x_N) = 0$. In a similar way, one can show that also the next Z_{N+j} polynomials are identically zero for every $j \leq N-1$. Indeed, to prove that Z_{N+j} is zero, it is sufficient to multiply the equations for Z_j to Z_{N+j} for the same coefficients as before to obtain the general relation

$$p_{n,n} Z_{N+j}(x_1, x_2, \dots x_N) = \sum_i w_i x_i^j L_n(x_i) - \int_{-1}^{1} x^j L_n(x) dx. \tag{3.20}$$

As before, the first term on the right hand side is identically zero for the same choice of x_i, while the second term is equal to $\langle L_j(x) | L_n(x) \rangle$ which is zero for any $j < n$. In conclusion, with $2N$ free parameters (the weights and the sampling positions) we have solved $2N$ equations granting a precision of the numerical integration of $\mathcal{O}(h^{2N})$.

3.3 Time-independent Schrödinger Equation

In the rest of the chapter, we review the most common and well-established methods to solve partial differential equations numerically. In particular, we specialize our presentation on the exact numerical solution of both time-independent and time-dependent Schrödinger equation.

The fundamental problem of quantum mechanics is to find the eigenvalues and eigenvectors of the Hamiltonian describing the system. Once the Hamiltonian is diagonalized, all information about the system can be computed. A general approach to achieving such result is to write the Hamiltonian matrix representation and solve it with the numerical methods presented in Chap. 2. Hereafter, we present the standard approaches to recast a quantum mechanical problem defined in a continuous space into a discretized version, essential to write the Hamiltonian in a discrete and finite matrix representation. The main approaches are indeed of two classes: the discretization of the space, introducing a lattice as reviewed in Sect. 3.3.1 and the introduction of proper basis functions used to expand the Hamiltonian, as shown in Sect. 3.3.2.

3.3.1 Finite Difference Method

The time independent Schrödinger equation to solve a single particle quantum mechanical problem defined in a three dimensional space in presence of a potential $V(\vec{x})$, can be formally written

$$H\,|\psi(\vec{x})\rangle = \left[\frac{\vec{p}^2}{2m} + V(\vec{x})\right]|\psi\rangle = \left[-\frac{\vec{\nabla}^2}{2m} + V(\vec{x})\right]|\psi\rangle = E\,|\psi\rangle\,, \tag{3.21}$$

where $\hbar = 1$ and $\vec{x}$, $\vec{p}$ are three dimensional vectors for the position and momentum, respectively. To solve numerically this equation, one can introduce artificially a lattice in space $\{\Delta\vec{x}\}$ to define a discrete eigenproblem and then check a posteriori the convergence of the solution to the continuous one in the limit $\Delta\vec{x} \rightarrow 0$. As for the integration presented in the previous chapter, the estimate of the error as a function of $\Delta\vec{x}$ is fundamental for an accurate discretization of the problem (see Appendix B). Hereafter, we will briefly review the simplest possible choice, that is to introduce a lattice with constant spacing h in each spatial direction. The first necessary ingredient to discretize Eq. (3.21) is to approximate the second order derivative by means of finite differences, at some order in h. Similarly to what is shown in Sect. B.3 for the first derivative, the discrete version of the second order derivative of a function $f(x)$ can be computed by means of the Taylor expansion of the function around a given point

$$f(x\pm h) = f(x)\pm hf'(x)+\frac{h^2}{2}f''(x)\pm\frac{h^3}{6}f^{(3)}(x)+\frac{h^4}{24}f^{(4)}(x)+\mathcal{O}\left(h^5\right). \tag{3.22}$$

Adding the two equations above and solving for the second derivative of the function we obtain

$$f''n = \frac{f_{n+1} - 2f_n + f_{n-1}}{h^2} + \mathcal{O}\left(h^2\right). \tag{3.23}$$

In conclusion, the discrete version of the time-independent Schrödinger equation at the fourth order in a one dimensional interval $[a : b]$ such that $h = (b - a)/N$ is given by

$$\begin{pmatrix} 2/h^2 + V_1 & -1/h^2 & 0 & \dots & 0 & 0 & 0 \\ -1/h^2 & 2/h^2 + V_2 & -1/h^2 & \dots & 0 & 0 & 0 \\ & & & \vdots & & & \\ 0 & 0 & 0 & \dots & -1/h^2 & 2/h^2 + V_{N-1} & -1/h^2 \\ 0 & 0 & 0 & \dots & 0 & -1/h^2 & 2/h^2 + V_N \end{pmatrix}$$

$$|\psi\rangle = E\,|\psi\rangle\,, \tag{3.24}$$

where $V_i = V(x_i)$ and we set $2m = 1$. The generalization of Eq. (3.24) to more dimensions and at higher orders is straightforward. As the reader has undoubtedly noticed, the matrix above is tridiagonal (and in general sparse); thus, it can be highly efficiently diagonalized, see Sect. 2.3. However, especially in dimensions larger than one, the size of the eigenvalue problem typically becomes intractable very fast, as it scales as the number of lattice points in a single dimension to the number of dimensions. For example, keeping one thousand points for each dimension, in three dimensions results into a matrix of the size of a billion, which is hard to attack even with the powerful methods reviewed in Chap. 2. For this reason, the introduction of more efficient discretization methods is highly recommended in such cases, such as the *finite elements methods* [129].

3.3.2 Variational Method

A second approach to treat high dimensional problems is to avoid the introduction of a lattice and perform the discretization using a finite set of basis functions. Indeed, it is always possible to choose a proper set of functions to expand the system wave function

$$|\psi\rangle = \sum_{i=1}^{K} a_i\,|\phi_i\rangle\,, \tag{3.25}$$

where $a_i = \langle\phi_i|\psi\rangle$, and $|\phi_i\rangle$ can be chosen by means of physical intuition or under the assumption that increasing their number K, any set of independent wave functions will eventually cover the full Hilbert space. As we will see later on, it is desirable that the functions $|\phi_i\rangle$ to be orthonormal, however, to provide the exact solution to the problem this is not a necessary condition. Having introduced the basis functions $|\phi_i\rangle$ it is straightforward to compute the matrix representation of the Hamiltonian in such basis, e.g. for a particle living in a one-dimensional external potential,

$$H_{i,j} = \langle\phi_i|\hat{H}|\phi_j\rangle = \int d\vec{x}\ \phi_i(\vec{x})^* \left(-\frac{\vec{\nabla}^2}{2m} + V(\vec{x})\right) \phi_j(\vec{x}); \tag{3.26}$$

and the overlap matrix

$$S_{i,j} = \langle\phi_i|\phi_j\rangle = \int d\vec{x}\ \phi_i(\vec{x})^* \phi_j(\vec{x}). \tag{3.27}$$

A choice of an orthonormal basis clearly results in $S = \mathbb{1}$, however, this is not true in general. Notice that this is the same approach exploited—together with other approximations—by the Hartree-Fock methods, one of the most powerful methods up to date used to compute molecular configurations [17, 134]; and to introduce models such as the Hubbard and Bose-Hubbard models to describe, e.g., electrons in solids and cold atoms in optical lattices respectively [96, 98].

It is now possible to compute the energy of any wavefunction $|\psi\rangle$ taking care of enforcing normalization, that is to compute

$$E = \frac{\langle\psi|\hat{H}|\psi\rangle}{\langle\psi|\psi\rangle}, \tag{3.28}$$

and to minimize such expression to find the ground state energy. If the wavefunctions $|\phi_i\rangle$ are the set of the Hamiltonian eigenfunctions, it is easy to show that the energy of any wavefunction is such that $E \geq E_0$, where E_0 is the system's ground state energy [135]. On the contrary, given a generic finite basis of K wavefunctions $|\phi_i\rangle$ it is possible to look for an approximation from above (the energy expectation value of any wave function is bigger or equal than E_0) of the ground state of the system within the spanned subspace. Once the minimization is performed, it is always possible to increase the number of basis functions to check for convergence.

The expression in Eq. (3.28) is minimized by any (normalized) wave function that satisfies the following equality (written in the chosen set of basis functions $|\phi_i\rangle$) with the minimal value of E:

$$\hat{H}\,|\psi\rangle = E\hat{S}\,|\psi\rangle\,. \tag{3.29}$$

If $S = \mathbb{1}$ this is once again an eigenvalue problem of dimension K that can be solved by means of standard methods presented in Chap. 2. On the contrary, it represents a *generalized eigenvalues problem* that can be solved recasting it in a standard eigenvalue problem as follows: the overlap matrix S is Hermitian by definition, and thus it can be diagonalized by means fo a unitary matrix W such that $W^\dagger \hat{S} W = D$ where D is some diagonal matrix of real numbers. Rescaling properly each eigenvectors it is then possible to find the orthogonal transformation such that $V^\dagger S V = \mathbb{1}$ and define the wave function $|\psi\rangle = V|\xi\rangle$ for which the eigenproblem of Eq. (3.29) can be rewritten as

$$\hat{H} V |\xi\rangle = E \hat{S} V |\xi\rangle \Rightarrow V^\dagger \hat{H} V |\xi\rangle = E \hat{V}^\dagger S V |\xi\rangle = E |\xi\rangle . \tag{3.30}$$

Thus, a standard eigenvalue problem can be defined for the operator $V^\dagger \hat{H} V$ and the wave function $|\xi\rangle$: once solved the solution for the original problem can be found by means of the relation $|\psi\rangle = V|\xi\rangle$.

We conclude this section with an example of such approach, that is, how to compute the solutions of anharmonic oscillator: a one-dimensional quantum particle subject to a third order potential $V(x) = x^2 + x^3$. In this simple case, it is natural to choose the solutions of the harmonic oscillators as the set of basis functions $|n\rangle$ and expand the potential in such basis. Clearly, the result is a matrix with the harmonic oscillator eigenenergies on the diagonal and off diagonal terms given by

$$H_{n,m} = \langle n|\hat{x}^3|m\rangle = \langle n|(\frac{a + a^\dagger}{2})^3|m\rangle , \tag{3.31}$$

where $a^\dagger, a$ are the creation and destruction operator. The Hamiltonian is then a band matrix with bandwidth seven which shall be diagonalized.

3.4 Time-dependent Schrödinger Equation

In this section, we review the most common approaches to solve the time-dependent Schrödinger equation building on the linear algebra methods that we have introduced so far: we aim to solve the problem of finding the evolved state $|\psi(t = T)\rangle$ of a quantum system according to

$$i\hbar \frac{\partial |\psi\rangle}{\partial t} = \hat{H} |\psi\rangle \tag{3.32}$$

given the boundary condition $|\psi(t = 0)\rangle = |\psi_0\rangle$. The possible applications of such numerical method are ubiquitous and range, for example, from the evolution of a qubit or spin one-half system, to the transport properties into quantum wires or the description of chemical reactions or light-harvesting processes.

3.4.1 Spectral Method

For time-independent Hamiltonians the most straightforward approach one can exploit follows directly the standard quantum mechanics prescription: to decompose the initial state in the Hamiltonian eigenfunctions $|E_i\rangle$ and compute exactly the time evolved state

$$|\psi(T)\rangle = \sum_i c_i \, e^{-\imath E_i T/\hbar} \, |E_i\rangle \,, \tag{3.33}$$

where $H\,|E_i\rangle = E_i\,|E_i\rangle$ and $c_i = \langle E_i|\psi_0\rangle$. This is indeed the most precise method and allows to propagate the system for virtually infinitely large times T, as the error mainly comes from the computation of the eigenstates $|E_i\rangle$. However, in this nice property lays also the seed of its limitation: the computation of the whole spectrum of the system is the limiting factor in terms of scalability. Indeed, despite the powerful methods introduced in Chap. 2, this approach is mainly limited to systems with small Hilbert space size, that is few-body discrete systems or low-dimensional continuous ones.

3.4.2 Split Operator Method

Another efficient approach to solve time-independent Schrödinger equation in real space is the *split operator method*. The first step, as for most of numerical solutions of partial differential equations, is to discretize the time variable, such that $T = n\,\Delta t$ (see also Appendix B): the formal solution of the Schrödinger equation in Sect. 3.4 for the wave function after one time step Δt is given by

$$|\psi(t+\Delta t)\rangle = e^{-i\hat{H}\Delta t}\,|\psi(t)\rangle \,, \tag{3.34}$$

where we set $\hbar = 1$. Consider now a generic one-dimensional Hamiltonian defined over a continuous space and composed by a kinetic and a potential part

$$\hat{H} = \hat{T} + \hat{V} = \frac{\hat{p}^2}{2m} + \hat{V}(\hat{x}). \tag{3.35}$$

The split operator method stems from the observation that the two parts of the Hamiltonian being functions of a single variable—either $\hat{x}$ or $\hat{p}$—are diagonal in the space and momentum representation respectively. Thus, even though the two parts do not commute, exploiting the Baker-Campell-Hausdorff relation [135] we can split the time-evolution operator that propagates the system from t tp $t + \Delta t$ as

$$e^{-i\hat{H}\Delta t} \simeq e^{-i\hat{V}\Delta t/2} e^{-i\hat{T}\Delta t} e^{-i\hat{V}\Delta t/2} + \mathcal{O}\left(\Delta t^3\right). \tag{3.36}$$

A simple algorithm can now be written to compute the time evolution: each terms in the r.h.s of Eq. (3.36) is diagonal either in space or momentum representation. The change of base between position and momentum representation is the Fourier transform $\mathcal{F}$, for which it exists an highly efficient algorithm, the Fast Fourier

Transform (FFT). The FFT scales as $\mathcal{O}(N \log N)$, thus clearly it outperforms a generic change of basis (matrix vector multiplication, $\mathcal{O}(N^2)$). Inserting the changes of bases in the Eq. (3.36) we obtain

$$e^{-i\hat{H}\Delta t}|\psi(x,t)\rangle \simeq e^{-i\hat{V}\Delta t/2}e^{-i\hat{T}\Delta t}e^{-i\hat{V}\Delta t/2}|\psi(x,t)\rangle \tag{3.37}$$

$$= U_V(x)\mathcal{F}^{-1}\mathcal{F}U_T(x)\mathcal{F}^{-1}\mathcal{F}U_V(x)|\psi(x,t)\rangle \tag{3.38}$$

$$= U_V(x)\mathcal{F}^{-1}\mathcal{F}U_T(x)\mathcal{F}^{-1}\left|\psi'(p,t)\right\rangle \tag{3.39}$$

$$= U_V(x)\mathcal{F}^{-1}U_T(p)\left|\psi'(p,t)\right\rangle \tag{3.40}$$

$$= U_V(x)\left|\psi''(x,t)\right\rangle = |\psi(x,t+\Delta t)\rangle\,. \tag{3.41}$$

In conclusion, provided that the proper change of basis is performed right before applying $U_T(p)$ or $U_V(x)$, all operators appear in their diagonal form. Thus, the split operator method requires to store only $\mathcal{O}(N)$ elements and its computational costs is given by the FFT, providing high speed-up when it is possible to apply it. Its precision can be enhanced considering higher terms in the expansion of Eq. (3.36) introducing an additional overhead to the algorithm computational cost. The split operator method can be straightforwardly generalized to dimensions larger than one, however, similarly to the spectral method, it can be hardly applied to systems living in more than two dimensions. Moreover, the split operator can also be applied to different Hamiltonians made of generic noncommuting terms. However, the transformation between different bases will differ from a Fourier transform, and thus the favorable scaling of the FFT algorithm cannot be exploited.

3.4.3 Partial Differential Equations Solvers

Whenever the two previous approaches cannot be exploited, one shall apply the direct integration of the Schrödinger equation. The main problem of the formal solution of the Schrödinger equation given by Eq. (3.34) is that to compute the exponential of the Hamiltonian operator one shall diagonalize it. However, as we have seen for the spectral method, diagonalizing the full Hamiltonian highly reduce the field of application of the method. Thus, the idea is to compute the relation at the right-hand side of Eq. (3.34) without computing explicitly the exponential of the matrix H. Exploiting the fact that Δt can be chosen small compared to any energy scale of the system, the ready solution to the problem is of course to expand the exponential in series of Δt: at first order one recovers the *Euler method*

$$|\psi(t+\Delta t)\rangle \simeq (\mathbb{1} - i\hat{H}\Delta t)|\psi(t)\rangle\,. \tag{3.42}$$

However, this rough approximation has the major problem that does not conserve the norm of the state. Indeed, the approximated time evolution operator is not unitary. This can be easily seen computing the norm of the time-evolved state after one time step at second order in Δt, $|\langle\psi(t+\Delta t)|\psi(t+\Delta t)\rangle|^2 = |1 - \Delta t^2\langle\Delta\hat{H}\rangle^2|^2$, where $\langle\Delta\hat{H}\rangle^2 = \langle\hat{H}^2\rangle - \langle\hat{H}\rangle^2$ is the (positive definite) variance of the Hamiltonian operator computed at time t.

To correct for such inconvenient artifact of the expansion, it is possible to impose unitarity of the truncated time evolution operator, simply expressing the time evolution operator as a function of the product of two adjoint operators. Indeed the following relation holds

$$e^{-i\hat{H}\Delta t} = (e^{i\hat{H}\Delta t/2})^{-1}e^{-i\hat{H}\Delta t/2} \simeq (\mathbb{1} - i\hat{H}\Delta t/2)^{-1}(\mathbb{1} - i\hat{H}\Delta t/2). \tag{3.43}$$

The right hand side of Eq. (3.43) is unitary by construction and thus solves the problem of the norm conservation, however it introduces the necessity of computing the inverse of a matrix. The solution of this problem, which goes under the name of the *Crank-Nicolson method* exploits our capability of solving systems of linear equations efficiently, as explained in Chap. 2. Indeed Eqs. (3.34) and (3.43) implies the relation

$$(\mathbb{1} - i\hat{H}\Delta t/2)\,|\psi(t+\Delta t)\rangle = (\mathbb{1} - i\hat{H}\Delta t/2)\,|\psi(t)\rangle \equiv |\phi\rangle\,. \tag{3.44}$$

Thus, to compute the evolved state $|\psi(t+\Delta t)\rangle$ one has to solve the system of linear equations defined by the above expression.

Finally, we will show that the methods presented above are only a specific example of numerical integration of the partial differential equation: as we have seen in Chap. 3 depending on the order of interpolation used, the precision of the method can increase drastically. Indeed, writing Schrödinger equation in its integral formulation

$$|\psi(t+\Delta t)\rangle = |\psi(t)\rangle + \int_t^{t+\Delta t} \mathcal{L}\,|\psi(t)\rangle\,, \tag{3.45}$$

where $\mathcal{L} = -i\hat{H}$, it is clear that solving the problem is equivalent to compute the integral on the right hand side of the equation. Thus, one obtains

$$|\psi(t+\Delta t)\rangle - |\psi(t)\rangle = \Delta t\,[w_1\mathcal{L}\,|\psi(t)\rangle + w_2\mathcal{L}\,|\psi(t+\Delta t)\rangle]\,. \tag{3.46}$$

where w_i are weights that can be chosen freely. The relation above can be rewritten as

$$\hat{A}\,|\psi(t+\Delta t)\rangle = \hat{B}\,|\psi(t)\rangle\,, \tag{3.47}$$

where $\hat{A} = \mathbb{1} - w_2\mathcal{L}\Delta t$ and $\hat{B} = \mathbb{1} - w_1\mathcal{L}\Delta t$, and the time propagator is defined as $\hat{A}^{-1}\hat{B}$. The trapezoidal rule prescribes $w_i = 1/2$ which indeed correspond to the Crank-Nicolson method. However, one could in principle make a different choice (e.g. to approximate the integral with a rectangle of height equal to one extrema of the function) and choose $w_1 = 1$, $w_2 = 0$, obtaining the Euler method, which as we have seen, is less precise than the latter.

Having identified the solution of the time-dependent Schrödinger equation with an integration of the Liouvillian operator $\mathcal{L}$ as in Eq. (3.46), it should be now clear how to improve the precision beyond the Crank-Nicolson using the Gaussian integration schemes presented in Sect. 3.2. Indeed, it is possible to introduce K intermediate function evaluations and compute the integral in Eq. (3.46) as

$$|\psi(t+\Delta t)\rangle - |\psi(t)\rangle = \Delta t \sum_{k=0}^{K} w_k \mathcal{L} |\psi(t+\tau_k)\rangle , \tag{3.48}$$

with $\tau_0 = 0, \tau_K = \Delta t$. We have seen in Sec. 3.2 that by means of Gaussian integration we can, in principle, achieve an accuracy of $\mathcal{O}(\Delta t^{2K})$. However, here an additional complication arise as in Eq. (3.48) $K-1$ additional variables have been introduced: the intermediate wave functions $|\psi(t+\tau_k)\rangle$, $k = 1, K-1$ are unknown (differently from previous scenario of the integration of a known function $f(x)$) and shall be computed together with $|\psi(t+\Delta t)\rangle$. The solution of this problem can be found exploiting the structure of the function to be integrated $\mathcal{L}_k \equiv \mathcal{L}|\psi(t+\tau_k)\rangle$, that is, exploiting the relation in Eq. (3.48) for each intermediate time steps, the Sect. 3.4 and the Taylor expansion in time of the wave function around the initial time t. Indeed, one can write the relation Eq. (3.48) for each time steps τ_k as

$$|\psi(t+\tau_1)\rangle - |\psi(t)\rangle = \Delta t T_{1,0}\mathcal{L}_0 \tag{3.49}$$

$$|\psi(t+\tau_2)\rangle - |\psi(t)\rangle = \Delta t (T_{2,0}\mathcal{L}_0 + T_{2,1}\mathcal{L}_1) \tag{3.50}$$

$$\vdots$$

$$|\psi(t+\tau_k)\rangle - |\psi(t)\rangle = \Delta t \sum_{j=0}^{k} T_{j,k}\mathcal{L}_k, \tag{3.51}$$

where $T_{i,j}$ is a transfer matrix which contains the to-be-determined coefficients that relate the first $k-1$ intermediate L_k to the k-th one. It is then possible to write a system of equations for the coefficients w_i (which relate the L_k with the final state), the transfer matrix coefficients $T_{i,j}$ and the intermediate time intervals τ_k such that the gaussian integration is performed. Hereafter we report the solution for $K = 2$, the *Runge-Kutta method* at fourth order. The Eq. (3.48) for $K = 2$ becomes

$$|\psi(t+\Delta t)\rangle - |\psi(t)\rangle = \Delta t (w_0\mathcal{L}_0 + w_1\mathcal{L}_1). \tag{3.52}$$

Recalling that $\mathcal{L} = \mathcal{L}(\psi, t)$ one can expand it at first order obtaining $\mathcal{L}_1 = \mathcal{L}_0 + \frac{\partial \mathcal{L}}{\partial \psi}\Delta\psi + \partial\psi\partial t\tau_1$. Inserting Eq. (3.49), recalling that $\tau_1 = c_1\Delta t$ for some coefficient c_1 one can rewrite Eq. (3.52) as

$$|\psi(t+\Delta t)\rangle = |\psi(t)\rangle + \Delta t(w_0+w_1)\mathcal{L}_0 + \Delta t^2 w_1(T_{1,0}\mathcal{L}_0\frac{\partial \mathcal{L}}{\partial \psi} + c_1\frac{\partial \mathcal{L}}{\partial t}). \qquad (3.53)$$

Equating the above expression term by term with the second order Taylor expansion of $\psi(t+\Delta t)$ one obtains the system of equations

$$\begin{cases} w_0 + w_1 = 1 \\ w_1 T_{1,0} = 1/2 \\ w_1 c_1 = 1/2 \end{cases},$$

whose solution provides the necessary coefficients to implement a Runge-Kutta solver at fourth order in Δt. Implementations of higher order approximations and adaptive Runge-Kutta methods can be found in most common platforms and programming languages.

3.5 Exercises

1 Consider a one-dimensional quantum harmonic oscillator defined by the Hamiltonian

$$H = \frac{\hat{p}^2}{2} + \frac{\hat{x}^2}{2}$$

and its eigenvalues E_k and eingenvectors $|\psi_k\rangle$ ($\hbar = m = \omega = 1$).

(a) Given the exact ground state $|\psi_0\rangle$, compute numerically the expectation value of the ground state energy $\langle\psi_0|H|\psi_0\rangle$ and compare it with the exact value E_0.
(b) Is the error due to the wave function discretization or to the approximation of the integral? *Hint: increase the integral approximation (Trapezoidal rule, Simpson's rule ...) for fixed wave function discretization and vice versa.*

2 Consider a one-dimensional quantum harmonic oscillator defined by the Hamiltonian defined in the previous exercise.

(a) Write a FORTRAN program to compute the first k eigenvalues E_k and eigenvectors $|\psi_k\rangle$.
(b) How would you rate your program in terms of the priorities we introduced in class for good scientific software development (Correctness, Stability, Accurate discretization, Flexibility, Efficiency), see Appendix B?

3 Given a time-dependent one-dimensional quantum harmonic oscillator defined by the Hamiltonian

$$H = \frac{\hat{p}^2}{2} + \frac{(\hat{q} - q_0(t))^2}{2};$$

with $q_0(t) = t/T$ and $t \in [0 : T]$. Given $|\psi_0\rangle = |n = 0\rangle$ (ground state of the Harmonic oscillator), compute $|\psi(t)\rangle$ for different values of T.

Part II

The Many-body Problem

Numerical Renormalization Group Methods 4

Simone Montangero

In this chapter, we introduce the renormalization group (RG) approach from a numerical perspective. While the analytical RG approach is a powerful tool routinely used to study different phenomena [23, 136, 137], hereafter, we concentrate on its application to numerically attack the many-body quantum problem. To set the stage, we start with the mean-field treatment of the many-body quantum problem and present its application to study the quantum Ising model in a transverse field. We then introduce the numerical RG approach, and finally, we review the density matrix renormalization group (DMRG) method in its original formulation to highlight its connection with the standard RG.

4.1 Mean Field Theory

Many-body quantum systems are by definition composed of many quantum degrees of freedom. Hereafter, for simplicity, we assume that each quantum degree of freedom is local and lives on a lattice site. The paradigmatic example is a chain of spins, but the same construction can be derived starting from continuous degrees of freedom in many distinct settings, e.g., atoms in optical lattices [96]. The system wave function $|\psi\rangle$ lives in the N-body Hilbert space $\mathcal{H}_N$ formed by the tensor product of all the local ones $\mathcal{H}_N = \mathcal{H}_1 \otimes \mathcal{H}_1 \otimes \ldots \mathcal{H}_1$. By definition, the wave function gives the amplitude of the probability of each possible system configuration $|\alpha_1 \alpha_2 \ldots \alpha_N\rangle$, where the index α_i labels the possible single-body configurations of

S. Montangero (✉)
Università di Padova, Padova, Italy
e-mail: simone.montangero@unipd.it

T. Felser, S. Montangero (eds.), *Introduction to Tensor Network Methods*,
Graduate Texts in Physics, https://doi.org/10.1007/978-3-032-17635-6_4

the i-th lattice site, $\{\alpha_i\}_1^d$ (e.g., if the system is composed by spin one-half systems, $|\alpha_1\rangle = |\uparrow\rangle$ and $|\alpha_2\rangle = |\uparrow\rangle$). Thus, a generic state is described by the N-rank tensor

$$|\psi\rangle = \sum_{\vec{\alpha}} \psi_{\alpha_1\alpha_2\ldots\alpha_N} |\alpha_1\alpha_2\ldots\alpha_N\rangle, \tag{4.1}$$

where $\vec{\alpha} = \alpha_1\alpha_2\ldots\alpha_N$, and $\mathcal{P}_{\alpha_1\alpha_2\ldots\alpha_N} = |\psi_{\alpha_1\alpha_2\ldots\alpha_N}|^2$ gives the probability to find the system in the configuration $\alpha_1\alpha_2\ldots\alpha_N$. Hereafter, the local Hilbert space dimension d can also be the result of a truncation of the local Hilbert space, e.g., while studying bosons in a lattice. Even though here for simplicity we consider the local Hilbert space dimension d uniform throughout the whole lattice, it could also be site dependent. Notice that the dimension of the N-rank tensor $\psi_{\alpha_1\alpha_2\ldots\alpha_N}$ increases exponentially with the system size.

The mean field approach is a very powerful method to get insights into the physics of many-body quantum systems, which relies on a strong approximation [15]. Within the mean field approximation, one assumes that quantum correlations shall be neglected, that is, the many-body wave function is constrained to be the product of N independent single-body wave functions,

$$|\psi^{\mathrm{MF}}\rangle = |\psi^1\rangle \otimes |\psi^2\rangle \otimes \cdots \otimes |\psi^N\rangle = \tag{4.2}$$

$$= \sum_{\alpha_1=1}^{d} \psi^1_{\alpha_1} |\alpha_1\rangle \otimes \sum_{\alpha_2=1}^{d} \psi^2_{\alpha_2} |\alpha_2\rangle \otimes \cdots \otimes \sum_{\alpha_N=1}^{d} \psi^N_{\alpha_N} |\alpha_N\rangle. \tag{4.3}$$

Thus, the object of study has been simplified from an exponential to a linear problem in the number of lattice sites N. An additional assumption (even though not strictly necessary) results in a further simplification: imposing translational invariance—all $|\psi^j\rangle$ are equal—recasts the original problem into a problem independent of the system size N. Indeed, the number of degrees of freedom (free parameters) is d^N in the original problem, Nd in the mean field scenario and only d in the translationally invariant one. It is then not surprising that, within the mean field approximation, efficient algorithms or even analytical solutions exist to very complex problems. However, the drawback of this powerful approach is the fact that the introduction of Eq. (4.3) is, in general, unjustified and uncontrolled. Indeed, independent checks of the mean field approximation are always needed, either via comparison with experimental results or by other means. Nevertheless, whenever quantum correlations do not play a major role, the mean field approach can be very powerful and, as we will see in the later sections, it can give a powerful insight into the physics of the system and can serve as a guide and starting point for further, more refined, analysis.

4.1.1 Quantum Ising Model in Transverse Field

We present the mean field approach using an example of a paradigmatic model, namely the quantum Ising model in a transverse field. The quantum Ising model represents one of the simplest nontrivial many-body quantum systems, and has played and plays a fundamental role in our understanding of correlated states of matter [97, 137]. As we will show in more detail in the last part of this book, this model has been instrumental to our understanding of critical phenomena in and out of equilibrium, and it is the first testbed for any numerical methods which aim to become a competitive one. The system has an analytical solution—a mapping to free fermions via the Jordan-Wigner transformation—that allows for precise benchmarking of any software developed to study numerically correlated many-body quantum system [97, 138–140]. The Hamiltonian of the model reads

$$\mathcal{H}^I = \sum_{i=1}^{N-1} H_{i,i+1} = -\sum_{i=1}^{N-1} \sigma_i^x \sigma_{i+1}^x + \lambda \sum_{i=1}^{N} \sigma_i^z; \tag{4.4}$$

where σs are Pauli matrices and represents a linear chain of N interacting spins $1/2$ in presence of an external field of intensity λ. The mean field analysis of this model starts from the definition of Eq. (4.3), where

$$|\psi^{\mathrm{MF}}\rangle = \otimes_{i=1}^{N} |\psi^1\rangle = \otimes_{i=1}^{N} \sum_{\alpha_i=1}^{d} A_{\alpha_i} |\alpha_i\rangle; \tag{4.5}$$

where for spin $1/2$ one has clearly $\alpha_i = \uparrow, \downarrow = 1, 2$ and $d = 2$. Under the assumption of Eq. (4.5), the computation of the system's energy can be readily done as

$$E^{\mathrm{MF}} = \langle\psi^{\mathrm{MF}}|\mathcal{H}^I|\psi^{\mathrm{MF}}\rangle = -\sum_{i=1}^{N-1} \left(\langle\psi^1|\sigma_i^x|\psi^1\rangle\right)^2 + \lambda \sum_{i=1}^{N} \langle\psi^1|\sigma_i^z|\psi^1\rangle. \tag{4.6}$$

The above quantity is extensive, thus it diverges at the thermodynamical limit, while the energy density $e = E/N$ is intensive and its limit for $N \to \infty$ can be calculated and is given by

$$e = -\left(\langle\psi^1|\sigma_i^x|\psi^1\rangle\right)^2 + \lambda\langle\psi^1|\sigma_i^z|\psi^1\rangle \equiv -r_x^2 + \lambda r_z, \tag{4.7}$$

where r_x and r_z are the two-level parametrization in the Bloch sphere. The expression for the energy density above is then a function of two variables r_x, r_z, which shall be minimized to compute the ground state energy density of the Ising model at the thermodynamical limit within the mean field approximation. This minimization can be performed analytically (even though more complex models might require numerical minimizations which can be performed with your preferred method [122]): the wave function normalization imposes the additional constraint $r_x^2 + r_y^2 = 1$, resulting in an expression function of the external magnetic field λ

$$\begin{cases} e = -1 - \lambda^2/4 & \lambda \in [-2:2] \\ e = -|\lambda| & \lambda \notin [-2:2]. \end{cases} \tag{4.8}$$

The expression above is not exact; however, it captures the physics of the system in the sense that it signals that there are two special points, $\lambda = -2, 2$ where the second derivative of the energy density becomes a discontinuous function: as we will see in Chap. 10 this signals the presence of a quantum phase transition. The quantum phase transition is indeed present; however, the mean field analysis is unable to compute the exact transition point and also to correctly describe the correlations occurring at the transition point. Indeed, by construction, the computation of any connected correlation functions $C_{i,j}$ between two observables $\hat{O}_i$ and $\hat{O}_j$ acting on two different spins will result to exactly zero:

$$C_{i,j} \equiv \langle \hat{O}_i - \langle \hat{O}_i \rangle \rangle \langle \hat{O}_j - \langle \hat{O}_j \rangle \rangle = \langle \hat{O}_i \hat{O}_j \rangle - \langle \hat{O}_i \rangle \langle \hat{O}_j \rangle \Rightarrow \tag{4.9}$$

$$C_{i,j}^{\mathrm{MF}} = \langle \psi^1 | \hat{o}_i | \psi^1 \rangle \langle \psi^1 | \hat{o}_j | \psi^1 \rangle - \langle \psi^1 | \hat{o}_i | \psi^1 \rangle \langle \psi^1 | \hat{o}_j | \psi^1 \rangle = 0. \tag{4.10}$$

In conclusion, mean field analyses are fundamental to get an insight of the physics of the system: whenever quantum correlations play a dominant role the mean field result fails to correctly describe the system, and thus calls for more advanced and sophisticated tools which we will introduce in the next sections.

4.1.2 Cluster Mean Field

A first possible correction to the mean field ansatz is cluster mean field. Indeed, whenever it is physically relevant to assume that the system ground state is characterized by short-range correlations, the mean field ansatz of Eq. (4.5) can be generalized to encompass them, thus greatly improving the result precision. A simple approach is to write a mean field ansatz of a cluster of M sites, i.e.,

$$|\psi^{\mathrm{MF}}_{\alpha_1\alpha_2\ldots\alpha_N}\rangle = \prod_{i=1}^{N/M} |\psi_i^M\rangle = \prod_{i=1}^{N/M} \sum_{\beta_i=1}^{d^M} A_{\beta_i} |\beta_i\rangle; \tag{4.11}$$

where the indexes β_i run on all possible states of each cluster. As in the previous section, starting from this ansatz, it is possible to find the cluster mean-field description of the ground state of the system under study. This analysis can also be useful to check the confidence of the results of the mean field analysis: increasing the cluster size and comparing the results, it is possible to estimate the errors and in some cases to extrapolate the results to the exact case $M = N$. Clearly, the exponential dependence of the number of states in Eq. (4.11) makes this approach inefficient as increasing the cluster size is exponentially expensive. However, a simple example might be helpful to clarify the scenarios where the cluster mean field will clearly outperform the mean field description. Let's consider the Hamiltonian

$$H_{XY}^{D} = \sum_{\langle i,j \rangle} h_{i,j} \left(\sigma_i^x \sigma_j^x + \sigma_i^y \sigma_j^y \right), \tag{4.12}$$

where $\langle i, j \rangle$ defines a disjoint cover of a graph (e.g., a two-dimensional square lattice). It is simply possible to solve exactly this problem, diagonalizing each two sites hamiltonian independently: the result is that the two sites form a Bell state (maximally entangled state) among the interacting pairs of sites. The mean-field treatment of the problem will result in the description provided in the previous section, neglecting the correlations present in the system ground state completely. On the contrary, a cluster mean-field expansion with $M = 2$ will result in the exact solution. Beyond this simple scenario, it shall be now clear that cluster mean-field descriptions encompass exactly M-party entanglement while neglecting the entanglement beyond the cluster size M. In conclusion, whenever the M-reduced density matrix is pure M-cluster mean field provides the exact result. Vice versa, it will provide only an approximate description with non-decreasing precision with M while the computational cost increases exponentially with M. We will see in the next sections how this intimate relation between the purity of the reduced density matrix, i.e., the number of populations different from zero, can be made explicit and exploited to build powerful approximation methods beyond mean-field.

4.2 Real-space Renormalization Group

Real-space renormalization group method is an approximation method based on a very powerful physical intuition [23]: the hypothesis that the ground state of a system is composed of low-energy states of the system's (non-interacting) bipartitions. Based on this assumption, it is indeed possible to introduce an algorithm that allows describing the ground state properties of many-body quantum systems with large sizes N, up to the thermodynamical limit corresponding to the fixed point of the renormalization flow [141].

The algorithm proceed as follows:

0. Consider a system composed of N sites that can be studied in an exact numerical way. Build the Hamiltonian $\mathcal{H}_N : \mathbb{C}^{d^N} \to \mathbb{C}^{d^N}$.
1. Diagonalize $\mathcal{H}_N$, finding its eigenvalues and eigenvectors $\mathcal{H}_N = \sum_{i=1}^{d^N} E_i |E_i\rangle \langle E_i|$, where the eigenvalues E_i are in increasing order. Consider the projector onto the lowest m eigenstates $P = \sum_{i=1}^{m} |E_i\rangle\langle E_i|$ which project the Hilbert space on the subspace spanned by the first m low-energy laying eigenstates. Compute the projected Hamiltonian $\tilde{\mathcal{H}}_N = P^\dagger \mathcal{H}_N P$ as well as any other needed operator representation in the projected space, i.e. $\tilde{O} = P^\dagger O P$.

2. Construct the Hamiltonian of a system of size $2N$ using the projected Hamiltonian $\tilde{\mathcal{H}}_N$ for each bipartition and the interaction among them, $\mathcal{H}_{2N} = \tilde{\mathcal{H}}_N \otimes \mathbb{1} + \mathbb{1} \otimes \tilde{\mathcal{H}}_N + \tilde{H}_{int}$. The interaction Hamiltonian can be obtained as $\tilde{H}_{int} = \tilde{A}_N \otimes \tilde{B}_N$ where $\tilde{A}$ ($\tilde{B}$) are the projected operator acting on each system bipartition $\tilde{A} = P^\dagger A P$ ($\tilde{B} = P^\dagger B P$).
3. Repeat the steps 1–2 until the desired system size is reached or convergence to the renormalization group fixed point is achieved. Notice that, at each step of the algorithm, the dimension of the described system is doubled ($N \rightarrow 2N$) while the dimension of the Hamiltonian representation is kept constant to m.

In the last forty years, this elegant and powerful idea has triggered an enormous amount of theoretical and numerical work and allowed us to enhance enormously our understanding of many-body quantum systems [136, 137, 142]. However, the underlying assumptions are not always true and indeed, there are important classes of physical systems that simply violate it. Hereafter, we review one of such examples, to present in detail the algorithm and to highlight the intuition behind the development of yet another powerful algorithm which overcame such issue, the Density Matrix Renormalization Group method introduced in the next section [27].

The problem we consider is one of the simplest problems in quantum mechanics, the free particle in an infinite box potential given by Eq. (3.24): $V(x) = 0$ in an finite interval $[a : b]$, while $V(x)$ is infinite overwise (we set $h = 1$ to simplify the notation). In this example, the scaling quantity is not the number of particles, but the number of points we introduce to discretize the space. Thus, in the first step we set $N = 2$ (two discrete positions, $|-1\rangle, |1\rangle$ in arbitrary units) and write the corresponding Hamiltonian

$$\mathcal{H}_2 = \begin{pmatrix} 2 & -1 \\ -1 & 2 \end{pmatrix}. \tag{4.13}$$

The Hamiltonian above can be easily diagonalized resulting in $E_{1,2} = 1, 3$, with groundstate $|E_1\rangle^{N=2} = (|-1\rangle + |1\rangle)/\sqrt{2}$. As sketched in Fig. 4.1a, this simple result is in agreement with the known solution of the infinite box potential under the assumption of the minimal discretization introduced above [143]. Following step one of the algorithm above, we now project out high-energy levels of the system. However—given there are only two levels—for this very first step, we keep all of them setting $m = 2$ and, consequently, $P = \mathbb{1}$. We then proceed to step two of the algorithm and build the Hamiltonian of the system with $N = 4$, that is,

$$\mathcal{H}_4 = \begin{pmatrix} 2 & -1 & 0 & 0 \\ -1 & 2 & -1 & 0 \\ 0 & -1 & 2 & -1 \\ 0 & 0 & -1 & 2 \end{pmatrix}. \tag{4.14}$$

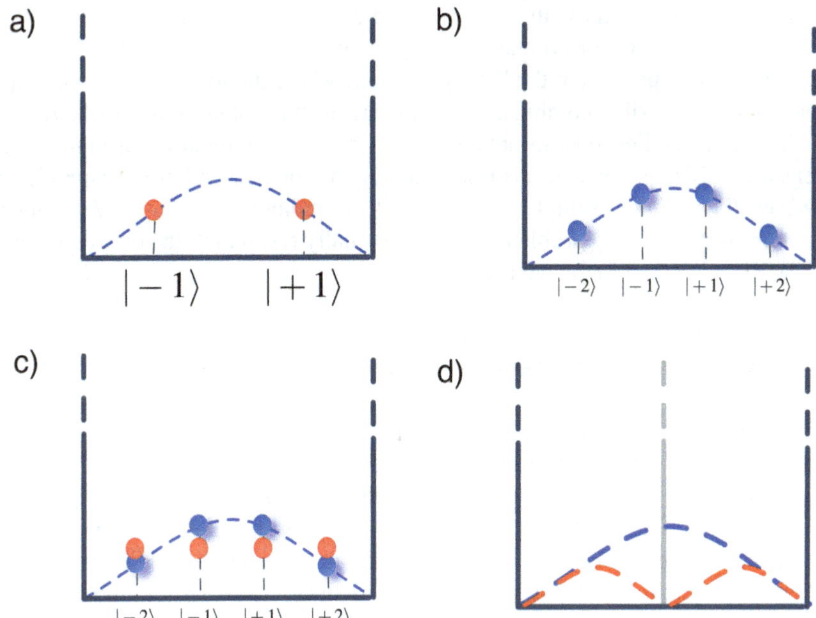

Fig. 4.1 Schematic representation of the infinite box potential (*black lines*) and its solution (*dashed blue line*). (**a**) Discretized amplitudes of probability in a minimal representation of a particle in the box, formed by two states (red circles) and (**b**) a box of double size, formed by four states (*blue circles*). (**c**) Comparison between $|E_1\rangle^{N=4}$ (*blue circles*) and $|E_1\rangle^{N=2} \otimes |E_1\rangle^{N=2}$ (*red circles*). (**d**) Large N solutions $|E_1\rangle^{2N}$ (*blue dashed line*) and $|E_1\rangle^{N} \otimes |E_1\rangle^{N}$ (*red dashed lines*)

This Hamiltonian is still exact due to the fact that previously all states have been kept ($m = 2$). The ground state of Hamiltonian(4.14) is $|E_1\rangle^{N=4} = (b|-2\rangle + a|-1\rangle + a|1\rangle + b|2\rangle)$ with $a \sim 0.3717$, $b \sim 0.6015$ (represented schematically in Fig. 4.1b), while the first excited state is $|E_2\rangle^{N=4} = (-a|-2\rangle + -b|-1\rangle + b|1\rangle + a|2\rangle)$. We keep again $m = 2$ states, and build the effective Hamiltonian which describes the low-energy physics of the system with $N = 4$ with only $m = 2$ states, using the projector $P = |E_1\rangle\langle E_1| + |E_2\rangle\langle E_2|$. We shall also update all relevant operators in the new truncated basis, and repeat the procedure which results in the ground state for a system of double size, and eventually to the thermodynamical limit. Although keeping only two states at each step ($m = 2$) is an extreme approximation, the procedure is correct it will result in an approximate description of the system properties. Increasing the number of kept states m, the approximation can be improved. However, for this precise example, the assumption on which RG is based on is violated, and thus it performs very badly. Indeed, the ground state in the case of $N = 4$ is composed almost only by the tensor product of two ground states of size $N = 2$ as shown in Fig. 4.1c and as can be verified computing the overlap between the two states $\langle E_1{}^{N=4} | E_1{}^{N=2} \otimes E_1{}^{N=2} \rangle \sim 0.97$. However, the overlap decreases rapidly with N: in the large N limit, the ground state of a system of size $2N$ and the tensor product of two ground states of size N differ drastically (see Fig. 4.1d): in particular, the latter has a node where the former has a maximum. Even worst, all possible combinations of the eigenvalues of systems of size N have a node in the middle, thus, to obtain $|E_1\rangle^{2N}$, a large combination of high-energy eigenstates $|E_i\rangle^N$ is needed, falsifying the assumptions on which RG is based on. However, this counterexample has the merit that pointed out that hard boundary conditions can be highly problematic and eventually resulted in the development of the DMRG presented in the next section [27].

4.3 Density Matrix Renormalization Group

The DMRG is a powerful modification of the original RG algorithm, where the truncation rule is improved resulting in a higher precision description of the final state at the price of slowing down the growth of the system size. Indeed, in the DMRG algorithm, the system size increases linearly instead than exponentially with the number of iterations. The original algorithm, developed to describe systems with nearest neighbor interaction at the thermodynamical limit (infinite DMRG) is composed of the following steps:

1. Consider the biggest system size N that can be diagonalized exactly with reasonable resources and regroup the N sites in two single sites in the middle and others in two groups of M sites as in Fig. 4.2a (the reason of this grouping will become clear later on). The system's Hamiltonian can be written as $\mathcal{H}_N = \mathcal{H}_{L+1} + \mathcal{H}_{int} + \mathcal{H}_{R+1}$, where $\mathcal{H}_{L+1}$ and $\mathcal{H}_{R+1}$ in the literature are referred at as the left and right enlarged blocks, and $\mathcal{H}_{int}$ is the interaction among them.

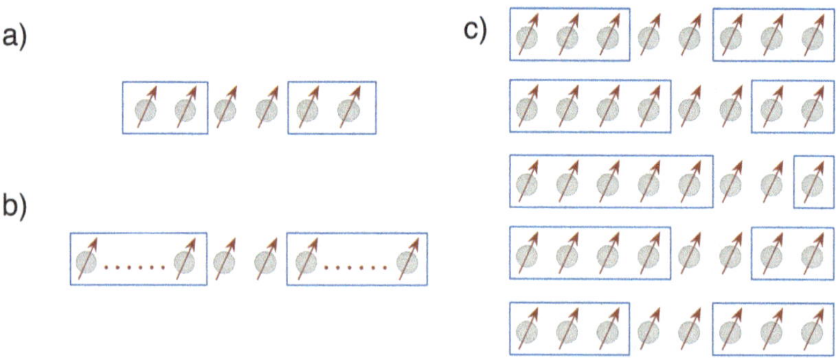

Fig. 4.2 (**a**) Sketch of step 1 of the INFINITE DMRG algorithm: (**a**) construction of the system Hamiltonian as left (right) block and two free sites. (**b**) a system of $2M + 2$ sites is represented by an effective Hamiltonian of dimension md. (**c**) Graphical representation of half-sweep of the FINITE DMRG algorithm: at each iteration, the number of sites included in one block is increased by one, while the other block decreases by one

The dimension of the Hilbert space of the system is $(dm)^2$, where d is the single site physical dimension and $m = d^M$ the dimension of the M grouped sites. The diagonalization of $\mathcal{H}_N$ returns the ground state expressed in the basis

$$|E_1^N\rangle = \sum \psi_{\beta_1\beta_2}|\beta_1\rangle|\beta_2\rangle$$

where $|\beta_i\rangle$ spans the basis of the left and right half sites $\beta_i = 1, \ldots, d^{M+1}$.

2. Compute the density matrix of the ground state

$$\rho_1^N = |E_1^N\rangle\langle E_1^N| = \sum_{\beta_1,\beta_2}\sum_{\beta_1',\beta_2'} \psi_{\beta_1,\beta_2}\psi^*_{\beta_1',\beta_2'}|\beta_1\rangle|\beta_2\rangle\langle\beta_1'|\langle\beta_2'|, \tag{4.15}$$

and the reduced density matrix of one half of the system

$$\rho_L = \mathrm{Tr}_R\rho_1^N = \left(\sum_{\beta_2}\psi_{\beta_1,\beta_2}\psi^*_{\beta_1',\beta_2}\right)|\beta_1\rangle\langle\beta_1'|. \tag{4.16}$$

3. Diagonalize $\rho_L = \sum_{i=1}^{md} w_i|w_i\rangle\langle w_i|$ and order the eigenvalues w_i (being ρ_L a density matrix they are non negative real numbers such that $\sum w_i = 1$) in descending order. If we assume the system to be left-right symmetric we also have that $\rho_L = \rho_R$. Define the projector $P = \sum_{i=1}^{m}|w_i\rangle\langle w_i|$ composed by only the first m eigenvectors of ρ_L.
4. The projector P defines a truncation of the Hilbert space from md to m states that can be used to compute the effective Hamiltonian of the system and all necessary operators in the reduced space, $\tilde{\mathcal{H}}_{L+1} = P^\dagger\mathcal{H}_{L+1}P$ and given that $\mathcal{H}_{int} = \sum_k c_k A_L^k \otimes B_R^k$ (where A_L^k and B_R^k acts on the left and right enlarged block respectively) the interaction term in the projected space can be computed applying the projector P on the left and right separately. We thus obtain an effective matrix describing the Hamiltonian for the system of N sites of dimension m instead than md.
5. The algorithm is iterated starting again from Step 1 provided that the Hamiltonian of the left and right block are replaced with the effective Hamiltonians computed in the previous step. The net effect is that one can describe a system of $N + 2$ sites with a Hamiltonian of size $(md)^2$ as depicted in Fig. 4.2b. At every step the size of the described system is incremented by two sites while keeping the computational resources constant.

The algorithm described above forms the core the DMRG approaches and differs from the RG approaches mainly from the truncation rule: indeed, the choice of keeping the states of the reduced density matrix with higher population can be shown to be equivalent to have the best possible approximated description of the system (in terms of observables and correlations) with the available resources. Indeed, the DMRG truncation rule optimize the overlap between the truncated and the exact system wave functions [144, 145]. For example, the computation of local observables acting on site i (belonging to the left enlarged block, and in general any observables with support only on the left block) for the exact wave function of a system of size N is

$$\langle \hat{O}_i \rangle = \langle \psi | \hat{O}_i | \psi \rangle = \mathrm{Tr}\{\rho_L\} \hat{O}_i = \sum_{i=1}^{md} w_i \langle w_i | \hat{O}_i | w_i \rangle. \tag{4.17}$$

The expectation value after the truncation step is $\langle \hat{O}_i \rangle_{\mathrm{AP}} = \sum_{i=1}^{m} w_i \langle w_i | \hat{O}_i | w_i \rangle$, thus the difference is $|\langle \hat{O}_i \rangle - \langle \hat{O}_i \rangle_{\mathrm{AP}}| \leq \epsilon \cdot C$, where C is an observable-dependent constant and $\epsilon = (1 - \sum_{i=1}^{m} w_i)$ is a non-increasing function with m. It is clear that in this basis, given the descending order of the reduced density populations, any other truncation choice would result in a higher error. In particular, if the matrix populations become exactly equal to zero for some $i > K$ (i.e., K is the Schmidt rank of the density matrix), no error is introduced by the truncation in the computation of the expectation value if $m \geq K$. A particular scenario when this can occur is once more whenever the exact system wave function is mean field, that is, $K = 1$ and thus $m = 1$ is sufficient to describe exactly the system properties. However, it should be clear by now that the DMRG can go beyond that, and exactly describe also systems beyond mean field provided that $m \geq K$. One can show that requiring to optimize the description of the entanglement present in the system (as quantified by the von Neumann entropy $S = \sum_i w_i \log w_i$) results exactly in the DMRG truncation prescription [144].

The considerations above are equivalent to state that the infinite DMRG algorithm results in the best possible approximation of the reduced density matrix of half of the system given the available resources (m states). However, the optimization concern only the bipartition of the system in two equal subsystems: it might be possible to have a better approximation of the global many-body states if we could optimize the reduced density matrix of all possible bipartitions of the system. This refinement procedure is provided by the finite DMRG algorithm, which aims to refine the representation of a system composed of N sites. The algorithm is defined as follows:

1. Build the representation of a system composed of $N = 2M + 2$ sites using the Infinite DMRG algorithm. At each step store the Hamiltonian (and operators) truncated representations $\tilde{\mathcal{H}}_j$, with $j = 1, \ldots, M$.
2. Start a new DMRG iteration taking care when building the new system Hamiltonian to keep the system size fixed to N. That is, $\mathcal{H}_L = \mathcal{H}_{M+2} + \mathcal{H}_{\mathrm{int}} + \mathcal{H}_M$. Notice that the left and right enlarged block now are different and represent a different number of sites. Obtain the truncated representation of the Hamiltonian $\tilde{\mathcal{H}}_{M+2}$.
3. Keep on iterating, increasing the size of the left block and decreasing that of the right block, keeping N constant. When the boundary is reached, reverse the process and keep iterating inverting the role of the left and right block as sketched

in Fig. 4.2c. At each iteration, the energy of the system shall decrease while the precision of the wave function representation increases. Once the free sites are in the middle for the second time, a DMRG *sweep* has been completed: additional sweeps will result in higher precision and improved convergence.

The described algorithms can be straightforwardly be generalized to more general scenarios, such that of Hamiltonians with space-dependent parameters or beyond the nearest neighbor interaction (regrouping physical sites in logical ones of larger size, even though this procedure is inefficient and thus always limited to short-range interactions). Finally, notice that Step 1 of the infinite DMRG algorithm can be modified in the sense that one can start with four sites ($M = 1$) and increase the system size without truncating until $d^M \leq m$. A step-by-step very clear and instructive example of the DMRG algorithm can be found in [146].

4.4 Problems

1. Given a Hamiltonian on a one-dimensional lattice with nearest neighbor interaction of the general form

$$H = \sum_{i,\alpha} h_i^\alpha + \sum_{i,\alpha} \lambda_\alpha h_i^\alpha h_{i+1}^\alpha$$

where $i = 1, \dots N$ runs on the lattice site and α on different operators; consider the mean field translational invariant ansatz

$$|\psi_{MF}\rangle = |\phi\rangle|\phi\rangle|\phi\rangle \dots |\phi\rangle$$

Compute the general mean field approximation of the ground state energy per site and specialize it for the quantum Ising Hamiltonian in transverse field in Eq. (4.4) for different values of λ.

2. Compute the mean field approximation of the ground state energy for the antiferromagnetic Heisenberg model

$$H = \sum_i \sigma_i^x \sigma_{i+1}^x + \sigma_i^y \sigma_{i+1}^y + \sigma_i^z \sigma_{i+1}^z$$

using $|\psi_{MF}\rangle$ and the staggered mean field ansatz

$$|\psi_{MFD}\rangle = |\phi_1\rangle|\phi_2\rangle \dots |\Phi_1\rangle|\Phi_2\rangle.$$

3. Given the quantum Ising Hamiltonian in transverse field of Eq. (4.4) on a one-dimensional lattice with nearest neighbor interaction:

(a) Compute the ground state energy as a function of the transverse field λ by means of the real-space RG.
(b) Compute the ground state energy as a function of λ by means of the infinite DMRG algorithm.
(c) Compare the resulting ground state energy density between them and with the mean field solution.

Tensor Network Methods

5

Timo Felser and Simone Montangero

In this chapter, we introduce the core concepts of tensor network methods: we first define the tensors themselves in Sect. 5.1 and the most common operations for their manipulations (Sect. 5.2) as the fundamental building-blocks. From that on, in Sect. 5.3, we introduce the principles of tensor network states and illustrate the most prominent tensor network geometries. Then, we present some of the most successful algorithms developed exploiting tensor networks to study the many-body problem (Sect. 5.4). The presented material, despite being far from covering the whole fast growing literature dedicated to the field, forms a solid starting point to enter in the field of tensor network methods. We conclude the chapter with a quick overview and pointers to the literature on further tensor network methods recently developed to push beyond our capability of simulating many-body quantum systems in Sect. 5.6.

It is worth mentioning that this chapter provides a comprehensive introduction for a foundational understanding of tensor network methods while selected topics of this chapter are discussed in more technical detail with respect to hands-on implementations and various concrete applications in the subsequent chapters of the book.

5.1 Tensors and Tensor Operations

5.1.1 Tensor Definition and Graphical Representation

We ficts we will use hereafter, and of the standard operations to act on them.

T. Felser (✉)
Tensor AI Solutions GmbH, Pfaffenhofen, Germany
e-mail: timo.felser@tensor-solutions.com

S. Montangero (✉)
Università di Padova, Padova, Italy
e-mail: simone.montangero@unipd.it

T. Felser, S. Montangero (eds.), *Introduction to Tensor Network Methods*,
Graduate Texts in Physics, https://doi.org/10.1007/978-3-032-17635-6_5

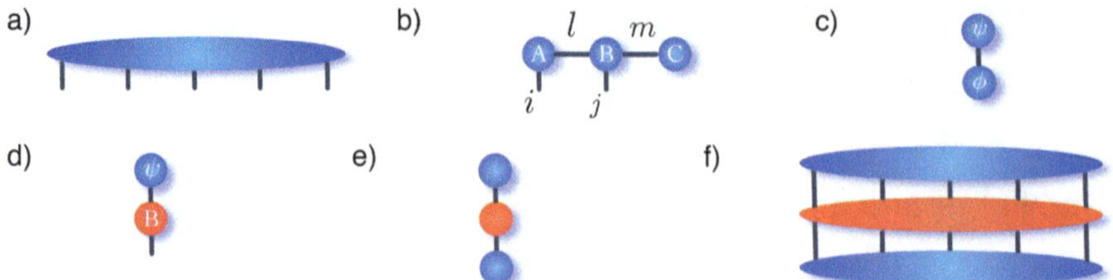

Fig. 5.1 Tensor graphical representations: (**a**) A n-rank tensor. (**b**) A tensor network of three tensors with tensor contraction equivalent to Eq. (5.5). (**c**) Wave functions scalar product $\langle\phi|\psi\rangle$. (**d**) Operator B acting on a wave function ψ. (**e**) Expectation value of an observable (*red tensor*) of the state represented by the wave function encoded in the *blue tensor*. (**f**) As (**e**) but for a N body wave function

Definition 5.1 (Tensor) A Tensor $\mathcal{T} \in C^{d_1 \times d_2 \times \cdots \times d_N}$ is an N-dimensional array with N indices that, in general, has the form

$$\mathcal{T}_{\alpha_1,\ldots,\alpha_N}, \tag{5.1}$$

and can be represented as shown in Fig. 5.1a. The number of indices N of a tensor is also referred to as its *rank* or its *order*.

Computationally, a single tensor $\mathcal{T}$ can be viewed as a generalization of vectors and matrices in linear algebra. In fact, a vector is equivalent to a rank-1-tensor, where the elements are addressed by only one index i_1. Moreover, a matrix forms a tensor of rank 2 with two indices i_1 and i_2 for its rows and columns, respectively. Equivalently, a tensor of order 3 can be considered as a three-dimensional cube of numbers with one index for each dimension. This concept of representing information can be extended to arbitrary higher orders of tensors. The Table 5.1 below reports the most common names of low rank tensors and typical examples.

For reasons of compactness, a graphical representation for the mathematical tensor notation was introduced for Tensor Networks: We represent tensors by circles (or similar closed objects) and the indices by lines or links that can connect two tensors with eachother. Each link connecting two different tensors shows a *contraction* over a matching index as explained in more detail in Sect. 5.2.1.

Table 5.1 Definitions of different rank-N tensors

Name	Example	Rank	Notation
Scalar	Constant	0	λ
Vector	Wave function	1	ψ_i
Matrix	Operator	2	$O_{i,j}$
$\vdots$	$\vdots$	$\vdots$	$\vdots$
rank-N tensor	N-body wave function	N	$\psi_{\alpha_1,\ldots,\alpha_N}$

5.1.2 Special Tensors

In practical applications, Tensor Networks use or consist of special tensors that fulfill certain mathematical constrains. The main ones commonly used in various Tensor Network computations are *isometries* and *disentanglers* which can be seen as generalisations of unitary matrices to higher order tensors.

5.1.2.1 Isometries

The *isometry* is frequently used in a large number of Tensor Network computations and is vital for many Tensor Network geometries to ensure an efficient implementation. Equally to $UU^\dagger = \mathbb{1}$ for a unitary matrix U, an isometry U becomes an identity when it is contracted with its complex conjugate:

$$\sum_{\{i_r\}\setminus i_k} (\mathcal{T})_{i_1,\ldots,i_R} (\mathcal{T}^\dagger)_{i'_1,\ldots,i'_R} = \delta_{i_k,i'_k}. \tag{5.2}$$

In this case, the tensor would be an *isometry* with respect to the link i_k. In general, an arbitrary tensor in a tensor network can be *isometrised* towards a link i_k or even a set of links $\{i_k\}$ by performing *QR decompositions* as described later on in Sect. 5.2.4. Further, we will see in more detail in Chap. 8 that a Tree Tensor Network is frequently using order-3-isometry tensors $\mathcal{T}_{[i_1,i_2,i_3]}$ for efficient computations.

5.1.2.2 Disentangler

A *disentangler*, in contrast, is a special kind of order-4-isometry tensor. In fact, a disentangler u_k can be viewed as a generalization of a unitary matrix U to a tensor of order 4, obeying the unitary condition:

$$\sum_{i_3,i_4} (u)^{i_1,i_2}_{i_3,i_4} (u^\dagger)^{i_3,i_4}_{i'_1,i'_2} = \delta_{i_1,i'_1} \delta_{i_2,i'_2}. \tag{5.3}$$

Thus, a disentangler u_k performs a unitary operation on two links i_1 and i_2. This particular isometry is a key element of some Tensor Network geometries, such as the MERA [147, 148], which we will show later on in Sect. 5.3 and the aTTN [149, 150], which is explained in detail in Chap. 9. In both applications, these disentanglers are intended to locally decouple, or *disentangle*, relevant degrees of freedom in a quantum system computation.

5.2 Tensor Manipulations

In this section, we introduce the basics operations on tensors that will be needed throughout the rest of the book to describe faithfully and efficiently the physical system of interests: they allow us to manipulate, compress and transform the entries of the tensors within the tensor networks.

5.2.1 Tensor Contraction

The tensor contraction is one of the most fundamental operations on tensors, especially since each tensor network in general is a network composed by different tensors connected via the contraction of different indexes (see Sect. 5.3 for more details). As the natural generalization of the standard matrix-matrix-multiplication, the tensor contraction sums over coinciding indices of two tensors following

$$C_{l,n} = \sum_{m} A_{l,m} B_{m,n} \equiv A_{l,m} B_{m,n} \tag{5.4}$$

where in the rightmost equality we introduced Einstein's notation: repeated indexes are implicitly summed. The key difference for the tensor algebra with respect to the matrix-multiplication is that each tensor A, B and C in Eq. (5.4) might now have several indices attached, so that $l = (l_1, ..., l_\lambda), m = (m_1, ..., m_\mu)$ and $n = (n_1, ..., n_\nu)$. Consequently, in this tensor algebra, the matrix-multiplication constitutes a special tensor contraction of two rank-2 tensors over one single coinciding link m. Furthermore, the contraction depicted in Fig. 5.1b is equivalent to

$$\sum_{l,m} A_{i,l} B_{l,j,m} C_m \equiv A_{i,l} B_{l,j,m} C_m. \tag{5.5}$$

The result here is a tensor with two indexes (a matrix) as all others have been contracted. Similarly, the computation of the scalar product between two wave functions, that is,

$$\langle \phi | \psi \rangle = \sum_{i} \phi_i^* \psi_i = \phi_i^* \psi_i, \tag{5.6}$$

can be depicted as shown in Fig. 5.1c, where we introduce the convention that tensors with indexes pointing upwards are the adjoint of the same tensor with indexes pointing downwards. Notice that the resulting object has no indexes (all indexes are contracted) and thus it is a scalar, as it shall be. The application of an operator to a wave function and the computation of an expectation value are similarly depicted in Fig. 5.1d and e respectively. Again, a check of the correctness of the results can be obtained by counting the final number of free indexes: it should match the expected one for the computed quantity. Notice that computing the scalar product of two N-body wave functions written in a product basis $(|\alpha_1\rangle \otimes |\alpha_2\rangle \otimes \cdots \otimes |\alpha_N\rangle)$ does not differ from what presented in Eq. (5.6). Indeed, in this case we have

$$\langle \phi_N | \psi_N \rangle = \phi^*_{\alpha_1,...,\alpha_N} \psi_{\alpha_1,...,\alpha_N} = \phi^*_{\mathbf{i}} \psi_{\mathbf{i}}, \tag{5.7}$$

where we have introduced the notion of global or fused index, that is, the operation of grouping and splitting indexes (introduced in the next section). Finally, one can recast operator operations in similar forms, e.g. the computation of the energy expectation value can be clearly written as

$$\langle\psi_N|\hat{H}|\psi_N\rangle = \psi^{*\beta_1,\ldots,\beta_N} H^{\alpha_1,\ldots,\alpha_N}_{\beta_1,\ldots,\beta_N} \psi_{\alpha_1,\ldots,\alpha_N} = \psi^{*\mathbf{i}} H^{\mathbf{j}}_{\mathbf{i}} \psi_{\mathbf{j}}. \tag{5.8}$$

We introduced here the notion of up (superscript) and down (subscript) indices for the sake of compactness and in alignment with the link direction in Fig.5.1f.

5.2.2 Tensor Contraction Complexity

Hereafter, we show in more details the computational costs of tensor contractions and why it is always important to work with tensors with minimal rank. Indeed, starting from the wave function scalar product depicted in Fig. 5.1c and defined in Eq. (5.6), it is straightforward to see that its computation requires $O(m)$ operations, where m is the dimension of the Hilbert space the wave function lives in, that is, the dimension of the index j. A slightly more complex contraction is represented in Fig. 5.1b and expressed in Eq. (5.5). A simple inspection returns the number of operations needed to compute it: for each element of the resulting matrix (two free indices) two nested sums shall be computed, resulting in an overall $O(m^4)$ operations (assuming all indices have dimension m). From these examples, it is easier to extract a general rule that can be used to compute the number of operations needed to perform a tensor network contraction: *the contraction complexity is given by the product of the dimensions of the free indices and the contracted ones*. Indeed, the product of the free indexes gives the number of coefficients of the resulting tensor to be computed, while the product of the dimensions of the contracted indices gives the order of the number of operations that are necessary to compute each of them. We shall mention that this estimate, although it is very useful to guide algorithm developments, in practice it gives only an estimation of the contraction scaling. Indeed, it is possible to decrease it by using advanced linear algebra manipulation algorithms, e.g., divide-and-conquer schemes for matrix-matrix manipulations [124]: For example, the Strassen algorithm for matrix-matrix multiplication provides a scaling of approximatively $O(n^{2.8})$ instead of $O(n^3)$ [151].

In conclusion, the rule above results in an exponential algorithmic complexity of any operation performed on an N-body wave function. Indeed, the computation of the scalar product in Eq. (5.7) is $O(m^N)$ either if performed as the contraction of two rank-N tensors or as viewed as a contraction of two rank-1 tensor (as in the latter case the dimension of the contracted index is m^N). As we will see in the following sections and in particular in Sect. 7.1.3, the introduction of tensor networks reduces in many cases the contraction complexity from exponential to a polynomial function, practically allowing the manipulation N-body wave functions for systems composed of hundreds of sites.

5.2.3 Index Fusion and Splitting

Tensors are nothing more than a structured way to organize information, that is, an ordered collection of numbers or variables. For example, to describe three dimensional relativists systems, the four space-time variables x, y, z, and t, typically are organized in a single vector to improve our representation and manipulation capabilities. However, if it turns out to be more convenient, one could arrange them in a 2×2 matrix. As we will see, the recasting of vectors in matrices and the generalizations of this operation, are indeed very useful in different settings. In general, a rank-n tensor can be rearranged in a tensor of a different rank following some given (invertible) rule: in the example given above one has

$$X_i = \begin{pmatrix} x \\ y \\ z \\ t \end{pmatrix} \Leftrightarrow O_{j,k} = \begin{pmatrix} x & y \\ z & t \end{pmatrix}; \tag{5.9}$$

where we have changed the way we organize the variables of interest, and no information is lost. Clearly, also the operations acting on the objects shall be changed accordingly. The transformation performed in Eq. (5.9), if applied to wave functions, is known as Liouville representation and, more formally, as Choi-Jamiolkowski isomorphism [152–157]. Notice that in transforming a matrix in a vector, we fuse two indexes in a single one. This index fusion operation can be generalized to any number of indexes simply as

$$\alpha_1 \ldots \alpha_n \equiv \mathrm{d} - \mathrm{coding}(\mathbf{i}); \tag{5.10}$$

where $\alpha_i = 1, 2, \ldots, d$ and the transformation is given by the d-nary coding of the global index $\mathbf{i}$. From now on, we will indicate the fusion operation as $\alpha_1 \ldots \alpha_n \succ \mathbf{i}$. The inverse operation is the splitting of indexes, $\mathbf{i} \prec \alpha_1 \ldots \alpha_n$. Finally, note that with the above definition, the last equality in Eq. (5.7) automatically holds. Similarly, the relation between the graphical representations in Fig. 5.1e and f is the operation of fusing all indexes into a single one.

5.2.4 Factorisation and Compression

The fusion operation allows one to recast any n-rank tensor in a matrix form. Indeed, it is always possible to choose a bipartition of n indexes in two disjoint groups and fuse together the indexes belonging to each group. That is,

$$T_{\alpha_1,\ldots,\alpha_N} \equiv T_{\mathbf{i},\mathbf{j}}, \tag{5.11}$$

where $\{\alpha_k, \alpha_j, ...\alpha_l\} \succ \mathbf{i}$, $\{\alpha_m, \alpha_n, ...\alpha_p\} \succ \mathbf{j}$, and $\{\alpha_k, \alpha_j...\alpha_l\} \cup \{\alpha_m, \alpha_n...\alpha_p\} = \{\alpha_1, \alpha_2, \ldots, \alpha_N\}$. Given that the tensor is now recast in matrix form, one can exploit the whole potential of linear algebra operations acting on matrices.

This allows us to factorise or decompose a given tensor using linear algebra operations, such as the *Singular Value Decomosition* (SVD) or the *QR-decomposition*. Thus, for properly factorizing a tensor, we transform it into a matrix, decompose the matrix and subsequently split the previously fused links back to their original form. Here, the QR-decomposition is a handy tool no only for decomposing tensors but also for creating isometries. Since it decomposes $T_{\mathbf{i},\mathbf{j}} = Q_{\mathbf{i},\mathbf{k}} R_{k,j}$ where Q refers to an orthogonal matrix with $Q^\dagger Q = \mathbb{1}$, the resulting tensor $Q_{\alpha_1,\ldots,\alpha_N}$ is an isometry after splitting the link $\mathbf{i}$ back and fulfills the unitary condition of Eq. (5.2).

The SVD, on the other hand—being the generalization of matrix diagonalization—is a key tool in practical applications of Tensor Networks. It allows to copmute entanglement properties in loop-free networks, such as the TTN which is described in Chap. 8 in more detail. In addition, the SVD can be used to compress a tensor or even a complete tensor network efficiently as we see in what follows. Indeed, given the tensor $T_{\mathbf{i},\mathbf{j}}$ it is always possible to decompose it in three tensors via the SVD, such that,

$$T_{\mathbf{i},\mathbf{j}} = \sum_{\mathbf{k}=1}^{d^N} S_{\mathbf{i},\mathbf{k}} V_{\mathbf{k},\mathbf{k}} D_{\mathbf{k},\mathbf{j}}, \tag{5.12}$$

where V is a diagonal and positive real matrix, while S and D are unitary matrices. The diagonal elements $V_{j,j} \equiv \lambda_j$ of the tensor V are the singular values of T which are real and positive, and thus they are bounded from below by zero and can be ordered. From now on, we assume them to be ordered in descending order, that is, $V_{\mathbf{1},\mathbf{1}} \equiv \lambda_1 > V_{\mathbf{2},\mathbf{2}} \equiv \lambda_2 > V_{\mathbf{3},\mathbf{3}} \equiv \lambda_3 \cdots > 0$. A central role in tensor network algorithms is played by the fact that, if $\lambda_j < \epsilon$, $\forall\, j > z$, it is possible to disregard some of the singular values λ_j. That is, whenever the singular values decay fast enough (or are exactly zero after some given value $\lambda_z = 0$) we can neglect them. Correspondingly, the matrices S, V can be truncated. This reduction introduces an error in the representation of the original tensor $T_{\mathbf{i},\mathbf{j}}$ which can be estimated in

$$||T_{\mathbf{i},\mathbf{j}} - \sum_{\mathbf{k}=1}^{\mathbf{m}} S_{\mathbf{i},\mathbf{k}} \lambda_{\mathbf{k}} D_{\mathbf{k},\mathbf{j}}|| = ||\sum_{\mathbf{k}=\mathbf{m}+1}^{d^N} S_{\mathbf{i},\mathbf{k}} \lambda_{\mathbf{k}} D_{\mathbf{k},\mathbf{j}}|| < \sum_{\mathbf{k}=\mathbf{m}+1}^{d^N} ||S_{\mathbf{i},\mathbf{k}} \epsilon D_{\mathbf{k},\mathbf{j}}|| < \epsilon\, \mathrm{C} \tag{5.13}$$

where C is a finite constant that can be estimated in terms of some matrix norm $||.||$ and their dimensions. In fact, with respect to the Frobenius norm $||T||_F^2 := \sum_{ij} |T_{ij}|^2$, the introduced error ζ by such a truncation equals the sum of all squared singular values that have been discarded

$$\zeta = \sum_{k=m+1}^{d^N} \lambda_k^2 \tag{5.14}$$

In particular, if ϵ—and thereby ζ—drops to zero (or equivalently to its numerical representation, typically of the order of 10^{-15} when working in double precision) truncating the tensor dimension to m introduces no error, drastically reducing the computational resources needed to describe the system of interest. It can be shown that in any case the SVD allows the best possible approximation of $T_{\mathbf{i},\mathbf{j}}$ by a compressed $\tilde{T}_{\mathbf{i},\mathbf{j}}$ in a smaller space (i.e. with a lower matrix rank) with respect to the Frobenius norm. Since this norm in Tensor Networks is linked to physical quantities, such as to the norm of the represented quantum many-body state, this compression technique allows us to efficiently and controllably reduce computational costs while keeping the most amount of relevant information in the computations.

5.3 Tensor Network States and Operations

As described above, a Tensor Network is a network composed by different tensors connected via the contraction of different indexes. Thereby, Tensor Networks decompose an exponentially large, N-dimensional Tensor into a set of smaller interconnected tensors that can be controlled and manipulated separately. In this way, we can represent a quantum many-body state $|\Psi\rangle$ as introduced in Eq. (4.1)—recalling that $|\Psi\rangle$ can be written as an N-dimensional tensor with the open links corresponding to the local Hilbert spaces of the system (see Fig. 5.2). The exact decomposition of such a tensor into a Tensor Network, i.e. connection of the single tensors in the network, can be realised in various ways giving rise to different *Tensor Network geometries* as we will describe later on in more detail. However, we can see each Tensor Network geometry similar to a mathematical graph with the key difference that a Tensor Networks can have open links, i.e. the *physical links*. In this

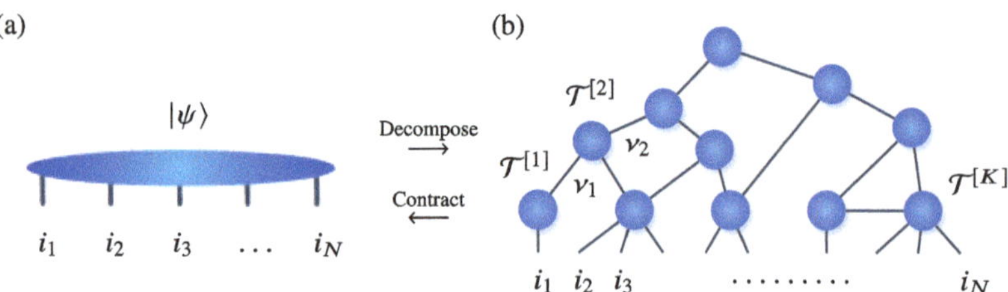

Fig. 5.2 A high-dimensional (rank-N) tensor $|\Psi\rangle$ provides an exact description of a quantum many-body wavefunction, encompassing all exponentially numerous coefficients (**a**). A possible Tensor Network representation of the wavefunction (**b**). The dimension of these bond-links ν_j connecting the tensors $\mathcal{T}^{[k]}$ can be adjusted to compress the information encapsulated within the network

context of *graph theory*, we can formally define a Tensor Network similar to a graph using the Definition from Ref. [150]:

Definition 5.3.1 (Tensor Network)
TN A Tensor Network shall be defined, as a tuple $\mathcal{G} = (\mathfrak{T}, \mathfrak{N})$ consisting of a set $\mathfrak{T} = \{\mathcal{T}\}$ of two or more tensors and a set $\mathfrak{N} = \{\nu\}$ of corresponding links. The set $\mathfrak{N}$ is further divided into *internal* links $\{\chi\}$ connecting the tensors with each other and *external* or *physical* links $\{i\}$ which are connected to one tensor only, thus having one open ending each.
Further, we denote:

- The *size* of the network K shall be defined as the total number of tensors within the tuple $\mathcal{G}$, i.e. the size $\mathcal{G}$ or $\mathfrak{T}$ respectively.
- The total number of physical links N, i.e. the size of $\{i\}$, is the *system size* [or *physical size*].
- The typical literature introduces $d_k = \dim i_k$ as the dimension of the physical link i_k. Depending on the network and description, the dimension of an internal link is typically referred to as χ, m, or D. [...]
- Every Tensor Network should, in theory, allow to be contracted to a single tensor $\Psi_{\{i\}}$ when contracting the constituting tensors over all internal links $\{\chi\}$, following

$$\Psi_{i_1,\dots,i_N} = \sum_{\chi_1=1}^{m_1} \sum_{\chi_2=1}^{m_2} \cdots \sum_{\chi_N=1}^{m_N} \left(\prod_{\kappa=1}^{K} \mathcal{T}^{[\kappa]}_{\{i\}_\kappa,\{\chi\}_\kappa} \right), \tag{5.15}$$

  where $\mathcal{T}^{[\kappa]}_{\{i\}_\kappa,\{\chi\}_\kappa}$ is the κ-th tensor in the set $\mathfrak{T}$ with the links $\{i\}_\kappa$ and $\{\chi\}_\kappa$ denoting the set of all physical and internal links respectively attached to the κ-th tensor. Noting that each internal link χ_κ occurs twice within the set of tensors $\mathfrak{T}$ and shall be contracted over, while each physical link occurs only once within $\mathfrak{T}$.
- Finally, we define the maximum dimension of all internal links as the *bond-dimension* of the network $m = \max_{\kappa=1,\dots,K}\{\dim \chi_\kappa\}$.

In what follows, we describe in more detail typical operations performed on tensor networks.

5.3.1 Tensor Network Differentiation

In many cases, we shall perform an optimization with respect to a given figure of merit of the entries of the tensors within a tensor network, for example when minimizing the expectation value of the energy to find a tensor network representation of the system ground state. Thus, the ability to extremize functions of the tensor network's coefficients is fundamental. The computation of the mean energy expressed in Eq. (5.8) is a quadratic function of two N-rank tensors:

$$\langle \psi_N | \hat{H} | \psi_N \rangle = \mathcal{H}(\psi^*_{\beta_1,\ldots,\beta_N}, \psi_{\alpha_1,\ldots,\alpha_N}) \equiv \mathcal{H}(\psi^*_{\mathbf{i}}, \psi_{\mathbf{j}}), \tag{5.16}$$

where we recall that being complex valued, $\psi^*_{\mathbf{i}}$ and $\psi_{\mathbf{j}}$ are independent variables as in complex analysis [158]. Thus, to find the extrema of the function $\mathcal{H}$ with respect to the entries of the tensors, the differential of the function shall be set to zero,

$$\frac{\partial \mathcal{H}}{\partial \psi_{\mathbf{j}}} \stackrel{!}{=} 0; \tag{5.17}$$

which given the expression in Eq. (5.8), is equivalent to the condition $\psi^*_{\mathbf{i}} H^{\mathbf{i}}_{\mathbf{j}} = 0$ as the variable appears linearly in the function.

In conclusion, not surprisingly, the derivative of any linear function of tensors (any function defined via a tensor network where each tensor appears only once) with respect to a particular tensor A, is given by the tensor network where the tensor A has been removed. The graphical representation of an example of such computation is reported in Fig. 5.3b.

5.3.2 Gauging

An operation that plays a prominent role in tensor network algorithms is the gauging, that is, to change the tensors entries by means of local transformations, without changing the global state to exploit some desired properties. Indeed, whenever two tensors are contracted, it is possible *to insert an identity operator between them* without changing the overall tensor network properties. Using the equivalence $\mathbb{1} = U^\dagger U$, that holds by definition for any unitary operator U, the

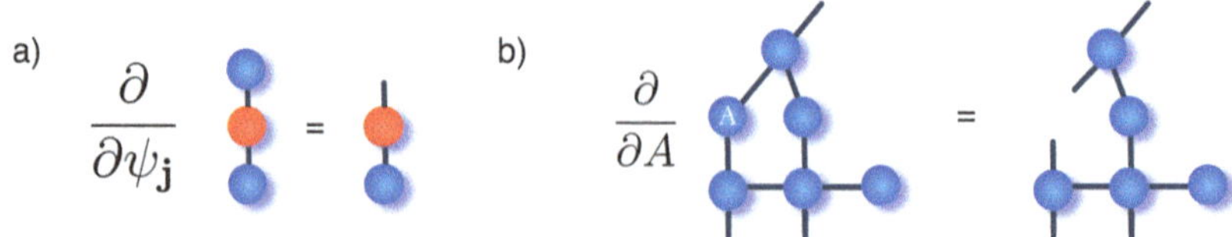

Fig. 5.3 Graphical representation of (**a**) the derivation of Eq. (5.17); (**b**) a generic partial derivation with respect to a tensor A embedded in a tensor network

tensor entries can be changed as

$$A_{\mathbf{i}} B_{\mathbf{i}} = A_{\mathbf{i}} \mathbb{1}_{\mathbf{i},\mathbf{j}} B_{\mathbf{j}} = A_{\mathbf{i}} U^{\dagger}_{\mathbf{i},\mathbf{k}} U_{\mathbf{k},\mathbf{j}} B_{\mathbf{j}} = \tilde{A}_{\mathbf{k}} \tilde{B}_{\mathbf{k}}; \tag{5.18}$$

with the net result that the tensor structure remains unchanged, but the tensors entries have changed and thus can satisfy some desired properties. Clearly, this freedom can be exploited at the level of the fused indices $\mathbf{i}, \mathbf{j}$ as well as at each single indexes α_j independently.

To avoid a possible confusion, we shall warn the reader that in the following (see Sect. 6.4) we will introduce gauge invariant tensor networks: despite the similar name and the clear connection (in both cases we exploit invariant properties of the tensor network under the action of local transformations), the two constructions are in general different and have different physical origins. Here, we are exploiting the invariance under transformations on auxiliary indexes, in the latter the invariance arises from the symmetries of the theory one is studying, and the gauge transformations act on the physical indexes.

Another powerful manipulation tool is the polar decomposition, that is, the property that every matrix can be decomposed in a unitary and nonunitary part (geometrically, any linear operation can be decomposed in a rotation and a scaling), that is:

$$A_{\mathbf{i},\mathbf{j}} = U_{\mathbf{i},\mathbf{k}} P_{\mathbf{k},\mathbf{j}}, \tag{5.19}$$

where U is a unitary matrix, and P a positive-semidefinite matrix.

As we will see, these operations allow to drastically reduce the computation time of some global contractions. For example, we will show that the computation of the norm of large tensor networks can be reduced to a simple small vector norm computation, exploiting the properties of unitary matrices.

5.3.3 Area Law and Entanglement Properties

The section discusses the impact of quantum entanglement on the representation of many-body states as tensor networks. While a general quantum wavefunction resides in an exponentially growing vector space, making an exact representation with TNs unfeasible, physical quantum states often adhere to entanglement bounds known as *area laws*. In contrast to the naive expectation that entanglement scales with the volume of the system, as well referred to as *volume law*, empirical evidence reveals that many physical states, especially in higher-dimensions, follow more restrictive upper entanglement boundaries [38, 159–163].

The concept of entanglement entropy S, measured by the Von Neumann entropy, is introduced as a means to quantify the amount of entanglement between two bipartitions of a quantum system. In that frame, the entanglement entropy of quantum states obeying an area law is expected to increase proportionally to the

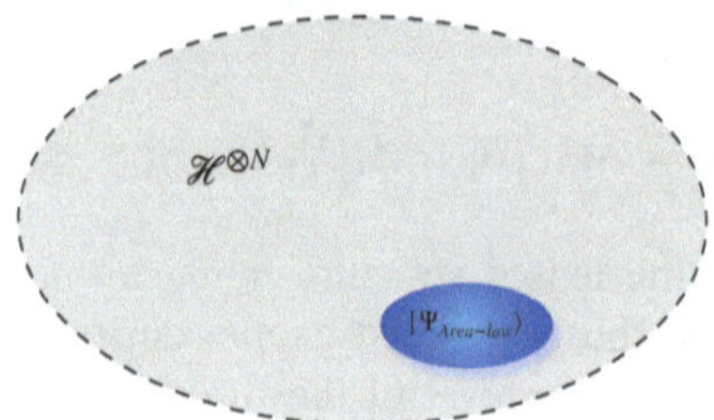

Fig. 5.4 The quantum many-body Hilbert space encompasses the entire spectrum of pure states represented by $|\Psi\rangle$. Within this exponentially large space, the states $|\psi_{Area-Law}\rangle$, which adhere to an area law scaling of entanglement entropy, occupy a small region within the overall manifold

boundary area of the bipartitions rather than the volume of the bipartitions, such that

$$S = O(\partial\mathcal{A}) \left(\sim L^{D-1}\right)$$

(the latter bracket stands for a D-dimensional system with equal dimensions L, i.e. $N = \prod_{i=1}^{D} L_i \overset{(L_i=L)}{=} L^D$). From an information theory perspective, where entropy is used as a measure of the information content inherent in a system [164–166], these area laws state that the relevant information to describe the physical state lies within a relatively small corner of the exponentially large Hilbert space (compare Fig. 5.4). Therefore, area laws of a quantum system often enable efficient decomposition of states into Tensor Network structures with minimal loss.

In TN representations the Von Neumann entropy can be assessed using the *Schmidt decomposition* [166] to formally decompose the TN-state $|\Psi\rangle$ into two bipartitions $|\Psi^{[\mathcal{A}]}\rangle$ and $|\Psi^{[\mathcal{B}]}\rangle$. In fact, every bipartite quantum state $|\Psi\rangle \in \mathcal{H}_{\mathcal{A}} \otimes \mathcal{H}_{\mathcal{B}}$ can be expressed through the Schmidt decomposition as follows:

$$|\Psi\rangle = \sum_{\alpha=1}^{\chi} \lambda_{\alpha}^{[\mathcal{A},\mathcal{B}]} |\Psi^{[\mathcal{A}]}\rangle \otimes |\Psi^{[\mathcal{B}]}\rangle \, ,$$

where $\lambda_{\alpha}^{[\mathcal{A},\mathcal{B}]}$ represents the *Schmidt-coefficients*, and $|\Psi^{[\mathcal{A}]}\rangle$ and $|\Psi^{[\mathcal{B}]}\rangle$ are orthonormal bases, also known as *Schmidt bases*, corresponding to the Hilbert spaces $\mathcal{H}_{\mathcal{A}}$ and $\mathcal{H}_{\mathcal{B}}$ of the bipartition $\mathcal{A}$ and $\mathcal{B}$ respectively. The entanglement entropy S between the bipartitions $\mathcal{A}$ and $\mathcal{B}$ is now given by the Schmidt coefficients as

$$S = -\sum_{\alpha=1}^{\chi} |\lambda_{\alpha}^{[\mathcal{A},\mathcal{B}]}|^2 \log\{|\lambda_{\alpha}^{[\mathcal{A},\mathcal{B}]}|^2\} \, .$$

In practical applications, the Schmidt decomposition can be carried out using a Singular Value Decomposition (SVD). The quantum state $|\Psi\rangle$ is first represented

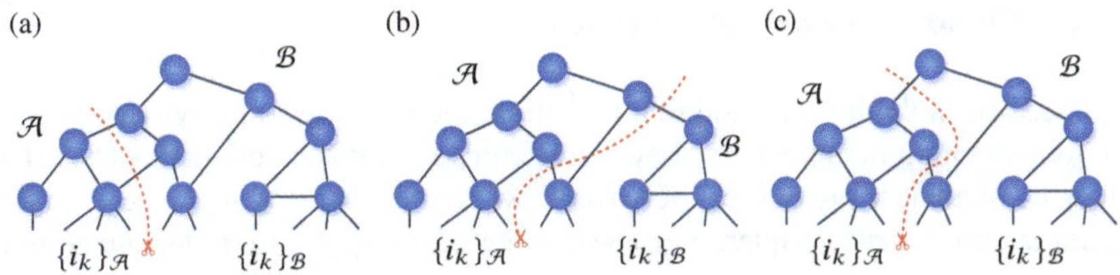

Fig. 5.5 Various tensor network bipartitions illustrating the links $\{i_k\}_{\mathcal{A}}$ (depicted in blue) and $\{i_k\}_{\mathcal{B}}$ (shown in orange), corresponding to the physical sites of the bipartitions $\mathcal{A}$ and $\mathcal{B}$ respectively. Each link cut provides an entropy contribution, capped at $\log m_{\ltimes}$, constrained by its bond dimension $m_{\ltimes}$

as a matrix $\Psi = \sum_{\{i\}} \psi_{(i_1,\ldots,i_k),(i_{k+1},\ldots,i_N)} |i_1, \ldots, i_k\rangle_{\mathcal{A}} |i_{k+1}, \ldots, i_N\rangle_{\mathcal{B}}$, with respect to the two bipartitions, and subsequently decomposed into $\Psi = U\sigma V^{\dagger}$. Here, U and V (with $U^{\dagger}U = \mathbb{1}$ and $VV^{\dagger} = \mathbb{1}$) span the orthonormal basis, while the non-zero, normalized singular values of σ are the Schmidt coefficients $\lambda_{\alpha}^{[\mathcal{A},\mathcal{B}]}$. Consequently, the minimal entropy $S = 0$ occurs when there is only one non-zero singular value ($\lambda_1 = 1$), signifying complete separability between the two bipartitions. Conversely, a higher $S > 0$ indicates increasing entanglement. Notably, the Von Neumann entropy is bounded by $S \leq \log \chi$.

Now, let Ψ denote a Tensor Network representation (as defined in Definition 5.3.1) with physical links $\{i_k\}$. The Subset $\{i_k\}_{\mathcal{A}}$ ($\{i_k\}_{\mathcal{B}}$) addresses the physical sites within the bipartition $\mathcal{A}$ ($\mathcal{B}$) (see Fig. 5.5). Bipartiting the TN state through cutting its internal links $\{\chi_{\ltimes}\}$ yields the following inequality with respect to the Von Neumann entanglement entropy:

$$S_{TN}^{[\mathcal{A},\mathcal{B}]} \leq \min_{\{\ltimes\}} \left[\sum_{\ltimes} \log m_{\ltimes} \right], \tag{5.20}$$

where $m_{\ltimes}$ represents the bond dimension of link $\chi_{\ltimes}$. This process breaks the network into sub-Tensor Networks, with each broken link contributing a maximum entropy of $\log m_{\ltimes}$ (bounded by its bond dimension $m_{\ltimes}$). The minimum over all possible cuts gives an upper bound for the network's entanglement entropy. Thus, a network with bond dimension $m = 1$ implies $S_{TN}^{[\mathcal{A},\mathcal{B}]} = 0$, indicating a product state without entanglement, aligning with mean-field theory. Increasing m allows the Tensor Network to faithfully represent states following the entanglement bounds in Eq. (5.20). Balancing computational efficiency and accurate representation involves scaling m or the number of cut links according to these entanglement bounds. Achieving this balance in higher dimensions poses a significant challenge for Tensor Network geometries.

5.3.4 Tensor Network Geometries

As described in the introduction to this section, there are several ways to represent a certain quantum many-body state with different Tensor Network geometries. The following shall just provide a brief overview over the most common geometries while the subsequent Chapters introduce some of them in a more technical note. In particular, Chap. 7 introduces the technical implementation of Matrix Product States (MPS) and different algorithms for it; Chap. 8 illustrates the hands-on implementation of Tree Tensor Networks (TTN); and Chap. 9 focuses on the implementation of the augmented TTN (aTTN).

In essence, for one dimensional systems, Matrix Product States (MPS) serve as the well-established Tensor Network structure for tackling equilibrium and, in many instances, out-of-equilibrium problems. On the other hand, the Multi-scale Entanglement Renormalization Ansatz (MERA) offers an efficient representation tailored especially for critical 1D systems [45, 147, 168, 169]. However, the development of Tensor Network algorithms for two or more dimensions is an ongoing endeavor, as achieving a balance between precision and scalability remains a significant challenge in simulating high-dimensional systems using Tensor Network methods [170, 171]. Notable Tensor Network representations for higher dimensions include Projected Entangled Pair States (PEPS) [172–176], mainly designed for two-dimensional systems, and Tree Tensor Networks (TTN) [42, 170, 177–179, 282, 353], which, in principle, can be defined for systems of any dimensionality.

5.3.4.1 Mean Field as Tensor Networks

As introduced in Sect. 4.1, the mean field approach is used to simplify the description of complex quantum systems by approximating the interactions between particles or quantum degrees of freedom. In this approach, each quantum particle or degree of freedom is treated as if it experiences an average or mean interaction field due to the presence of all other particles. This approach can be reinterpreted as a tensor network algorithm: We can model the (single-body) mean field approximation introduced in Sect. 4.3 as a tensor network composed by the product of N single rank tensors for each quantum particle, i.e. physical site. The task is then to find the entries of the tensors that minimize the expectation value of the energy computed within the mean-field ansatz. In Chap. 7, we describe this approach of mean field calculation as tensor network in more detail leading up to the technical introduction of MPS as a generalisation of the mean field approach.

5.3.4.2 Matrix Product States

The Matrix Product State (MPS) stands as one of the most successful tensor network frameworks, initially introduced by Ostlund and Rommer [47] and shown to be equivalent to a DMRG description [25]. It has found routine application in the study of one-dimensional quantum systems, both in equilibrium and out-of-equilibrium scenarios. As an extension of the mean-field ansatz, the MPS incorporates additional links between different local sites to represent correlations

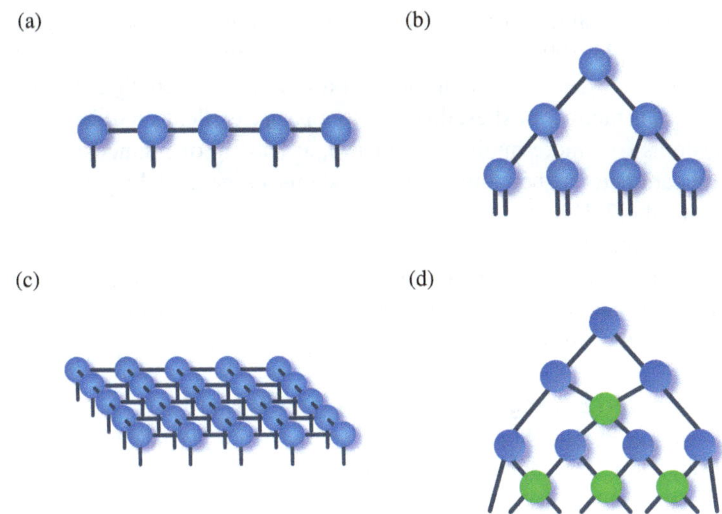

Fig. 5.6 Most frequent tensor network geometries for representing quantum many-body wavefunctions: the MPS (**a**), the TTN (**b**) and the MERA (**c**), and the PEPS (**d**)

in the system. Consequently, it consists of several tensors $\mathcal{T}^{[k]}$, one for each physical site k, which are each connected to their nearest neighbours by internal, auxiliary links as illustrated in Fig. 5.6a. Due to this train-like structure it is as well referred to as *Tensor Trains* in the mathematical community [183]. Unlike the cluster mean-field ansatz introduced in Sect. 4.1.2, MPS exhibits translational invariance and the flexibility to interpolate between mean-field and exact many-body representations based on the *auxiliary dimension* of the internal links.

MPS has become a cornerstone in the realm of Tensor Networks, with various algorithms developed over the last two decades. With a numerical complexity of $O(m^3)$ for prominent minimisation and time-evolution algorithms, it is the primary computational tool for one-dimensional problems in equilibrium [184, 185], and in many cases even out-of-equilibrium [53, 186]. While MPS can be extended to two-dimensional systems (2D-DMRG) [188, 189] with optimized scaling for practical applications in cylindrical configurations, its utility diminishes for higher-dimensional systems. In dimensions greater than one, MPS requires exponentially large bond dimensions to compensate for its inherent limitations, hindering its ability to faithfully represent quantum many-body states as system sizes increase.

In Chap. 7, we describe the MPS ans its technical implementation in more detail with hands-on examples and exercises.

5.3.4.3 Tree Tensor Networks

The Tree Tensor Network (TTN), in contrast to the MPS, is composed out of a hierarchical tensor network structure. On one edge of this spectrum there are binary trees as depicted in Fig. 5.6b and commonly referred to as Tree Tensor

Networks, where each tensor has rank three, the minimal rank necessary to have a tree structure [42, 44, 190, 191, 282]. Due to its hierarchical structure and to the fact that each couple of tensors is connected via a link-path of length $2 \log N$ at most with N being the number of sites, this structure is an excellent candidate to describe critical systems (see more on that in Part IV), at least in one dimension [44, 192]. Its generalization to higher dimensional systems or general loop-free graphs is straightforward [42, 150, 191, 193, 194].

In Chap. 8, the TTN and its implementation for high-dimensional systems is presented in more details featuring a formal definition, typical structures, its optimisation, the calculation of observables and most critically the practical hurtles when extending this Tensor Network to higher dimensions.

5.3.4.4 Projected Entangled Pair States (PEPS)

Similar to Matrix Product States (MPS) in one dimension, Projected Entangled Pair States (PEPS) provide an approximation of the full state vector for two-dimensional quantum systems: Each physical site corresponds to a tensor, and these tensors are connected in a grid configuration that mimics the underlying physical lattice structure (see Fig. 5.6c). This results in a Tensor Network that contains "loops" or non-local gauge redundancies. Consequently, PEPS represents an intuitive and potentially powerful approach for describing two-dimensional quantum many-body wavefunctions. However, PEPS is associated with a high numerical complexity, typically scaling as $O(m^{10})$ for finite-sized PEPS [39, 174], and lacks the capability for exact expectation value calculations [195]. The contraction of the complete PEPS for such calculations scales exponentially with the average system length L for a finite square lattice with $N = L \times L$ sites [196]. Consequently, optimizing the PEPS ansatz computationally often results in typical bond dimensions of around $m \sim 10$ which is insufficient for systems with higher local dimensions, such as those encountered in higher-order spin systems, Lattice Gauge Theories, or fermionic Hubbard-like systems. Furthermore, PEPS is primarily tailored for two-dimensional problems with open boundary conditions, making its application to systems with periodic boundary conditions or higher dimensionality practically infeasible.

5.3.4.5 Multi-scale Entanglement Renormalization Ansatz (MERA)

The Multi-Scale Entanglement Renormalization Ansatz (MERA) made its debut in 2007 as an efficient representation for the analysis of critical one-dimensional (1D) systems [45]. In contrast to Matrix Product States (MPS), MERA excels in handling the entanglement entropy in critical 1D systems, extending beyond the area law. Its intricate structure comprises alternating layers of isometries and disentanglers (as shown in Fig. 5.6d), allowing for efficient contraction. However, optimizing MERA comes with a complexity of at least $O(m^7)$ or higher, depending on the specific configuration [147].

The concept underlying MERA can also be extended to two-dimensional (2D) systems [148]. Unlike Projected Entangled Pair States (PEPS), 2D MERA can calculate expectation values exactly. However, its computational complexity skyrockets to $O\left(m^{16}\right)$, rendering it impractical for high-dimensional systems in real-world applications. Consequently, the utilization of 2D MERA has waned in recent years due to growing competition from PEPS and Tree Tensor Networks (TTN). While in principle, MERA can be applied to higher-dimensional systems, the numerical scaling becomes increasingly unfavorable with dimensionality growth. Noteworthy applications of MERA are found in quantum gravity [168, 169, 197–200], where it has connections to the geometry of space, such as Anti-de Sitter (AdS) or Conformal Field Theory (CFT).

5.3.4.6 Augmented Tree Tensor Network

The *augmented Tree Tensor Network* (aTTN) combines the ideas of an ordinary TTN with the strengths of a MERA. By placing disentanglers—as used in the MERA—only at the bottom of a TTN structure, it can combine the benefits of low computational scaling of the TTN with a better representation of complex quantum systems especially for high dimensional systems. In Chap. 9, we introduce this Tensor Network geometry and describe in more technical detail how it augments the TTN approach in a way that encodes the area law and boosts computations in higher dimensions.

5.3.5 Tensor Network Operators

In this section, we briefly show how to describe operators acting on many-body states in Tensor Network representation. We show how operators can be represented in the Tensor Network language as general *Tensor Product Operators* (TPO)

These TPOs $\mathcal{T}$ are a means of representing general operators in a decomposed Tensor Network form. This concept was introduced in Ref. [170, 201] where the operators can be expressed as:

$$(\mathcal{T})^{\{i_j\}}_{\{i'_j\}} = \sum_{\{\gamma_\nu\}} \prod_j (\mathrm{t}^{[j]})^{\{\gamma_{\nu'}\}}_{i_j, i'_j} \,. \tag{5.21}$$

In this expression, each tensor $\mathrm{t}^{[j]}$ acts locally on the site i_j and is connected through the links $\gamma_{\nu'}$ to other tensors within the TPO. Consequently, the entire TPO operates on the physical sites i_j. Unlike Tensor Network states, which have one physical link for each physical site i_j, TPOs possess two physical links, denoted as i_j and i'_j, for each local site they address. This TPO formalism allows us to describe various entities, including local observables ($\mathcal{T}_i$), interaction components ($\mathcal{T}_{i_j} = \mathcal{H}_p$) of a Hamiltonian ($\mathcal{H} = \sum_p \mathcal{H}_p$), string observables ($\mathcal{T}_{i_j}$), as well as more complex structures like a *Matrix Product Operator* (MPO) [185, 203]—which are explored in more detail in Sect. 7.1.4—or a *Projected Entangled Pair Operator* (PEPO) [204–206]. Figure 5.7 provides visual representations of various TPOs frequently encountered in the Tensor Network analysis of typical quantum

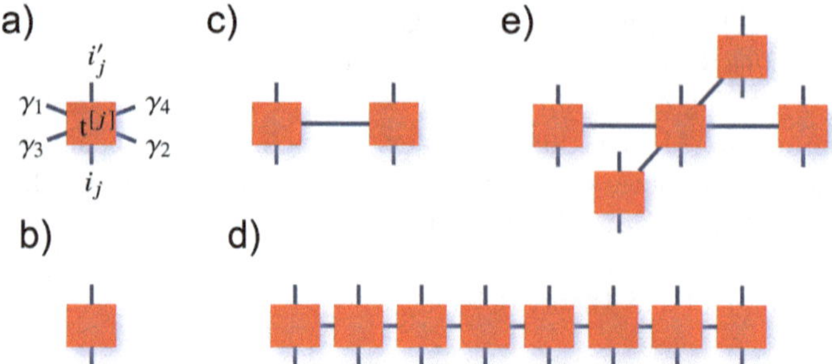

Fig. 5.7 Each Tensor Network Operator (TPO) is constructed from local tensors $t^{[j]}$ that act on physical sites i_j corresponding to different local Hilbert spaces (**a**). TPOs for Tensor Network calculus include a local operator (**b**), a two-body operator or correlator (**c**), a star operator simulating a potential interaction in a two-dimensional system (**d**), and a Matrix Product Operator (MPO) frequently employed for modeling a Hamiltonian in MPS analysis (**e**)

many-body systems. To ensure computational efficiency, it is crucial to maintain a reasonable number of links and manage their internal bond-dimension κ effectively.

A TPO, or a collection of TPOs, serves as a useful tool for representing for instance the Hamiltonian $\mathcal{H} \in \mathscr{H}$ of a quantum system in a manner suitable for Tensor Network calculations. The computation of the expectation value $\langle \mathcal{H} \rangle_\psi$ for the complete Hamiltonian $\mathcal{H} \in \mathscr{H}$ can be prohibitively expensive, even when employing a Tensor Network representation. However, it is often the case that the complete Hamiltonian can be expressed as a product of interactions, i.e., $\mathcal{H} = \sum_p \mathcal{H}_p$. Consequently, each interaction term $\mathcal{H}_p$ can be effectively described as a Tensor Product Operator (TPO). In Sect. 8.2.3, we provide an illustrative example that demonstrates this representation of the Hamiltonian using a set of TPOs. This example showcases the optimization procedure for a two-dimensional Tree Tensor Network.

5.4 Tensor Network Algorithms

Tensor Networks find extensive applications not only in Quantum Physics but also in various other domains. In the following sections, we will briefly scratch on some of the most notable algorithms and challenges addressed through the utilization of Tensor Networks.

5.4.1 Ground State Computation

Determining the ground state of an interacting quantum many-body system often presents a formidable challenge for numerical techniques. In the context of Tensor Networks, this task can be carried out with high efficiency, i.e. maintaining a reasonably low numerical complexity necessary for energy minimization. Given the Hamiltonian $\mathcal{H}$, we optimize the variational parameters of the Tensor Network wavefunction ψ to discover the system's ground state by minimizing the energy:

$$E = \langle \psi | \mathcal{H} | \psi \rangle \, . \tag{5.22}$$

The notable advantage of Tensor Networks lies in the fact that each individual tensor within the network can be treated as an independent variable. Consequently, the fundamental idea is to minimize one or a limited subset of tensors at a time. This approach transforms the global, exponentially large optimization problem into a series of local minimizations within a substantially smaller subspace, the size of which is determined by the chosen bond dimension of the network. Building upon this concept, various optimization techniques can be applied, such as eigenvalue minimization, gradient descent methods, or imaginary time evolution. In Sect. 8.1.4, we will explore a more detailed technical illustration of this approach using Tree Tensor Networks.

5.4.2 Time Evolution

Especially in one dimensional systems, Tensor Networks are frequently deployed to investigate non-equilibrium dynamics by evolving a state in time under a unitary Hamiltonian evolution. In particular, there are two established methods for performing such a time evolution numerically. The first to be mentioned would be a class of algorithms, such as the Time-Evolving Block Decimation (TEBD) [53] and the tDMRG [186], which are based on decomposing the unitary evolution by means of a Suzuki-Trotter decomposition into a set of local operators to be applied to the network. Another remarkable method is the Time-Dependent Variational Principle (TDVP) [54, 207] which exploits mathematical concepts of differential calculus to compute a time evolution based on the geometry of the network. While these techniques have a broad range of applications such as performing various quenches [208–211], controlled dynamics [212] or, using imaginary time, quantum annealing, these applications are currently limited to either small system sizes in one-dimension or short time-scales.

5.4.3 Open Systems Dynamics

Another noteworthy application of Tensor Networks involves the representation of mixed states in open systems [213–217]. Tensor Networks can indeed be utilized to investigate steady states or analyze dissipative time evolutions governed by a Lindblad master equation. The concept behind this application is to efficiently decompose the exponentially large density matrix of a mixed many-body state via Tensor Networks, rather than representing the wave function of a pure state. The dynamics prescribed by the underlying master equation can then be implemented through direct integration [214, 216] or employing stochastic methods [217]. This avenue of research is relatively recent but holds promise as a competitive approach for simulating and validating dissipation in emerging quantum technologies, such as quantum computers.

5.4.4 Machine Learning

Tensor Networks (TN) have emerged not only as a numerical method in quantum physics: They are gaining increasing relevance in the realm of applied mathematics and in particular in machine learning [218–226]. Here, TNs excel at efficiently representing vast amounts of information in a compact form, making them versatile for recognizing patterns and classifying data. Given their initial development aimed at addressing large-dimensional linear quantum systems, Tensor Networks (TNs) offer a straightforward means of computing quantities that provide valuable insights into the data learned for machine learning applications [227].

In Chap. 13 we illustrate in more detail the ideas behind machine learning and how tensor networks can be used as an efficient tool for artificial intelligence. In this chapter, we highlight some of the applications of tensor network machine learning ranging from classifying proton-proton collisions in data from LHCb at CERN [227] up to autonomous driving [228].

5.5 Measurements and Accessible Information

Although Tensor Networks are proficient in efficiently representing a quantum state by capitalizing on its inherent entanglement structure, it may not always be straightforward to extract the embedded information from a general Tensor Network geometry. Indeed, the ability to access this information poses a critical challenge in designing Tensor Network configurations. Ultimately, our objective is to extract meaningful information about the state from the Tensor Network. To address this vital point, let's explore the typical quantities of interest used in the analysis of quantum systems with Tensor Networks.

5.5.1 Expectation Values

Calculating the expectation values $\langle\Psi|O|\Psi\rangle$ of an observable O stands as perhaps the most vital means of gleaning information about a quantum many-body state from a Tensor Network representation. Naturally, one of the foremost observables of interest in the analysis of a quantum system is its Hamiltonian, denoted as $\mathcal{H}$. As introduced in Sect. 5.3.5, $\mathcal{H}$ can be expressed in the form of *Tensor Network Operators* (TPOs), enabling the computation of the energy expectation value $\langle\mathcal{H}\rangle_{\psi}$ in practical applications.

In addition to $\mathcal{H}$, there are typically three categories of observables that pique our interest: *Local observables*, denoted as $O = O^{[s]}$, which operate on a single local site s and are frequently employed to discern local order parameters; *correlators*, expressed as $O = O^{[s_1,s_2]}$, which act on two sites and are essential for investigating long-range order; and more expansive *string observables*, given by $O = O^{[s_1,s_2,\ldots,s_\omega]}$, which act on ω sites within the system and are crucial, for instance, in detecting topological properties.

The ability to compute such expectation values exactly or the necessity to approximate them depends on the geometry employed. A notable example of the latter is the PEPS, which is not *efficiently contractible*, meaning that the network cannot be contracted with itself in polynomial time relative to its size.

In Sect. 8.1.5, we show the calculation of such observables in more technical detail for a TTN as hands-on example.

5.5.2 State Overlap and Fideltiy

Comparing two physical states, denoted as $|\psi^{[1]}\rangle$ and $|\psi^{[2]}\rangle$, we can evaluate their overlap $\langle\psi^{[1]}|\psi^{[2]}\rangle$ or their fidelity $|\langle\psi^{[1]}|\psi^{[2]}\rangle|^2$, as measures for assessing the similarity between the two states. In practice, this calculation involves contracting the two networks, one that represents $|\psi^{[1]}\rangle$ and one that represents $|\psi^{[2]}\rangle$, over their shared physical links. Tensor Networks that can be efficiently contracted can readily provide this information, while for networks like PEPS, this contraction can become computationally challenging as the system size increases.

Furthermore, the norm of a state, denoted as $\langle\psi|\psi\rangle$, can be computed as a special case of the state overlap computation using a similar approach. Additionally, for carrying out this computation, the Tensor Network's gauge freedom can often be exploited to further reduce the computational costs [170, 185].

5.5.3 Entanglement Properties

Since entanglement plays a pivotal role in quantum systems and their Tensor Network representations, we are typically interested in entanglement properties of quantum states. In that context, we can calculate for loop-less Tensor Networks the entanglement of certain bipartitions of the system that bipartitite the Tensor Network into two sub-networks. To gauge the amount of entanglement encapsulated within a Tensor Network for any given bipartition, one can contract one of the corresponding sub-networks. This contraction allows us to reconstruct the spectrum of the reduced density matrix, enabling measurements of both the *Von Neumann entropy* and all *Rényi entropies* [42, 229, 230].

However, depending on the Tensor Network's geometry, contracting an arbitrary sub-system can become exceedingly challenging. In particular, *loop-less* Tensor Networks excel in this regard (see Chap. 8 for more details). By employing a suitable isometrization technique, they can be directly expressed in terms of the Schmidt decomposition, rendering it computationally straightforward to extract the Von Neumann entropy (as elaborated in Sect. 8.2.2).

5.6 Further Developments

The tensor network ansatzes and algorithms introduced so far form the basic tools to study mainly one-dimensional finite closed many-body quantum systems. However, starting from the tools introduced in this chapter, tensor network methods have been extended—with different levels of developments and success—to encompass many other scenarios. Indeed, for example, TN formulations to address directly the thermodynamical limit [52,202] or to work in the momentum or hybrid space [231–233] have been introduced. TN methods to compute thermodynamical properties of many-body quantum systems have also been presented [234, 235] Moreover, has already mentioned, the extension to finite temperature and to study open Lindbladian dynamics has been introduced by means of MPDO or LPTN ansatzes and by means of quantum trajectories [51,236–241].

Other extensions of TN methods go in the promising direction of describing systems defined in the continuum, e.g., quantum field theories. A more rigorous formalization has followed the first attempts to combine second quantization and TN methods in the so-called continuous MPS [242,243]. Alongside, condensed matter systems such as Wigner crystals has been efficiently described using an adaptive local basis [244].

Finally, worth mentioning are the increasing number of applications of TN methods to different fields beyond condensed matter and quantum science. On top of the already mentioned applications to the study of lattice gauge theories for high-energy physics (see Sect. 6.4); the application of TN methods to quantum chemistry problems is fast increasing: investigations are ongoing to individuate the most promising approach and to benchmark them against standard approaches [245–248]. The theoretical construction underlying the idea of the MERA tensor network has been exploited in some quantum gravity theory. Indeed, it is now conjectured that the MERA realizes some features of the CFT/AdS correspondence [249–252] and some numerical calculations—which could support the thriving theoretical activities along these lines—have been already performed, see e.g. [253,254].

Last but not least, there are many possible applications of tensor network methods to classical computer science problems. Indeed, being a class of tools to efficiently perform manipulation, compression, and optimization of a large amount of data, different applications are being explored: from image compression to machine learning and data analytics [221,224].

5.7 Software

To date, there are plenty software platforms available which allow entering the world of tensor networks at different levels. Some of them are ready-to-go suites which allow starting to investigate the model of interest without almost any knowledge of the details of the program running under the hood. The itensor and the TNT libraries, allow one to conveniently manipulate tensors and write TN algorithms without the need of low-level programming [255–257]. The TNT libraries also includes a DMRG/MPS code, similarly to other implementations freely available such as—among others—the ALPS project and the openMPS code [258, 259]. It is also possible to freely download specific implementations of TN codes tailored to attack the electronic structure problems for quantum chemistry [247, 260, 261].

A ready-to-go, simple but reliable and flexible (with an arbitrary number of Abelian symmetries embedded) DMRG and t-DMRG code has been distributed since 2006 by one of the authors of this book and his collaborators [262]. That software has undergone a variety of upgrades and now has matured in an open source project, the Quantum TEA [1], available for downloading and preinstalled in the Italian Supercomputer center CINECA, ready for HPC GPU-ready calculations. Quantum TEA offers a growing number of ready-to-go capabilities, from equilibrium and out of equilibrium simulations of many body quantum systems in one and two dimensions, the emulation of quantum computation compatible with standard quantum algorithm software (e.g., Qiskit) and tensor network machine learning tasks.

5.8 Problems

1. Define a class of objects to describe general n-rank tensors with the basic operations acting on them: initialization and contraction.
2. Include in the previous exercise the functions acting on a general tensor of indexes fusion, reshaping, SVD and compression.
3. Prove mathematically that an SVD enables to approximate any arbitrary matrix M by another matrix $\tilde{M}$ with lower rank $\tilde{r} < r$ with a minimal deviation introduced in its Frobenius norm $||M||_F^2 := \sum_{ij} |M_{ij}|^2 = \mathrm{tr}\{MM^\dagger\}$.

Symmetric Tensor Networks

6

Simone Montangero

As well known, symmetries play a fundamental role in physics: their mathematical description has been used heavily in many branches of physics to understand the main properties of the systems of interest, simplify their description, and improve the numerical performances of numerical codes employed to describe them. The most straightforward scenario in quantum mechanism—familiar to any physicist—where symmetries can be exploited, is that of a system described by a Hamiltonian invariant according to a given transformation g. Elementary quantum mechanics shows that the Hamiltonian is degenerate (i.e., different eigenstates with the same energy exist that behave differently under the action of the transformation), and the Hamiltonian and the operator g generating such transformation commute and can be simultaneously diagonalized [135]. That is, the system's eigenstates can be labeled according to a multiplet of labels which uniquely identify each system's eigenstate: the typical example being the eigenstates of any system with spherical symmetry (e.g., the Hydrogen atom), whose eigenstates can be labelled via three indexes n, l, m; the first one identifying the energy level, the others the system angular momentum and its projection on the quantization axis and thus the corresponding spherical harmonic [135]. Finally, Noether's theorem states that a system invariant with respect to a given continuous transformation has a conserved continuous quantity [263, 264].

In many-body physics, symmetries are combined with the additional degree of freedom introduced by the spatial extension of the system. Indeed, they might appear as global symmetries (treating the whole system as a single one, e.g., reflection symmetry, translational invariance, etc.), as global point-like symmetry (a global symmetry which results as invariance under the action of a combination of local observables, as the global magnetization, the total charge or the rotation of

S. Montangero (✉)
Università di Padova, Padova, Italy
e-mail: simone.montangero@unipd.it

T. Felser, S. Montangero (eds.), *Introduction to Tensor Network Methods*,
Graduate Texts in Physics, https://doi.org/10.1007/978-3-032-17635-6_6

spins around their own axes) and gauge symmetries (independent local symmetries as Gauss' law). Hereafter, we show how symmetries can be exploited to boost the performance of tensor network algorithm and to address specific symmetry sectors.

In this chapter, we first recap the most important elements of group theory, the mathematical description of symmetries and the theoretical ground of the formulation of symmetric tensor networks. Then, we introduce the reader to symmetric global point-like and gauge invariant tensor networks. This chapter is meant to be an introduction to the subject where technicalities are hidden as much as possible: the readers interested on the details of the technical aspects and of the implementation of symmetric TN, are invited to read Refs. [32,59,79] on which this chapter is mostly based on.

6.1 Elements of Group Theory

The statements reported in the introduction of this chapter can be introduced starting from the mathematical definition of a group $\mathcal{G}$: a set of elements g_i equipped with a multiplication operation (any composition rule between two elements of the set) which satisfies the conditions: (1) of being closed under multiplication, (2) the associative property holds, (3) the identity exists in the set and (4) the inverse of each element in the group exists in the set. If the multiplication is commutative the group is said to be *Abelian*, non-Abelian otherwise. Groups can be as simple as the identity group (composed only by one element, the identity) or composed of infinite elements. The number of groups elements is the *order* of the group. From the previous very abstract definition, it is possible to build a group-multiplication table which reports the result of all possible multiplication of the group elements. The multiplication table uniquely characterizes the group: two groups with the same multiplication table are said to be *isomorphic* , that is, there is a unique one-to-one correspondence between the elements of the group. A slightly more general property between groups we will need soon is that of homomorphism: two groups are *Homomorphic (group)* if it exists a correspondence one-to-many among them that is, for each element of the first group A_i^1 we can associate a set of elements in the second group $\{A_{i'}^2\}_{i'=1,\ldots m_i}$ such that if $A_i^1 A_j^1 = A_k^1$ then the product of any elements in the correspondent sets $\{A_i^2\}$, $\{A_j^2\}$ belongs to $\{A_k^2\}$.

Group theory is related to the symmetries introduced before as, defining some operations on the system (translation, rotations, etc.), it is possible to combine (multiply) them. If they form a group, it is possible to build the corresponding multiplication table. However, the previous abstract construction would be of hard use in quantum mechanics if not connected with the standard description that a physicist has of symmetry operations. The bridge is given by the theory of group representation: a representation of a group is any group of concrete mathematical entities (hereafter square matrices) homomorphic to the original group. If every matrix is different, then the two groups are isomorphic, and the representation is said to be faithful or true. The dimension of such matrices is called *dimensionality* of the representation. Finally, from a representation, it is possible to build an equivalent one by means of a similarity transformation S, i.e., $A' = S^{-1}AS$. Indeed, the group multiplication table is preserved under the action of a similarity transformation, and all representations related by a similarity are said to be *equivalent*.

Given any representation, it is always possible to build another representation by doubling the matrix dimension and defining its elements as a block-matrix with each block composed by the original matrix, i.e., $A^2 = A^1 \oplus A^1$: this is indeed a valid representation (preserve the multiplication table) but is *reducible*. A similarity transformation might conceal the block structure, and thus, to check for the reducibility of a representation, one shall look for the existence of a similarity transformation that brings all the elements of the group to the same block structure. If this is not possible, the representation is said to be the *irreducibile* representation (irrep), as it cannot be reduced to representations of smaller dimensionality.

In conclusion, it is possible to associate a matrix to any element in a group g, in such a way that according to the standard matrix multiplication, the group matrix multiplication table is reproduced. Moreover, it can be proved that any group representation by matrices (with non-zero determinant) is equivalent to a representation by unitary matrices $U(g)$ [265]. That is, a symmetry is identified by a unitary representation $U(g)$ of a group $\mathcal{G}$ which commutes with the Hamiltonian: $[U(g), H] = 0 \; \forall \, g \in \mathcal{G}$ [265]. As a consequence, simultaneously diagonalizing the Hamiltonian and the $U(g)$, it is possible to find an eigenbasis of the Hamiltonian such that each energy eigenstate can be labeled also via the eigenvalues of the group unitary representations (commonly referred as quantum numbers). Abelian groups, on which we will concentrate hereafter for the sake of simplicity, have one-dimensional unitary (complex) irreps. Specifically, as the group is Abelian (i.e. the elements commute and thus they can all be simultaneously diagonalized) such irreps in their diagonal form are phase factors $W^{[\ell]}(g) = e^{i\varphi_\ell(g)}$, where the phase φ depends on the quantum number ℓ and group element g, and $U(g)|\psi_\ell\rangle = W^{[\ell]}(g)|\psi_\ell\rangle$.

6.2 Global Pointlike Symmetries

In this section, we introduce one of the most common scenarios where a global symmetry is present and that can be exploited to build symmetry-invariant tensor networks: a system with a constant number of particles, such as atoms (either bosons or fermions) hopping and interacting on a lattice. To boil down the presentation to one of the simplest non-trivial scenarios where this construction can be exploited, we consider the one-dimensional Hamiltonian:

$$H = -t \sum_i (c_i^\dagger c_{i+1} + c_{i+1}^\dagger c_i) \tag{6.1}$$

where $c_j^\dagger$ (c_j) are creation (annihilation) operators at the site j of the lattice, t is the tunneling matrix element, m the mass of the particles, and the sum runs over the lattice sites. The Hamiltonian of Eq. (6.1) clearly conserves the total number of particles in the lattice. Indeed, it can be easily checked that it commutes with the total particle number operator $\hat{\mathcal{N}} = \sum_i n_i$, where $n_i = c_i^\dagger c_i$, that is, $[H, \hat{\mathcal{N}}] = 0$. However, the Hamiltonian of Eq. (6.1) has also another important property: it is invariant under the transformation

$$c_i \rightarrow e^{i\varphi} c_i, \quad \varphi \in [0, 2\pi], \tag{6.2}$$

indeed the two phases cancel out due to the presence, in each term, of a creation and an annihilation operator (for which $c_i^\dagger \rightarrow e^{-i\varphi} c_i^\dagger$ holds), and the Hamiltonian remains invariant. Notice that the invariance holds because the transformation is defined with a constant φ, independently from the lattice site. On the contrary, if we allow a site-dependent φ_i the Hamiltonian is not invariant anymore: an invariant Hamiltonian can be built by adding an additional operator, as we will see in the next section.

It shall be clear by now the connection with group theory revised in the previous section: the Hamiltonian of Eq. (6.1) is invariant under the action of the transformation defined in Eq. (6.2), that is, the unitary representation of an Abelian group (two different phase rotations commute). In particular, Eq. (6.2) defines a $U(1)$ symmetry, the invariance in under rotation in the complex plane, for which the group parameter is the rotation angle ϕ, the labels of the irreps are $\ell \in \mathbb{Z}$, and the phase $\varphi_\ell = \phi\ell$.

There is, however, an additional step to be carefully identified before moving forward: as we have seen before, the transformation has to be applied to all lattice sites at the same time, with the same phase. This springs from the fact that we moved from a single-body system described in Sect. 6.1, to a many-body setting: the many-body Hilbert space is the tensor product of the single-body ones, thus we have that irreps assume the form $U(g) = \bigotimes_j W_j(g)$, where $W_j(g)$ is the local representation of group element g at site j, and it does not depend explicitly on j. We refer to this particular kind of symmetries as *global pointlike* symmetries. In particular, Abelian global pointlike symmetries act on the many-body wave function $|\Psi\rangle = \sum_{\vec{\alpha}} \psi_{\alpha_1\alpha_2\ldots\alpha_N} |\alpha_1\alpha_2\ldots\alpha_N\rangle$, as

$$U(g)|\Psi\rangle = \bigotimes_j W_{\alpha_j}(g)|\Psi\rangle = \sum_{\vec{\alpha}} \psi_{\alpha_1\alpha_2\ldots\alpha_N} \prod_j e^{i\varphi_{\ell_j}} |\alpha_j, \ell_j\rangle, \tag{6.3}$$

where we have explicitly introduced the quantum number ℓ_j which labels each local basis state $|\alpha_j\rangle$ according to how it transforms under $W(g)$. It should now become clear that as it is possible to label each single-body state according to the quantum number ℓ, it is also possible to label and characterize each many-body basis state $|\alpha_1\alpha_2\ldots\alpha_N\rangle$ according to a global quantum number (or *charge sector*). Indeed, from Eq. (6.3) follows straightforwardly that

$$\prod_j e^{i\varphi_{\ell_j}} |\alpha_j, \ell_j\rangle = e^{i\sum_j \varphi_{\ell_j}} \prod_j |\alpha_j, \ell_j\rangle. \tag{6.4}$$

That is, every many-body basis state transforms under the action of the global pointlike symmetry as the single-body basis states, but with a phase which is the sum of all local phases $\varphi_{\ell_1,\ldots,\ell_N} = \sum_j \varphi_{\ell_j}$. In particular, for $U(1)$ global pointlike symmetries, we have $\varphi_{\ell_1,\ldots,\ell_N} = \phi \sum_j \ell_j \equiv \phi N$, and the many-body quantum number N characterize the global charge sector of the state, in this case the total number of particles in the system.

The tensor network ansätze introduced in Chap. 5 can be improved by exploiting the symmetric properties introduced above. If the Hamiltonian conserves the total number of particles, i.e., is invariant under $U(1)$ global pointlike symmetry, it exists an eigenbasis that simultaneously diagonalizes $\hat{H}$ and $\hat{N}$. This means that we can label the system's eigenstates according to N and look for, e.g., the ground state for fixed particle number (charge sector) N or the time evolution of an initial state with a well-defined particle number. Then, by definition, these states are composed by a superposition of states belonging to a single charge sector, that is,

$$|\Psi^N\rangle = \sum_{\vec{\alpha}\in\mathcal{U}_N} \psi_{\alpha_1\alpha_2\ldots\alpha_N} |\alpha_1\alpha_2\ldots\alpha_N\rangle; \quad \mathcal{U}_N \equiv \left\{\{\alpha_i, \ell_i\}_i \middle| \sum \ell_i = N\right\}, \tag{6.5}$$

where in the definition of the set $\mathcal{U}_N$ we have again explicitly introduced the quantum numbers ℓ_i. Our aim is to write a tensor network ansatz to describe states defined in the previous equation. These states, and in general any a state which is invariant under the action of the symmetry, according to Eq. (6.3) obey the relation

$$U(\phi)\big|\Psi^N\big\rangle = e^{i\phi N}\big|\Psi^N\big\rangle. \tag{6.6}$$

We look for a tensor network ansatz which obeys, by construction, to the condition in Eq. (6.5). A possible solution has been put forward in [266] and consists in forming a tensor network with *symmetric invariant tensors*, that is, where each tensor composing the network is invariant under the application of the symmetry transformation on each index. To introduce such an object, we label the basis states according to the charge sector ℓ_j they belong to. Moreover, we identify possible different states belonging to the same charge sector with a degeneracy index τ_j, that is, we replace the original state index α_j with a couple of indexes $\alpha_j \prec \{\ell_j, \tau_j\}$. Given a generic tensor $S_{\alpha_1,\ldots,\alpha_n} \equiv T_{\{\ell_1,\tau_1\},\ldots,\{\ell_n,\tau_n\}}$ the invariance condition reads, as depicted in Fig. 6.1,

$$S_{\{\ell_1,\tau_1\},\ldots,\{\ell_n,\tau_n\}} \overset{!}{=} \bigotimes_j W^{\diamond_j}_{\ell_j,\tau_j} S_{\{\ell_1,\tau_1\},\ldots,\{\ell_n,\tau_n\}} = e^{i\phi\sum_j(-1)^{\diamond_j}\ell_j} S_{\{\ell_1,\tau_1\},\ldots,\{\ell_n,\tau_n\}} \tag{6.7}$$

where $\diamond = 1, -1$ specify if the representation on the link has to be inverted or not, such that $WW^{-1} = \mathbb{1}$. Given that the condition in Eq. (6.7) has to be fulfilled for every ϕ, it implies that the elements of the tensor S can be different from zero only when $e^{i\phi\sum_j(-1)^{\diamond_j}\ell_j} = 1$, that is, when $\sum_j(-1)^{\diamond_j}\ell_j = 0$. This corresponds to the fact that the sum or irreps on each index has to sum to the identical irrep. In other

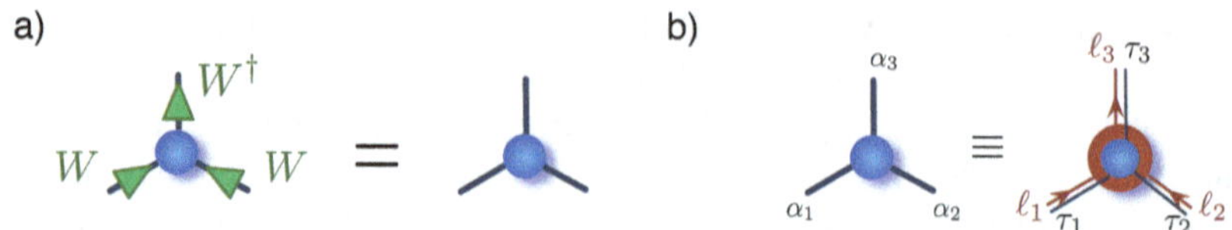

Fig. 6.1 (**a**) The symmetric tensor condition of Eq. (6.7): the tensor remains invariant after the application of the group generators $W^\diamond$ on each index. (**b**) The class of symmetric tensors satisfying such condition can be represented by splitting the index α into a structure index ℓ (containing no variational parameters) and degeneracy index τ (containing variational parameters). The arrows on the charge indexes recall the representation (direct W or inverse $W^\dagger$) under which the tensor is invariant. The degeneracy indexes can be equipped with an orientation as well via a QR decomposition as described in Chap. 5

words, for abelian symmetries, the sum of incoming and outcoming charges has to be zero: the overall charge is conserved passing through the tensor. The condition just stated can be recast in a specific form of the tensor S, separating the structural part dictated by the charge conservation condition, and the *degeneracy tensor* which contains the non-zero elements which can be used to perform variational algorithms. In conclusion, in the presence of an abelian symmetry, a symmetric tensor can be written as

$$S_{\{\ell_1,\tau_1\},\ldots,\{\ell_n,\tau_n\}} \equiv T_{\tau_1,\ldots,\tau_2}\delta_{\sum_j(-1)^{\diamond j}\ell_j,0}, \tag{6.8}$$

where the *structural tensor* $\delta_{\sum_j(-1)^{\diamond j}\ell_j,0}$ is a Kronecker delta and imposes the symmetry condition: the tensor can thus be written in block-diagonal form, and each operation acting on it can be done block by block, resulting in an improved performance of all algorithms [32, 267–269].

A simple example might help to clarify the theoretical construction presented above. We consider a tensor network composed of two lattice sites, each of them equipped with a Hilbert space whose local basis $\alpha_j = 0, 1$ labels the number of particles in each lattice site. We can build an MPS for the two sites, composed of two $d \times m$ tensors $S_{\alpha_j,\beta}$ depicted in Fig. 6.2a. If we exploit the charge conservation, we first split the global indexes in the charge and degeneracy indices, $\alpha_j \prec \{\ell_j, \tau_j\}$ and $\beta \prec \{m, \upsilon\}$, where $\ell_i = 0, 1$. The symmetric condition implies

$$S_{\alpha_1,\beta} = T_{\tau_1,\upsilon}\delta_{\ell_1,m}, \quad S_{\alpha_1,\beta} = T_{\tau_2,\upsilon}\delta_{\ell_2,-m}, \tag{6.9}$$

which, specialized for the sector of zero particles in the system (i.e. $\tau_i = \upsilon = 0$), results in a single variational parameter for the overall many body state! Although this might appear confusing at first glance, we should remember that we have constrained the system to be in the sector of global charge zero. Being this equivalent to the set of states with zero particles (composed only by the state $|00\rangle$), it is not surprising that the symmetric ansatz results in a single variational parameter: the amplitude of such state (the phase is global and can be removed). However, it would

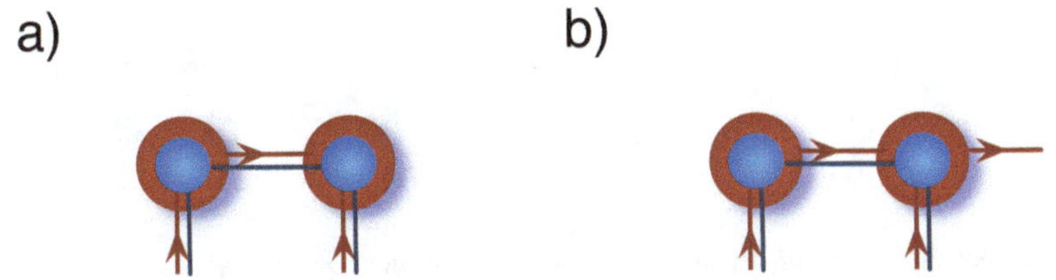

Fig. 6.2 (**a**) Symmetric MPS of two sites with structural (*directed red links*) and degeneracy link (*black lines*) (**b**) Symmetric MPS with symmetry charge selector (*open directed red link*)

be desirable to have the possibility of changing the overall symmetry sector, for example to study the set of states with a single particle. The selection of different symmetry sector can be promptly done slightly modifying the symmetric tensor network ansatz, that is, adding a link which can work as charge selector, as depicted in Fig. 6.2b. Notice that this structure is equivalent to the first two tensors of an MPS with N particles, i.e., whenever we cut a symmetric tensor network, the resulting free index brings the label of all possible charge sectors (number of particles) present in the partition. In our example, as each lattice site can have at most one particle: in two lattice sites the only possible states are those with overall up to two particles: As for system subsets the conservation of the number of particles does not hold, we shall indeed keep all the possibilities. In conclusion, proceeding as before, one obtains that $S_{\alpha_1,\beta}$ remains unchanged, while the new rank-three tensor for the second lattice site becomes

$$S_{\alpha_2,\beta_1,\beta_2} = T_{\tau_2,\upsilon_1,\upsilon_2}\delta_{\ell_2+m_1+m_2,0}. \tag{6.10}$$

It can be easily seen now that the structural tensor imposes that non-zero element appear when $m_2 = 0, 1, 2$ for which $\ell_2 = m_1 = 0$, $\ell_2 = 1, m_1 = 0$ or $\ell_2 = 0, m_1 = 1$, and $\ell_2 = m_1 = 1$ respectively. That is, the originally 4×4 tensor S is now replaced by a block diagonal tensor with three blocks (one for each charge sector): two non-degenerate and the charge-one sector of degeneracy two. One can keep on this exercise by adding more lattice sites and exploring the full power of such construction.

Finally, it shall be clear to the reader how to construct a global symmetric tensor network ansatz with charge symmetry selection: an additional open *charge selector link* is added and then used to project over the desired overall system charge sector. We conclude this introduction to symmetric tensor network by showing that indeed the construction fulfills the condition of Eq. (6.6). In Fig. 6.3 we depict such demonstration: the global pointlike operator is applied to the physical indexes, an identity $\mathbb{1} = WW^{-1}$ is inserted on each internal index of the network and the charge selector index. The invariance of every single tensor under the action of the symmetry group of Eq. (6.7) results in the demonstration of the thesis. In the presence of a charge selector, a global phase appears $\mathcal{N}$ being the charge of the sector selected by the charge selector link.

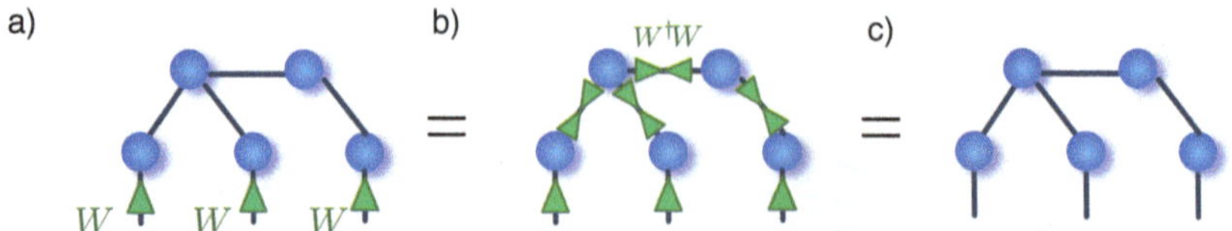

Fig. 6.3 Symmetric tensor networks: (**a**) The transformation $\bigotimes_j W_{\alpha_j}(g)$ is applied to every physical index of the TN. (**b**) The identity $W^{\dagger}W = \mathbb{1}$ is inserted in each internal index of the network and (**c**) given that each tensor satisfies the symmetric tensor condition of Eq. (6.7), the whole tensor network results invariant under the action of the group

We conclude this section stressing the fact that the construction introduced here for Abelian global symmetries, can be readily generalized to non-Abelian ones, and also to combinations of different symmetries for systems presenting more than one global symmetry, as the conservation of the total spin and the total number of particles. However, despite this generalization is somehow straightforward, it becomes quickly highly technical and thus we refer the interested reader to more specialized literature, which should be smoothly addressable after the introduction presented here [266, 270, 271].

6.3 Quantum Link Formulation of Gauge Symmetries

In the previous section, we have seen how we can introduce a tensor network ansatz which, by construction, respects global symmetries. However, there is another very relevant class of symmetries that would be desirable to exploit, that is, *gauge symmetries*. In this section, we define gauge symmetries and show that, similarly to the global symmetries, it is possible to construct a gauge invariant tensor network ansatz. To do that, we introduce briefly the *quantum link formulation* of lattice gauge theories as it can readily be encoded in the tensor network language. The discussion on the comparison between Wilson and quantum link formulation of lattice gauge theories goes beyond the scope of this book, the interested reader can find it in more technical literature [272, 273]. However, since tensor network methods require finite local Hilbert space, we consider it as one of the methods to truncate the local Hilbert space dimensions, an operation that shall be performed by any tensor network approach to lattice gauge theories. Notice that quantum link models in the limit of large representations of the link degree of freedom converge to the standard Wilson representation [91]. Finally, we end the section with a quick overview of the potential application of gauge invariant tensor networks and of other possible approaches to attack this extremely challenging problem. Once more, we specialize our presentation to Abelian symmetries in one dimension and a simple Hamiltonian to get rid, at this introductory stage, of the unnecessary technicalities, and to present the central concepts as clearly as possible. Extension to higher dimensions, more complex theories, and non-Abelian symmetries are possible and, despite being technically more challenging, are based on the ideas presented hereafter. A more detailed formal introduction to these concepts is presented in Chap. 11.

The main point of lattice gauge theories stems from the promotion of the global invariance of Eq. (6.2) to a local one, that is,

$$c_i \to e^{i\varphi_i} c_i, \quad \varphi_i \in [0 : 2\pi], \tag{6.11}$$

where the phase factor φ depends on the lattice index i and clearly $c_i^\dagger \to e^{-i\varphi_i} c_i^\dagger$. Notice that now we are requesting the Hamiltonian to be invariant under the action of each unitary representation of the local group $U_i(g)$, not only of the product of all N of them (one for each lattice site). In this sense, we are requesting the system to obey a much stronger constraint with respect to that introduced in the previous section: not only the total number of particles shall be conserved, but also the dynamics at each lattice site are highly constrained. As we will see, this is equivalent to imposing Gauss' law in QED (and its generalization in other theories). Consequently, exploiting this extensive number of symmetries can bring a large gain in terms of numerical efficiency.

Applying the transformation defined in Eq. (6.11) to the hopping Hamiltonian of Eq. (6.1), one obtains

$$H = -t \sum_i (e^{i\varphi_{i,i+1}} c_i^\dagger c_{i+1} + e^{i\varphi_{i+1,i}} c_{i+1}^\dagger c_i), \tag{6.12}$$

where $e^{i\varphi_{i,i+1}} = e^{i(\varphi_{i+1} - \varphi_{i+1})}$. As one might have expected, the Hamiltonian is clearly not invariant unless $\varphi_i = \varphi \, \forall i$, that is, in the case of global point-like symmetry studied in the previous section. A possibility to build a simple invariant Hamiltonian is then to add, an object that transforms in such a way that the phases in Eq. (6.12) cancel out. This line of reasoning leads to the introduction of new operators, the *parallel transporters* $\mathcal{U}$ which are responsible for the matter-fields coupling in the theories of fundamental interactions [274, 275]. The idea is to define an operator with support on the link between two lattice sites that transforms under the action of the group, for an $U(1)$ gauge symmetry, as

$$U_i(g) \mathcal{U}_{i,i+1} U_{i+1}(g)^\dagger = \mathcal{U}_{i,i+1} e^{-i\varphi_{i,i+1}}; \tag{6.13}$$

and use them to cancel the unwanted phases appearing in Eq. (6.12). Recalling that the phases φ in $U(1)$ symmetric systems are related to the charge ℓ on the lattice sites, we can look for their conjugate variable $E_{i,i+1} = -i\partial/\partial\varphi_{i,i+1}$ which obeys to the commutation relations

$$[E_{i,i+1}, \mathcal{U}_{i,i+1}] = \mathcal{U}_{i,i+1}; \quad [E_{i,i+1}, \mathcal{U}_{i,i+1}^\dagger] = -\mathcal{U}_{i,i+1}^\dagger, \tag{6.14}$$

and zero otherwise. It can be shown, that for theories such as QED, the charge ℓ and the field $E_{i,i+1}$ are indeed the electric charge and fields respectively, while for more complex theories such as QCD, the charges are related to the number of quarks present on the link, while the field represents the gauge fields [78, 275, 276].

In the quantum-link formulation of lattice gauge theories, the link operators $\mathcal{U}$ and the field operator are given by spin operators $S_{i,i+1}$, such that

$$\mathcal{U}_{i,i+1} \equiv S^{+}_{i,i+1}; \quad \mathcal{U}^{\dagger}_{i,i+1} \equiv S^{-}_{i,i+1}; \quad E_{i,i+1} \equiv S^{z}_{i,i+1}. \tag{6.15}$$

One can readily show that the commutation relations given in Eq. (6.14) are indeed satisfied with this choice. It shall be evident that the link operators $\mathcal{U}, \mathcal{U}^{\dagger}$ are the raising and lowering operators of one gauge field quanta, whose value is given by the expectation value of S^z. It is now possible to write a gauge invariant Hamiltonian in the quantum link formulation, that is,

$$H_t = -t \sum_{i,i+1} (c_i^{\dagger} S^{+}_{i,i+1} c_{i+1} + c^{\dagger}_{i+1} S^{-}_{i,i+1} c_i); \tag{6.16}$$

where he tunneling dynamics of a matter field is accompanied by a spin flip on the link between the two lattice sites. This is indeed reminiscent of what happens in electrodynamics (either classical or quantum) where, as depicted in Fig. 6.4, if an electron hops the electric field shall change accordingly not to violate Gauss' law. In this discrete one dimensional version, the familiar classical Gauss' law $\rho = \nabla \vec{E}$ can be expressed as

$$c_i^{\dagger} c_i = S^{z}_{i+1,i} - S^{z}_{i,i-1} \Rightarrow n_i = \Delta S^{z}_{i} \tag{6.17}$$

that is, the number of charges on the link shall be equal to the difference of the electric field entering and exit the lattice site. The condition in Eq. (6.17) can be rewritten introducing a local operator $G_i = c_i^{\dagger} c_i - \Delta S^{z}_{i}$ and imposing that its expectation value on each lattice site is identically zero, that is, imposing that the

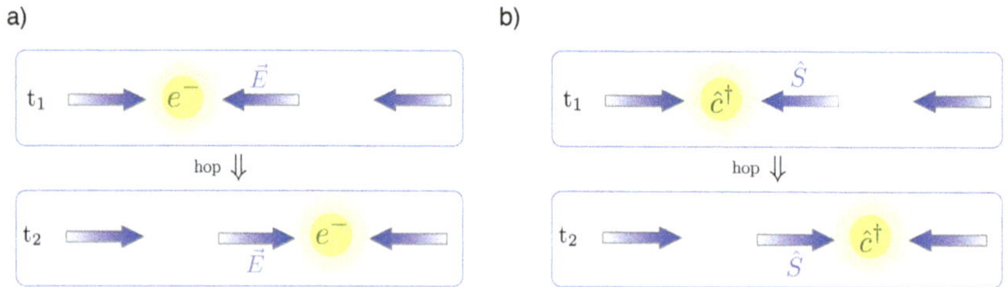

Fig. 6.4 (**a**) An electron hops from one lattice site at time t_1 to the next one at time t_2, the electric field between the two sites shall change accordingly. (**b**) The correspondent process in the quantum link description, when the matter hops from one site to the other, the spin on the link changes its state under the action of Hamiltonian (6.16) and in particular of $\hat{S}^{+}$ or $\hat{S}^{-}$ (i.e., for spin one-half it flips sign)

physical state has to fulfill the generalized Gauss' law

$$G_i|\psi\rangle = 0. \tag{6.18}$$

All states that do not satisfy Eq. (6.18) are unphysical states and form the gauge-variant space that shall be neglected. Finally, notice that the operator G_i are the generator of the group symmetry. Thus, one can reverse the construction presented here and start defining the generators of the symmetry group G_i according to the desired symmetry the theory shall fulfill: starting from Eq. (6.18), it is then possible to write the correspondent Abelian or non-Abelian gauge theory [78].

6.4 Lattice Gauge Invariant Tensor Networks

Before presenting the gauge invariant version of tensor networks, we shall make an additional formal step, rewriting the quantum link formulation in terms of other operators as this will be instrumental in the following theoretical construction. The necessary step is to reformulate the spin operator S_i defined on a link in terms of particles that live on the neighbouring lattice sites, which are conventionally referred at as *rishons*. The central idea, is to map a spin degree of freedom on a double well system with particles hopping between them. Once more, an example—depicted in Fig. 6.5—helps to give the intuitive idea: a spin 1/2 particle has two possible states, up and down, and they are formally equivalent to the possible states of a single particle hopping in a double-well potential in second quantization formalism. Indeed, we have that

$$\hat{S}^{+}_{i+1,i} = \hat{r}^{\dagger}_{i,L}\hat{r}_{i+1,R}; \qquad \hat{S}^{z}_{i+1,i} = \frac{1}{2}(\hat{n}^{r}_{i+1,R} - \hat{n}^{r}_{i,L}); \tag{6.19}$$

where $\hat{r}^{\dagger}_{i,\diamond}$ ($\hat{r}_{i,\diamond}$) are the rishon creation (annihilation) operators in the left or right ($\diamond = L, R$) well of the double well potential between the i-th and $i+1$-th sites and $\hat{n}^{r}_{i,\diamond} = \hat{r}^{\dagger}_{i,\diamond}\hat{r}_{i,\diamond}$ the correspondent number operator. The rishon fields are completely

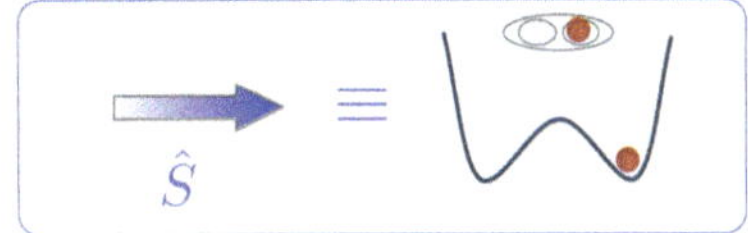

Fig. 6.5 Rishon representation of the gauge degree of freedom: a spin one-half is formally equivalent to a particle (a rishon) that hops in a double well potential. The spin quantization axis can be defined orthogonal to the symmetry axis of the double well potential: the spin pointing to the left (right) represents the occupied left (right) well. Upper diagram is a view from above of the two well potential, used in Fig. 6.6 to graphically depict different spin ice states via rishon states. The generalization to higher spin representation (higher number of rishons per link) can be obtained increasing the number of rishons in the double well

arbitrary, i.e. they can be either bosonic or fermionic as they appear always in pairs, ensuring that the gauge fields are bosonic operators. Stricktly speaking, the introduction of the bilinear representation in terms of rishons of the parallel transporter $\mathcal{U}$ can be formulated also without introducing the spin representation. Finally, the introduction of the rishons introduces an additional gauge symmetry,

$$\hat{N}^r_{i,i+1}|\psi\rangle = (\hat{n}^r_{i,L} + \hat{n}^r_{i+1,R})|\psi\rangle = N^r_{i,i+1}|\psi\rangle, \tag{6.20}$$

that is, the total number of rishons $N^r_{i,i+1}$ is conserved in each link: in the example of Fig. 6.5 we have indeed one rishon per link. This additional gauge symmetry will play a major role in the following tensor network formulation.

With the introduction of the rishon representation of the gauge field, we are now ready to start building the gauge invariant tensor network ansatz. We aim to embed the gauge constraint of Eq. (6.18) exactly in the tensor structure: to achieve this goal we exploit the fact that the spin operators $S_{i+1,i}$ written in terms of the rishons can be split in commuting operators which have support only on disjoint dressed sites. Hereafter, we will explicitly construct such an ansatz for a simple case and then present the general result. The example we consider is a simplified version of the *quantum spin ice* model, a paradigmatic model to study frustrated magnetism [102, 277–279]: It is the quantum counterpart of the classical spin ice, which is defined over a two-dimensional lattice, where a spin lives on each link. Assuming that the spin can have only two possible configurations $\{+, -\}$ (pointing in the two directions of the link), spin ice models looks for the lowest energy configuration of a given Hamiltonian (whose particular expression is not relevant here) under the constraint that the number of spins pointing towards each lattice site is equal to the number of spins pointing outwards. Two exemplary states fulfilling such condition are depicted in Fig. 6.6a together with their rishon representation (Fig. 6.6b): it should be clear by now that it introduces a gauge constraint that can be written in the form of Eq. (6.18): the magnetic flux through the contour line around every lattice site shall be null. The quantum spin ice is the quantum version of such a model, where the classical variables are replaced with quantum ones, namely spin one-half represented by the standard Pauli matrices. Here, the quantum spins are the physically relevant quantity (they are not a representation of physical gauge fields), however, they play exactly the role of the gauge fields,

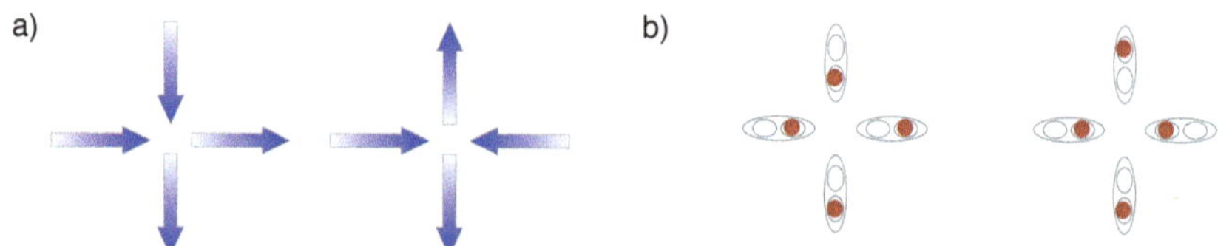

Fig. 6.6 (**a**) Two possible configuration of spin ice ($S = 1/2$) systems: an equal number spin points in- and out-ward every lattice site. (**b**) The same two states for quantum spin ice in their rishon representation, depicted with the same graphical notation as in Fig. 6.5

and everything we said before can be applied straightforwardly. Moreover, notice that differently from Eq. (6.16), in this model (and in related ones, e.g., quantum dimer models) there are no matter fields: the Hamiltonian is only a function of spin variables, and thus the gauge invariant tensor network construction is simpler than that necessary in theories with gauge field and matter coupling. However, despite the simplification in the detailed calculations, all the necessary steps to attack the general cases are present. For the same reason, we consider the one-dimensional version of the quantum spin ice, which despite being physically not exciting and possibly of little interest, it is a perfect example for our purposes.

The gauss law for the one-dimensional spin ice model can be written as

$$G_i|\psi\rangle = (\sigma^z_{i,i-1} + \sigma^z_{i+1,i})|\psi\rangle = 0; \tag{6.21}$$

indeed, out of the possible four spin configurations of the two spins living on the links connected to the i-th site, only the two configurations $|+,-\rangle$ and $|-,+\rangle$ obeys the condition (6.21). Equivalently, using Eq. (6.19) one can write the quantum spin ice Gauss' law in the rishon representation as

$$G_i|\psi\rangle = (n^r_{i,R} - n^r_{i,L})|\psi\rangle = 0, \tag{6.22}$$

assuming that the total number of rishons in each link $N^r_{i,i+1}$ is not only constant but equal on every link (in our example indeed $N^r_{i,i+1} = 1$). Finally, of all four possible states on the site $\left|n^r_{i,R}, n^r_{i,L}\right\rangle \in \{|0,0\rangle, |1,0\rangle, |0,1\rangle, |1,1\rangle\}$, only the two states $\mathcal{H}^1_G \equiv \{|0,0\rangle, |1,1\rangle\}$ are gauge invariant. Thus, one can define the gauge invariant local basis of the i-th site accordingly, reducing the local Hilbert space dimension d from four to two.

However, when building the composite Hilbert space of two or more sites, one shall take into account the constraint on the total number of rishons per link $N^r_{i,i+1} = 1$. Indeed, when composing two local Hilbert spaces $\mathcal{H}^2 = \mathcal{H}^1_G \otimes \mathcal{H}^1_G$, four possible states appear but only two satisfy condition in Eq. (6.20), namely the gauge invariant space (with respect to the local number of rishons, not the original one defined by Eq. (6.18)) is $\mathcal{H}^2_G = \{|00\rangle \otimes |11\rangle, |11\rangle \otimes |00\rangle\}$. The construction of the gauge invariant space of two sites can be achieved applying the corresponding projector to $\mathcal{H}^2$, which can be written as

$$\hat{P}_{N^r_{i,i+1}} = \delta_{n^r_{i,L}+n^r_{i+1,R}, N^r_{i,i+1}}. \tag{6.23}$$

It shall now be clear how it is possible to build quantum link representation of the gauge invariant space of L lattice sites: starting from the standard tensor product of the local Hilbert spaces in the computational basis, the Gauss' law can be enforced applying the projector that selects the gauge invariant states satisfying Eq. (6.17). This projector is local in the sense that lives on the left and right rishon Hilbert space with the same index, where the operators $c^r_{i,R}, c^r_{i,L}$ have support. Notice that this is possible only because we split the spin degree of freedom in two independent ones with the constraints given in Eq. (6.20). Finally, the application of the projectors $P_{N^r_{i,i+1}}$ enforce the condition Eq. (6.20). In the following, we show that this theoretical construction is valid in general and that can be straightforwardly encoded in tensor network language.

The gauge constraint in Eq. (6.17) can be imposed keeping only the gauge invariant states among all that generated by the tensor product of the local basis composed by the two rishons $|i_R\rangle$, $|i_L\rangle$ (and, if present, also the basis of the matter degree of freedom $|i_c\rangle$), that is, the gauge invariant basis can be written as

$$|g\rangle_i = \sum_{i_R, i_c, i_L} K^g_{i_R, i_c, i_L} |i_R\rangle |i_c\rangle |i_L\rangle, \tag{6.24}$$

where the linear operator K_i is an isometry such that $K_i K_i^\dagger = \mathbb{1}$, and $K_i^\dagger K_i = P_{G_i}$ where P_{G_i} is the projector on the gauge invariant subspace for the i-th site. Consequently $K_i P_{G_i} = K_i$, a property that we shall use to conveniently write the tensor network ansatz. Similarly, we can enforce the constraint of Eq. (6.20) applying the projector given in Eq. (6.23) on each link. Of course, to obtain a gauge invariant many-body state in the quantum link representation, we shall apply such operators to every site and links, resulting in the operators

$$K = \bigotimes_i K_i; \quad P = \bigotimes_i P_{G_i}; \quad P_{N^r} = \bigotimes_i P_{N^r_{i,i+1}}. \tag{6.25}$$

We can enforce the link and the gauge constraint applying first P_{N^r} and then K to a generic many-body wave function, obtaining

$$K P_{N^r} |\psi\rangle = K P P_{N^r} |\psi\rangle = K P_{N^r} P |\psi\rangle = K P_{N^r} K^\dagger K |\psi\rangle = P^G_{N^r} |\psi^G\rangle \tag{6.26}$$

as it can be shown that $[P, P_{N^r}] = 0$, $P^G_{N^r} = K P_{N^r} K^\dagger$, and $|\psi^G\rangle$ is a generic gauge invariant many-body state (i.e. written in the local gauge invariant bases $|g\rangle_i$). We can now introduce the tensor network ansatz: an MPS state with $|g\rangle_i$ to account for $|\psi^G\rangle$ in Eq. (6.26) and an MPO (without variational parameters) for the projector $P^G_{N^r}$ as depicted in Fig. 6.7. Moreover, it can be shown that the MPO has a very compact and diagonal representation [59]. Finally, starting from the introduced gauge invariant ansatz, one can straightforwardly apply the machinery introduced in

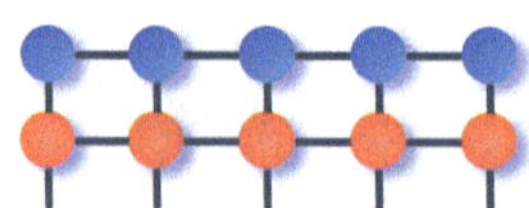

Fig. 6.7 Gauge invariant tensor network: a MPS (*blue tensors*) contains the variational parameters, while the non-variational MPO $P^G_{N^r}$ (*red tensors*) constraints the state in the gauge invariant subspace

Chap. 5 to perform equilibrium and out-of-equilibrium investigation of lattice gauge theories.

The aforementioned approach has been recently applied to investigate one and two dimensional Abelian and non-Abelian lattice gauge theories, studying in and out of equilibrium phenomena, such the ground state properties, the string breaking, the Schwinger mechanism, and scattering of mesons [55, 59, 79, 280, 281]. Notice that the construction presented here can be generalized with little effort to higher dimensional systems [181].

We conclude this introductory chapter mentioning that there are alternative formulations of lattice gauge theories compatible with tensor networks. Indeed, for specific group choices, other formulations of discrete gauge theories have also been introduced - see, e.g., refs. [55–58, 58–77, 114]. In particular cases such as the study of QED in one-dimension (the Schwinger model) it is possible—integrating the out the gauge degree of freedom—to map exactly the lattice gauge theory to a system of long-range interacting spins. This equivalent spin model can be studied exploiting again tensor network methods, providing also quantitative results at the continuum limit [56, 58, 62–64, 66, 75]. Interestingly, it has also been shown [62]—at least for the Schwinger model—that also at the continuum limit the population of the spin representation of the gauge fields decays exponentially with the square of the charge sector, corroborating the possibility that already with small spin representation of the quantum link formulation, one can correctly describe the main features the low-energy physics of lattice gauge theory.

A more technical and in-depth presentation of the concepts presented in this chapter can be found in Chap. 11.

6.5 Problems

1. Check that $[H, \hat{\mathcal{N}}] = 0$, and find other operators that commute or not commute with $\hat{\mathcal{N}}$. Check that [G,H]=0.
2. Write explicitly the tensor in Eq. (6.10) and extend the construction to a third site. Generalize numerically this construction to L sites.
3. Exploiting the relation of Eq. (6.26) evaluate numerically the dimension of the gauge invariant Hilbert space of the one-dimensional spin one sector as a function of the number of sites.
4. Compute the operator K and P_{N^r} for the two-dimensional quantum spin ice model.

Part III

Implementation of Tensor Networks

Matrix Product States 7

Timo Felser and Simone Montangero

After having introduced the core mathematical concepts of tensor networks in Chap. 5, this chapter aims to introduce the concept and hands-on implementation of tensor network computations, in particular as an exploration into frequently used algorithms and related topics. Concretely, we present some of the most successful algorithms developed exploiting tensor networks to study the many-body problem. We first reformulate the mean-field approach using the tensor notation. Then, we introduce the MPS and the reformulation of the DMRG we presented in the previous chapter in this new formalism. This shift of point of view has been accompanied by an explosion of new classes of variational tensor network states and algorithms, to accommodate different needs and describe more efficiently several physical phenomena. Hereafter, we present some algorithms to simulate the ground state properties and the time evolution of one-dimensional many-body quantum systems; and review the generalization of the MPS to the most straightforward hierarchical structure, the tree tensor network (TTN)—which is described in depth for hands-on implementations in the subsequent Chap. 8—and the algorithms for its optimization.

7.1 Ground States via Tensor Networks

In this section, after reformulating the mean field approach in terms of tensor networks, we introduce the most common and successful tensor network state ansatz, the *matrix product state*. Then, we present some algorithms to perform the

T. Felser (✉)
Tensor AI Solutions GmbH, Pfaffenhofen, Germany
e-mail: timo.felser@tensor-solutions.com

S. Montangero (✉)
Università di Padova, Padova, Italy
e-mail: simone.montangero@unipd.it

T. Felser, S. Montangero (eds.), *Introduction to Tensor Network Methods*,
Graduate Texts in Physics, https://doi.org/10.1007/978-3-032-17635-6_7

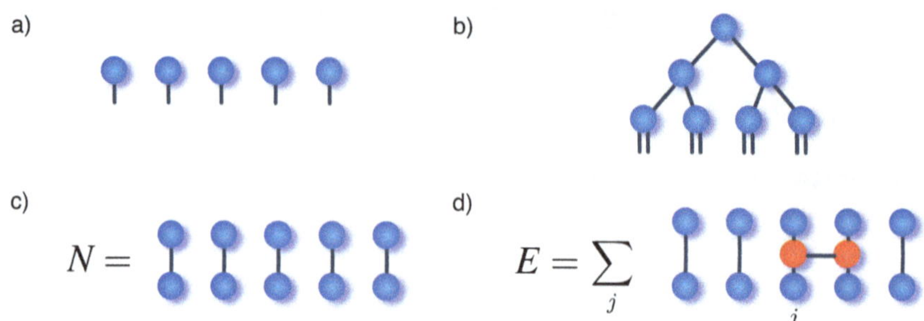

Fig. 7.1 (**a**) Tensor network representation of the mean field ansatz, (**b**) of the real-space renormalization group, aka tree tensor networks. (**c**) Computation of the norm of the state and (**d**) Energy expectation value E within mean-field approximation

variational optimization on this class of states to search for ground state properties of many-body Hamiltonians. Finally, we present the generalization of such approaches to general loop-free tensor networks, concluding the section with an overview on more general and complex looped tensor networks which are the focus of the modern research in TN methods.

7.1.1 Mean Field

Being equipped with the tensor network representation introduced in the last sections, we can now reinterpret the powerful mean field approach introduced in Chap. 4 as a tensor network algorithm. Indeed, the (single-body) mean field approximation introduced in Sect. 4.3 can be represented by a tensor network composed by the product of N single rank tensors, as depicted in Fig. 7.1. The task is then to find the entries of the tensors that minimize the expectation value of the energy computed within the mean-field ansatz. Notice that we can comfortably relax the translational invariant condition, i.e., the condition on all tensors ψ^1 of equation Eq. (4.5) to be equal, to accommodate for, e.g., boundary effects and/or non-translational invariant Hamiltonians. Thus, under the assumption that the system of interest is described by a nearest-neighbor Hamiltonian (as, for example, in the case of the Ising model in transverse field defined in Eq. (4.4)) the energy expectation value is given by the sum of $N-1$ terms as depicted in Fig. 7.1d.

Notice that each term of the Hamiltonian can be represented as a tensor network itself, as depicted in Fig. 7.1d (red tensors): indeed, in the case of the Ising Model and many other Hamiltonians, the explicit form of each tensor can be analytically found. The simplest example, in the case of zero transverse field, each red tensor in figure is exactly the correspondent Pauli matrix, and the dimension of the auxiliary index k is one. Alternatively, any two-site operator $H_{\alpha_i,\alpha_j}^{\alpha'_i\alpha'_j}$ can be decomposed in two smaller tensors via a singular value decomposition as show in Eq. (5.12), where the indexes have been fused as $\alpha_i, \alpha'_i \succ \mathbf{i}$ and $\alpha_j, \alpha'_j \succ \mathbf{j}$. As we will see later on, this is one of the simplest examples of a Matrix Product Operators defined over two sites (MPO) which can be used to described and manipulate efficiently operators acting on many-body quantum systems. The computation of each contraction of the tensors trivially results in the norm of each local wave function $\left|\psi_j^1\right\rangle$, except for the case where the Hamiltonian term is present. In the later sections, we will see how this simplification can also be introduced in more complex tensor networks, taking care of adequately gauging the tensors.

Finally, the minimization on the tensors entries shall be performed, taking into account that the normalization condition should hold. A formal way of doing such procedure, is to introduce the constraint using a Lagrange multiplier [125], that is, to build the Lagrangian

$$\mathcal{L}(\psi_1^1, \dots \psi_N^1, \psi_1^{1*}, \dots \psi_N^{1*}) = \langle\psi|\mathcal{H}|\psi\rangle - \lambda(\langle\psi|\psi\rangle - 1) \equiv E - \lambda(\mathcal{N} - 1). \quad (7.1)$$

We can search for the minima of the Lagrangian, by iterating over the different variables ψ_j^1 and ψ_j^{1*}. Each of them is then defined by the differential condition (hereafter for easy of notation we assume nearest neighbour interacting Hamiltonians and normalized single-site wave functions $\left|\psi^1\right\rangle$, and omit the superscript 1):

$$\frac{\partial \mathcal{L}}{\partial \psi_j^*} = H_{\alpha_{j-1},\alpha_j}^{\alpha'_{j-1}\alpha'_j} \psi_{\alpha_{j-1}}^* \psi_{\alpha'_{j-1}} \psi_{\alpha'_j} + H_{\alpha_j,\alpha_{j+1}}^{\alpha'_j\alpha'_{j+1}} \psi_{\alpha_{j+1}}^* \psi_{\alpha'_{j+1}} \psi_{\alpha'_j} - \lambda\psi_{\alpha_j'} = 0 \quad (7.2)$$

Finally, after computing the *effective Hamiltonian* $\tilde{H}_{\alpha_j}^{\alpha'_j} = H_{\alpha_{j-1},\alpha_j}^{\alpha'_{j-1}\alpha'_j} \psi_{\alpha_{j-1}}^* \psi_{\alpha'_{j-1}} + H_{\alpha_j,\alpha_{j+1}}^{\alpha'_j\alpha'_{j+1}} \psi_{\alpha_{j+1}}^* \psi_{\alpha'_{j+1}}$, the condition above can be expressed as

$$\tilde{H}_{\alpha_j}^{\alpha'_j} \psi_{\alpha'_j} = \lambda\psi_{\alpha'_j}, \quad (7.3)$$

that is, an eigenvalue problem for the effective Hamiltonian $\tilde{H}$. Solving it by means of the numerical methods introduced in Chap. 2, provides the new entries for the tensor ψ_j: by inspection, the ground state of the effective Hamiltonian provides also the minimal energy for the overall Hamiltonian. Moreover, due to the Hermitian property of the Hamiltonian, the adjoint tensor ψ_j^* solves automatically the problem $\frac{\partial \mathcal{L}}{\partial \psi_j^1} = 0$. Thus, there is no need to iterate on the adjoint part of the tensor network.

In conclusion, to solve a mean field problem within the tensor network paradigm corresponds to the following algorithm:

1. Define a tensor network composed by N independent tensors, each of them represented by a vector of dimension equal to the problem local dimension d. Fill the tensors with random entries and/or with a guess for the system ground state. If necessary, enforce normalization.

2. Iterate over each tensor ψ_j computing the effective Hamiltonian $\tilde{H}$ and solving the correspondent eigenproblem. Update the tensor ψ_j.
3. Sweep over the whole tensor network until convergence or the desired precision is reached.

We stress that despite the fact that the above-sketched algorithm results in a mean-field description of the systems of interest, thus limited by construction, it plays a key role in the development of tensor network algorithms as it provides an operative way to look for ground states of Hamiltonians. Indeed, it contains most of the concepts employed in a large class of tensor network algorithms.

7.1.2 Graphical Tensor Notation

From the equations presented in the previous sections, it is clear that the expressions needed for the tensor network manipulations became quickly overcrowded by indexes which are somehow cumbersome and prone to typos. This is indeed the reason why we introduced the graphical tensor notation in Sect. 5.1. However, the explicit formulas remain hard to read and risk to hide their physical and operational content. Moreover, the standard notation does not convey explicitly some important information, as the information on the gauging. Thus, hereafter we introduce a compressed notation that tries to mitigate these issues, possibly at the cost of being unconventional with respect to standard notation in mathematics and other fields where tensors are used.

The first assumption we make is that tensor networks are oriented, that is, a direction of reference $\mathcal{D}$ has to be chosen. Hereafter, we choose the down-up direction with respect to the page. The second assumption is that no links of the graph representing indexes are allowed to be orthogonal to such direction. This is always possible as single tensors edges can be moved freely. Moreover, in case there are two tensors contracted with an edge orthogonal to $\mathcal{D}$, a virtual tensor between them can always be introduced, effectively introducing two links which have a different direction without moving the original tensors, see Fig. 7.2a. As a consequence, each link of every tensor in the graph defines a direction (from the tensor) which has a non-zero projection, either parallel or anti-parallel, with respect to $\mathcal{D}$: we represent the formers as superscripts, while the latter as subscripts in the tensor notation. In general, we adopt the convention that the direction $\mathcal{D}$ corresponds to ket states labeled via subscripts and states defined in the dual space (bra states) are labeled via superscripts. Thus, superscripts contract only with subscripts and vice-versa. Moreover, going in the Liouville representation is equivalent to lowering superscripts and similarly for the reverse transformation (see Fig. 7.2b):

$$L(\rho_i^j) = L(\rho_i^j\,|i\rangle\,\langle j|) \equiv \rho_{i,j}\,|i,j\rangle\rangle = \rho_{i,j}\,|i\rangle\,|j\rangle = \rho_{i,j}\,. \tag{7.4}$$

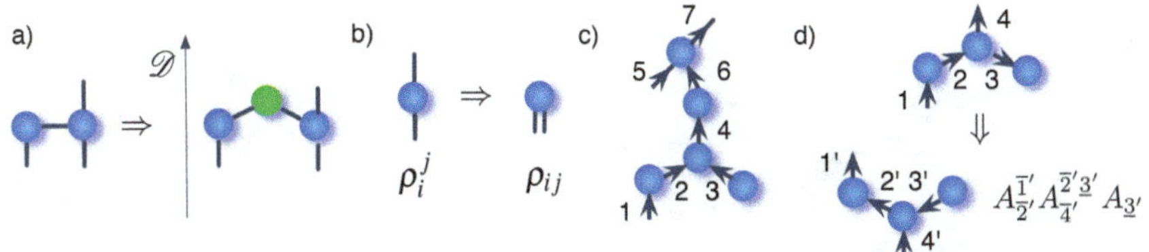

Fig. 7.2 Tensor network notation: (**a**) Virtual tensor (*green*) introducing a direction in the link. An auxiliary tensor containing the singular values of the left-right bipartition of the system has been explicitly introduced (*green* rank-2 tensor). (**b**) Liouville representation of a density matrix. (**c**) Directed and labeled tensor networks with respect to a reference direction $\mathcal{D}$ (**d**) Adjoint operation acting on a directed tensor network with the correspondent tensor notation. Notice that the subscripts lower than the superscripts implicitly implies that the tensor entries shall be conjugated

Hereafter, we number the links of the tensor network starting from bottom to top (with respect to $\mathcal{D}$) of the wave function and left to right, and we write explicitly only the index number, see Fig. 7.2c. According to our graphical notation, performing the adjoint of a tensor, corresponds to swapping super- with subscripts and performing the complex conjugate of the tensor entries. Notice, that all tensors belonging to a tensor network representing a wave function in the adjoint space will have the superscript smaller than the subscript, thus we can uniquely identify adjoint tensors.

In conclusion, the expression in Sect. 7.1.1 can be rewritten as

$$\frac{\partial \mathcal{L}}{\partial \psi^{j'}} = H^{j-1,j}_{j-1',j'} \psi^{j-1'} \psi_{j-1} \psi_j + H^{j,j+1}_{j',j+1'} \psi^{j+1'} \psi_{j+1} \psi_j - \lambda \psi_j \tag{7.5}$$

which despite being still a complex formula, it is more compact than the original one. Notice also that the Hamiltonian, as well as any operator acting on physical indexes, can be recognized by the fact that super- and sub-scripts have the same number.

Importantly, the introduced notation might generate some confusion as ψ_j is typically the j−th element of the vector ψ. Hereafter, whenever we want to specify the single entries of the tensor and some confusion might arise, we will write

$$[\psi]_j. \tag{7.6}$$

Alternatively, it is always possible to go back to the more complete notation reintroducing the Greek name of the index ψ_{α_j}. Finally, it will be necessary to introduce a directed tensor network graph, for example, to depict gauged or symmetric tensor networks. Whenever we specify the internal direction of the link, we underline (overline) depending on which direction the arrow points, that is either $\bar{j}$ or $\underline{j}$. Thus, the adjoint operation of an oriented tensors equals to swapping lower and upper indexes, see Fig. 7.2d.

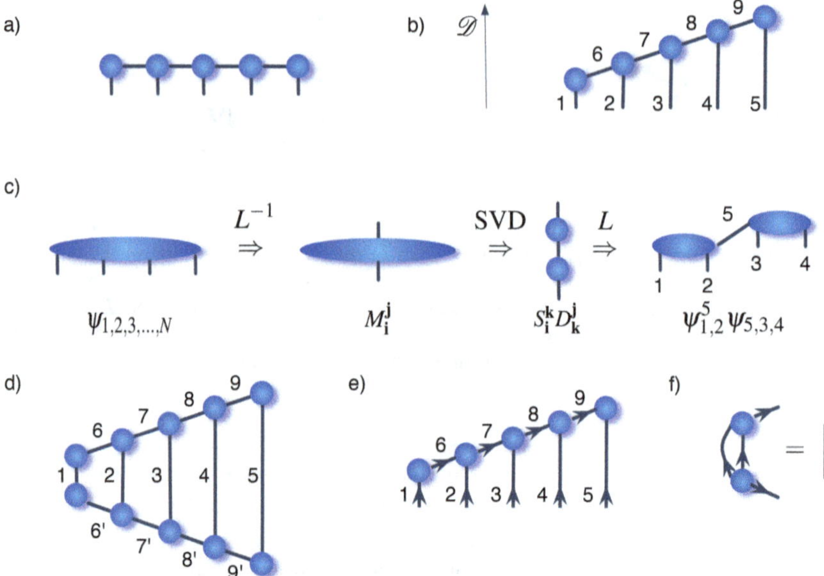

Fig. 7.3 (**a**) Matrix product state ansatz for $N = 5$ and (**b**) its ordering. (**c**) The first iteration of the recasting of a many-body wave function in MPS form. (**d**) The computation of the norm of an MPS state. (**e**) A gauged MPS state with respect to the rightmost tensor. (**f**) Isometric condition: the contraction in Eq. (7.9) of the two tensor is equal to an identity

From now on, we will use the notation introduced here, starting from the MPS ansatz we will introduce in the next section.

7.1.3 Matrix Product States

The Matrix Product State (MPS) is one of the most successful tensor networks introduced so far. It is by now used routinely to study one-dimensional quantum systems in- and out-of-equilibrium. The MPS has been introduced by Ostlund and Rommer [47] and has been recognized to be completely equivalent to a DMRG description [25]. The MPS is defined as

$$\left|\psi^{\rm MPS}\right\rangle = A_1^{N+1} A_{N+1,2}^{N+2} \cdots A_{N+j-1,j}^{N+J} \cdots A_{N-1,2N-2}^{2N-1} A_{N,2N-1} \left|1, 2 \ldots N\right\rangle \tag{7.7}$$

as depicted in Fig. 7.3b, where all indexes below and equal to N are refereed as *physical indexes* while the others *auxiliary ones*. The state is a clear generalization of the mean-field ansatz, as an additional link is present among each different local wave functions $|\psi^1\rangle$. As for the case of cluster mean field introduced in Sect. 4.1.2, these additional links allow representing the correlations between different sites. However, contrary to the cluster mean field ansatz, the structure is translationally invariant, and it allows one to interpolate between the mean field representation and the exact representation of the many-body state, depending on the *auxiliary dimension* of the newly introduced index. Indeed, disregarding momentarily, the problem of efficiency and memory constraints, starting from the exact many-body wave function $\psi_{1,2,3,\ldots,N}$, its MPS representation can be built with a subsequent series of singular value decomposition as follows:

1. Group all indexes in two groups splitting the tensor in the middle: $1, \ldots N/2 \succ \mathbf{i}$ and $N/2 + 1, \ldots N \succ \mathbf{j}$, and singular value decompose the matrix $M_{\mathbf{i}}^{\mathbf{j}} = L^{-1}(\psi_{\mathbf{i},\mathbf{j}})$. The resulting matrices $M_{\mathbf{i}}^{\mathbf{j}} = S_{\mathbf{i}}^{\mathbf{k}} D_{\mathbf{k}}^{\mathbf{j}}$ (we absorbed the singular values by contracting the matrix V with one of the other matrices) are square matrices of dimension $d^{N/2}$.
2. Repeat the above operation on each matrix separately, splitting the remain physical index in two at each iteration and grouping the auxiliary indexes according to the sketch in Fig. 7.3c.
3. Iterate until each matrix contains only one single physical index and two auxiliary ones. The resulting matrices form the MPS ansatz and define, by construction, the $A_{N+j-1,j}^{N+J}$ matrices.

As can be seen from point 1 above, the described algorithm allows us to map a many-body wave function in an MPS, but it scales exponentially with the number of constituents in the system N. However, suppose that in the process the resulting singular values are all precisely zero apart from the first one, that is, that the dimension of the auxiliary indexes is one. What one unveils is that the original wave function can be expressed exactly in a MPS with auxiliary dimensions one: it is an exact mean-field state. On the contrary, if the state cannot be represented as a mean-field state (i.e., it contains correlations), the auxiliary dimension has to be bigger than one.

The MPS approach, as for all other tensor network algorithms, stems from the assumption that the system of interest can be described with an auxiliary dimension bigger than one (i.e., beyond mean field) but not exponential in N, and thus it is possible to perform an efficient computation of the system's properties. Thus, $\mathbf{k} = 1, \ldots m$ with m which plays the role of the cut dimension in RG methods (see Chap. 4), and all results shall be checked for convergence in m. Extrapolation in $1/m$ allows extending the results virtually to the exact case, even though the arising errors shall be treated with due care.

The variational algorithm for ground state search within the MPS ansatz follows step by step the algorithm presented in the previous section, with an additional care that shall be taken with respect to the state gauging. We start introducing such a concept and then present the whole algorithm. The main idea stems from the fact that the MPS ansatz has some redundant degrees of freedom that can be used to simplify the calculations, gaining in efficiency and precision. The very first example of such possibility arise whenever the norm of the MPS state is computed

$$\begin{aligned} \mathcal{N} &= \langle \psi^{\text{MPS}} | \psi^{\text{MPS}} \rangle \\ &= A_1^{N+1} \dots A_{N+j-1,j}^{N+J} \dots A_{N,2N-1} A_{N+1'}^{1} \dots A_{N+J'}^{N+j-1',j} \dots A^{N,2N-1'} \end{aligned} \tag{7.8}$$

as depicted in Fig. 7.3d. This contraction can be done efficiently: indeed, starting from one extreme (e.g., the leftmost tensors), upper and lower tensor are contracted ($O(dm^2)$ operations). Then, the resulting tensor is contracted with the other two, one after the other with $O(dm^3)$ operations. The tensor structure is now effectively shorter by one physical site, and thus repeating the same procedure with the tensors on the right still to be contracted, the norm can be computed with $O(Ndm^3)$ operations. Despite the fact that the cost needed is linear in N, avoiding these operations could be highly desirable, also because similar computations appear in every step of the minimization algorithm. The solution comes from the gauging as depicted in Fig. 7.3f: indeed, if we gauge the MPS in such a way that each tensor obeys the isometric condition

$$A_{N+j-1,j}^{N+J} A_{N+j-1',j}^{N+J'} = \mathbb{1}, \tag{7.9}$$

the contraction in Eq. (7.8) is analytically equivalent to the contraction of the last tensor on the right with itself. This is what is typically referred as an MPS in the right gauge and is represented in Fig. 7.3e. Equivalently, there exists the left gauge when the arrows in the auxiliary dimension in the figure are reversed (and correspondingly the condition in Eq. (7.9)). The preparation of a gauged MPS starting from a not gauged one is possible using the tensor manipulation operation introduced previously: by means of a singular value or a QR decomposition, any tensor can be transformed into a product of a unitary and a generic tensor. The unitary tensor redefines the original one while the latter can be contracted into the next tensor in the chain. Repeating this procedure through all the chain, it is possible to enforce the isometric condition to all but the last tensors. In particular,

$$A_{N+j-1,j}^{N+J} \rightarrow A_k^l \rightarrow S_k^j V_j^j D_j^l \rightarrow \tilde{A}_{N+J-1,j}^l K_l^{N+J} \tag{7.10}$$

where now the tensor $\tilde{A}$ obeys the isometric condition and the tensor K shall be contracted into $A_{N+j,j}^{N+J+1}$ before performing its isometrization. A similar procedure can be performed using the QR decomposition [25, 32]. Moreover, by slightly varying the MPS tensor structure, i.e., adding a rank two tensor for each auxiliary link (as shown in Fig. 7.2a), it is possible to always keep explicitly stored the singular values: this gauge allows for an easy control of the entanglement present in the system and parallelization of the variational algorithms [25].

Finally, the algorithm to find the mean-field description of the ground state of a Hamiltonian can straightforwardly be generalized to the MPS tensor structure. Indeed, the algorithm is defined by the following steps:

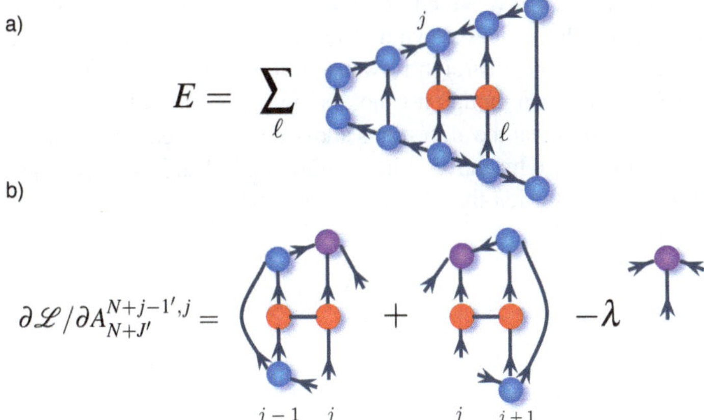

Fig. 7.4 (**a**) Energy expectation value of an MPS state gauged with respect to the tensor in position j. (**b**) Computation of the derivative of the Lagrangian in Eq. (7.1) with respect to $A^{N+j-1',j}_{N+J'}$ for an MPS gauged with respect to the j-th tensor. The condition $\partial\mathcal{L}/\partial A^{N+j-1',j}_{N+J'} = 0$ results in an eigenvalue problem for the tensor $A^{N+J}_{N+j-1,j}$ (highlighted in *purple*)

1. Initialize the MPS ansatz at given auxiliary bond dimension m following the preferred strategy, among which: (a) with random entries, (b) via a guess obtained via analytical formulas, e.g., via perturbation theory or simpler system parameters, (c) grow an MPS (performing an infinite DMRG) until system size N is reached.
2. Define the Lagrangian in Eq. (7.1) and find its extremum with respect to all tensors appearing in the network iterating sequentially on each of them, keeping the MPS gauged with respect to the optimized tensor at each step. For each tensor, impose the condition $\partial\mathcal{L}/\partial A^{N+j-1',j}_{N+J'} = 0$ computing the corresponding effective Hamiltonian and solving the eigenvalue problem for $A^{N+J}_{N+j-1,j}$. Here, the importance of gauging become evident as, as shown in Fig. 7.4, the computation is highly simplified, and it reduces a generalized eigenvalue problem into a standard one, for which numerical methods to solve them are faster and more stable [25, 32].

The computational cost of all operations included in this algorithm is upper bounded by the eigenvalue problem, whose solution can be computed via Lanczos algorithm. Thus, the algorithm complexity scales with the computation of the application of the matrix to the vector, i.e., the tensor contraction represented in Fig. 7.4b, that is, $\mathcal{O}(m^3)$.

The algorithm above optimizes each tensor singularly. Thus it is typically referred as a single tensor update. This is not the only possible choice, as one could solve the eigenvalue problem which arises first contracting two tensors and eventually separating the solution again with an SVD. This second choice is the two tensors update strategy, which strictly resembles the DMRG approach, and plays a prominent role whenever symmetries and thus different charge sectors are considered: it allows one a straightforward strategy to optimize the open charge sectors, see Sect. 6.2, even though it is not the only possibility [32].

Finally, we mention that the algorithms above can be generalized to study also the excited states of the system. Indeed, the Lagrangian in Eq. (7.1) can be generalized to include also the condition that the state which extremize it, is orthogonal to one or more given states $|\phi_k\rangle$:

$$\mathcal{L}_{ex} = \mathcal{L} - \sum_k \mu_k(\langle\phi_k|\psi\rangle). \tag{7.11}$$

Defining $|\phi_0\rangle$ as the ground state found extremizing the Lagrangian $\mathcal{L}$, thus allows to find the first excited state of the system (or in case of degenerate ground state, an orthogonal ground state with respect to $|\phi_0\rangle$). Repeating the procedure including an increasing number of states in principle allows computing the whole spectra. However, this approach remains limited to the first low energy levels due to the growing difficulty of convergence and high request of computational resources.

The MPS ansatz is not the unique choice of tensor networks which might be used to extend the mean-field state. Hereafter, we briefly review other possibilities which have been introduced, classifying them as loop-free or looped networks as the different topology introduces a fundamental difference between them: the formers can be used together with algorithms equivalent to that introduced in this section, while the latter requires special attention and much more careful treatment due to higher computational costs and additional technical difficulties.

7.1.4 Matrix Product Operators

A straightforward generalization of the MPS ansatz is the Matrix Product Operator (MPO) ansatz [283]: it represents the tensor network arising from the application to operators of the concepts and methods introduced above for states. Indeed, to represent an operator the number of physical indexes has to be doubled, one-half for the space it is acting on, and the second half for the dual space. Apart from that, the analysis presented in the previous paragraphs can be straightforwardly extended. In particular, it is easy to show that a MPO is equivalent to a MPS provided that the local dimension is squared: they are linked by a Liouville transformation and a fusion of the two indexes as follows

$$B_1^{3N-1,N+1} \dots \lambda_{N+J,N+J+1} B_J^{N+J+1,3N+J-2,N+J+2} \dots B_N^{3N-2,4N-2} \tag{7.12}$$

$$B_{1,3N-1}^{N+1} \dots \lambda_{N+J,N+J+1} B_{J,3N+J-2}^{N+J+1,N+J+2} \dots B_{N,4N-2}^{3N-2} \tag{7.13}$$

as depicted in Fig. 7.5a and b, where the B tensors are represented in red and the λ tensors containing the singular values in green. Finally, the original operator can be

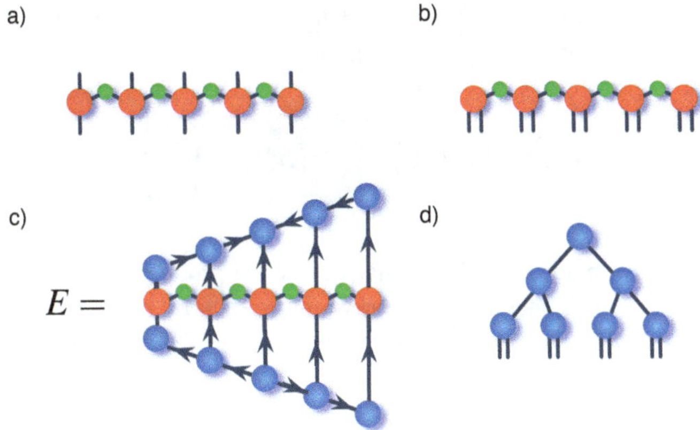

Fig. 7.5 (**a**) Matrix product operator graphical representation with the tensor B (λ) of Eq. (7.12) represented in red (*green*). (**b**) Graphical representation of the Liouville representation given in Eq. (7.13) of the MPO. (**c**) Computation of the energy expectation value via MPO and MPS tensor networks. (**d**) Tree Tensor Network graphical representation

represented via a wave function

$$|\psi_{MPO}\rangle = B_{\mathbf{1}}^{N+1} \ldots \lambda_{N+J,N+J+1} B_{\mathbf{J}}^{N+J+1,N+J+2} \ldots B_{\mathbf{N}}^{3N-2} \, |\mathbf{1}, \mathbf{2} \ldots \mathbf{N}\rangle \,, \tag{7.14}$$

where $J, 3N + J - 2 \succ \mathbf{J}$.

A very useful application of the MPO ansatz is its capability of representing Hamiltonians in a compact form. Indeed, having the MPO expression of a Hamiltonian allows computing the energy expectation value by a very simple-to-code and handle tensor network, as represented in Fig. 7.5c. The minimization procedure introduced in Sect. 7.1.3 can be readily applied starting from that tensor network. This approach is particularly relevant to treat long-range interacting Hamiltoniains, for which the algorithm presented in the previous section becomes pretty fast highly involuted, practically preventing its application in most interesting cases.

In the following, we briefly introduce how to recast Hamiltonians in compact MPO form, while we refer the interested reader to relevant literature for the complete mapping into MPO form of long-range Hamiltonians [40, 284]. We consider nearest neighbor Hamiltonians, $\hat{H} = \sum_{j=1}^{N-1} \hat{h}_j \otimes \hat{k}_{j+1}$, where $\hat{h}, \hat{k}$ are operators acting on the local Hilbert space of dimension d. Under this assumption, it is easy to verify that the MPO expression given in Eq. (7.12) corresponds to the product of matrices whose elements are matrix themselves. To show how this works, it is instructive to start with a simple example, where $N = 2$. In this case, the Hamiltonian becomes $\hat{H} = \hat{h} \otimes \hat{k} = h_1^4 k_2^5 = H_{1,2}^{4,5}$ which we aim to write it in the MPO form of Eq. (7.12), that is for two sites, $B_1^{4,3} B_2^{3,5}$. Equating the two previous relations, it is simple to identify the conditions for the equality to hold: the index 3, i.e. the auxiliary index between the two MPO matrices, shall play no role as it can have only one value, and $B_1^4 = h_1^4$ and $B_2^5 = k_2^5$. To generalize the relation for $N > 2$ we define the products of two operator-valued vectors

$$[B_1^4]^3 = \left(\mathbb{1}, \hat{h}, 0\right) \qquad [B_3^5]_3 = \begin{pmatrix} 0 \\ \hat{k} \\ \mathbb{1} \end{pmatrix}; \tag{7.15}$$

whose elements are matrices and such that their scalar product is $H_{1,2}^{4,5}$. It can be now easily checked that the general MPO expression for a nearest neighbour Hamiltonian $\hat{H} = \sum_{i=1}^{N-1} \hat{h}_i \otimes \hat{k}_{i+1}$ is formed by the product of $N-2$ matrices of the form

$$[B_J^{3N+J-2}]_{N+J+1}^{N+J+2} = \begin{pmatrix} \mathbb{1} & h_i & 0 \\ 0 & 0 & k_{i+i} \\ 0 & 0 & \mathbb{1} \end{pmatrix}, \tag{7.16}$$

and in first and last position the vectors of Eq. (7.15). Thus, single-operator nearest neighbour Hamiltonians can be represented with an MPO of auxiliary dimension (the dimension of the vectors in Eq. (7.15)) $m = 3$. The generalization to Hamiltonians of the form $\sum_{i=1}^{N-1} \sum_{k=1}^{p} \hat{h}_i^k \otimes \hat{k}_{i+1}^k$ (e.g., the Heisenberg model) is straightforward, increasing the dimension of the MPO bond dimension $m = p + 2$. Finally, as any two-body operator can be rewritten as a sum of at most d^2 terms via a SVD, the maximal needed dimension of a MPO r-range interacting Hamiltonians is $m = d^2 r + 2$ [284].

A second very useful application of MPO is their possible use to describe efficiently many-body quantum systems at finite temperature and, in general, the density matrix of open many-body quantum systems. While with this approach it is possible to study the out-of-equilibrium dynamics, as we will show in the next section [40], it is also possible to search directly for the steady state of a Liouvillian evolution via a variational algorithm [285, 286]. Indeed, by writing the density matrix of the system in an MPS form and the Liouvillian $\mathcal{L}$ in its superoperator MPO form, it is possible to search for the lowest eigenvector of $\mathcal{L}$ or of $\mathcal{L}^\dagger \mathcal{L}$ to find the steady state of the system by adapting the algorithm sketched in the previous section. While the latter choice (minimize $\mathcal{L}^\dagger \mathcal{L}$) ensures the operator to be Hermitian and semi-positive, the former provides an improved numerical performance at the price of a necessary more careful fine-tuning of the convergence [285, 286].

Finally, as said before, MPO can encode general operators which can be exploited to develop different tasks and more specialized algorithms. We will review in some details one of such application in Sect. 6.4, where we present an explicit MPO form for the projectors onto the gauge-invariant subspace of an abelian gauge symmetry.

7.1.5 Further Tensor Network Geometries

In the previous sections, we have shown how to optimize a tensor network, in particular MPS, to represent equilibrium and out-of-equilibrium properties of given many-body Hamiltonians. However, the MPS are not the only possible tensor network that can be used as a variational ansatz to describe such properties of many-body quantum systems. In the following we briefly touch on two different groups of tensor network geometries: *loop-free* networks and *looped* networks.

7.1.5.1 Loop-free Tensor Networks

Indeed, any loop-less graph can be used to define the correspondent tensor network: the variational algorithm presented in this section for MPS can be applied almost straightforwardly. However, general graphs will eventually result in computational costs which scale at least as the higher rank tensor present in the network (see Sect. 5.2.2). Thus, at constant auxiliary dimension, a convenient balance has to be found: higher rank tensors encodes more information, lower rank ones result in more favourable computational costs. On one edge of this spectrum there are binary trees as depicted in Fig. 7.5d and commonly referred at as Tree Tensor Networks, where each tensor has rank three, the minimal rank necessary to have a non-trivial structure [42, 44, 190, 191]. The network is composed starting from a top rank-two tensor, and then additional layers are created adding a rank-three tensor for each free link. This construction results in a doubling of the number of tensors and of physical indices at each layer. Eventually, after $\log_2 N$ levels, the tensor network can readily accommodate the description of N physical sites. The network is then defined as

$$\left|\psi^{TTN}\right\rangle = \prod_{k=1}^{\log_2(N/2)} \prod_{j=1}^{N/2^k} \Lambda^{j+f_2}_{2j-1+f_1,2j+f_1}, \tag{7.17}$$

where $f_1 = \sum_{l=1}^{k-1} N/2^{l-1}$ and $f_2 = \sum_{l=1}^{k} N/2^{l-1}$. Due to its hierarchical structure and to the fact that each couple of tensors is connected via a link-path of length $2 \log N$ at most, this structure is an excellent candidate to describe critical systems (see more on that in Part IV), at least in one dimension [44, 192]. Its generalization to higher dimensional systems or general loop-free graphs is straightforward [42, 191, 193].

Notice that under the assumption of translationally invariant tensors (all tensors at each level are equal) and that each tensor is constrained to be an isometry, i.e. $\Lambda^{j+f_2}_{2j-1+f_1,2j+f_1} \Lambda^{2j-1+f_1,2j+f_1}_{(j+f_2)'} = \mathbb{1}$, this tensor structure is equivalent to the real space renormalization group described in Chap. 4. However, it can be shown that relaxing the isometry constraint the precision of the resulting ground state energy remains constant with N at constant bond dimension [32, 44].

Finally, any loop-free tensor network can be represented as rank-three tensor networks using a sequence of SVD, which decompose a rank-n tensor in two rank-$n-1$ tensors. The price to pay is that the auxiliary dimension connecting the two

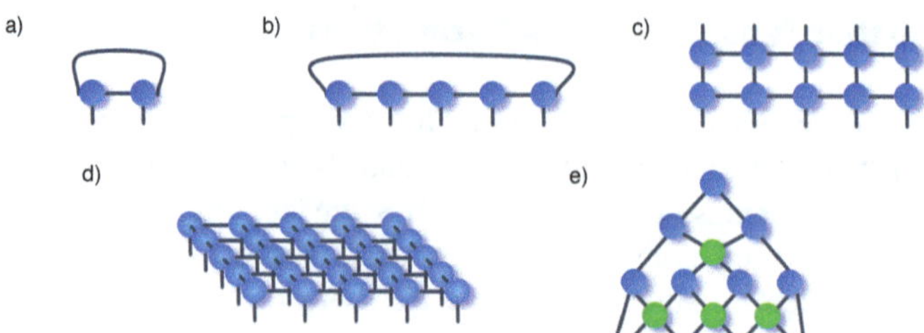

Fig. 7.6 (**a**) Infinite MPS (**b**) MPS with periodic boundary conditions (**c**) Locally Purified Tensor Network (**d**) Projected Entangled Pairs State (**e**) Multi-scale Entanglement Renormalization Ansatz composed by unitaries (*green tensors*) and isometries (*blue tensors*)

new tensors scales as the maximum of the products of the remaining dimensions of each tensor. However, one can truncate the smallest singular values of each SVD, thus introducing also an additional compression.

To recall the reader, this group of loop-free tensor networks will be discussed in form of the tree tensor networks (TTN) in great detail in Chap. 8 with hands-on examples and implementation advice for one- and higher-dimensional systems.

7.1.5.2 Looped Tensor Networks

Even though loopless tensor networks are generally speaking easier to implement and computationally more efficient, there are some scenarios where it is more natural, or it appears to be more convenient, to choose tensor networks with loops. Due to their different topology, such structures present additional difficulties when optimization algorithms are implemented, either due to a less favorable scaling or because they are prone to more numerical instabilities. Thus, for most of them, (if one wants to go beyond a proof-of-principle numerical experiment) the implementation requires additional care and higher expertise. A detailed presentation of such tools go beyond the scope of this book. Hereafter, we briefly recall some of the most common and successful looped tensor networks (also sketched in Fig. 7.6) and refer the reader interested to take this challenge to the relevant literature.

One of the most straightforward scenarios where looped tensor networks appear quite naturally is the infinite MPS (iMPS). Assuming that the system of interest is translationally invariant, one can start from the hypothesis that the entries of all tensors of the MPS ansatz of Eq. (7.7) are equal (i.e., a structure as $AAAAA$) and thus it is possible to work directly at the thermodynamical limit. In such case, the problem can be recast into the minimization of the tensor structure depicted in Fig. 7.6a and standard minimization techniques or imaginary-time evolution can be applied [25, 30]. Given the simplicity of the ansatz, the algorithm can be efficiently and easily implemented obtaining accurate results. However, the drawback of such method is that it cannot be applied to finite systems and thus it hardly compares with experimental results, and it cannot be applied to cases where translational invariance is broken, e.g., in the presence of imperfections, disorder or presence of trapping potential or boundaries. Moreover, being defined only at the thermodynamical limit, finite-size scaling (see Chap. 10) cannot be applied, preventing the use of a very powerful method to characterize the properties of critical systems. The iMPS ansatz can also be naturally extended and improved using a hybrid assumption, that is, to assume that the tensors are equal only with a periodicity bigger than one, i.e., structures as $ABABAB$ or $ABCABCABC$, similarly to the spirit of cluster mean-field presented in Sect. 4.1.2.

A second scenario where a looped tensor network has been introduced is to describe one-dimensional systems with periodic boundary conditions. Indeed, in such a case, it is natural to introduce an ansatz of the form

$$\left|\psi_{\text{PBC}}^{\text{MPS}}\right\rangle = A_{2N,1}^{N+1} A_{N+1,2}^{N+2} \cdots A_{N+j-1,j}^{N+J} \cdots A_{N-1,2N-2}^{2N-1} A_{N,2N-1}^{2N} \left|1, 2 \ldots N\right\rangle, \tag{7.18}$$

as depicted in Fig. 7.6b. Once again, it is possible to apply straightforwardly the ideas presented in the previous sections to this ansatz. However, the additional link that has been introduced, has the disadvantage to increase the scaling of the algorithm and hindering the gauging, practically discouraging its use unless no other solution is at hand and more sophisticated strategies are applied [287]. An alternative approach is based on the re-ordering of the lattice sites from $1, 2, 3, \ldots N$ to $1, N, 2, N-1, \ldots$. Thus, it is possible to map the system with periodic boundary conditions to one with open boundaries. Finally, one can use standard approaches either mapping two neighbour physical sites into a logical one at the price of increasing the local dimension from d to d^2 or, preferably, writing the resulting next nearest neighbour Hamiltonian in a convenient MPO.

Another looped tensor network has been introduced to describe one-dimensional open many-body quantum systems and to cope with the problem of the positivity of the density matrix: the Locally Purified Tensor Network (LPTN) stems from imposing the positivity of the density matrix, writing it as $\rho = XX^\dagger$ [288]. Writing the operator X as an MPO results in a tensor network as depicted in Fig. 7.6c. On such structure, it is possible to apply the strategies presented before, and a generalization to the complete Liouville operator of the time-dependent DMRG presented in the next section [241]. Moreover, being almost one-dimensional, LPTNs profit from the gauging unlike most other looped tensor network.

Finally, other remarkable TN classes, specialized for different scenarios, have been introduced such as Projected Entangled Pair States (PEPS) to simulate two-dimensional MBQS (Fig. 7.6d) [289, 290], the Multiscale Entanglement Renormalization Ansatz (MERA) to study hierarchical scale-invariant systems (Fig. 7.6e) [291–293], the branching MERA [294], the weighted graph states [295], the entangled-plaquette states [296], the string-bond states [297], and the hyperinvariant tensor networks [298]. Covering all these possibilities goes beyond the scope of this book, however, we refer the interested reader to the relevant literature.

7.2 Time Evolution via Tensor Networks

In the previous sections, we have seen how it is possible to find the best tensor network approximations of equilibrium states. In this section, we show how to investigate out of equilibrium properties of many-body quantum systems by means of solving time-dependent Schrödinger equation via tensor networks. The first step in this direction is typically the decomposition of the many-body time-evolution operator via the *Suzuki-Trotter decomposition* in such a way that the exponentially large operator is decomposed and approximated as the product of few-body operators [262,299]. Hereafter we focus on the case of the dynamics generated by time-independent Hamiltonians. However, the following presentation can be straightforwardly extended to the case of time-dependent Hamiltonians [300, 301].

As for most numerical solutions of time dependent problems, the starting point is to discretize the time axis and thus to recast the time-evolution operator in the form of products of propagators for a small time step Δt as in Eq. (3.34). Thus, the task to be solved is to efficiently apply multiple times the operator $\hat{U} = e^{-i\hat{H}\Delta t/\hbar}$ to the tensor network. This challenge is in general double-faced: on the one hand, the tensor structure of the state shall remain invariant at every step, otherwise the algorithm cannot be iterated. On the other hand, the dimensions of each tensor index shall remain constant (or increase up to a given threshold) to maintain the algorithm efficiency. We will see later on how this is possible in many interesting scenarios, starting form the decomposition of the operator $\hat{U}$ in terms of two sets of commuting operators having disjoint support. For example, for one dimensional nearest neighbour interacting systems, the Hamiltonian operator can be divided in odd an even terms

$$\hat{\mathcal{H}} = \sum_{i=1}^{N-1} \hat{H}_{i,i+1} = \sum_{j=1}^{N/2} \hat{F}_{2j-1,2j} + \sum_{j=1}^{N/2-1} \hat{G}_{2j,2j+1} \tag{7.19}$$

such that

$$[F_i, F_j] = [G_i, G_j] = 0 \qquad \text{and} \qquad [F_i, G_j] \propto \delta_{i,j}. \tag{7.20}$$

Finally, making use of Baker-Campbell-Hausdorff formula, the time evolution operator can be approximated as

$$\begin{aligned} \hat{U} = \exp\left(\frac{-i\sum_i \hat{F}_i \Delta t}{2\hbar}\right) \exp\left(\frac{-i\sum_i \hat{G}_i \Delta t}{\hbar}\right) \exp\left(\frac{-i\sum_i \hat{F}_i \Delta t}{2\hbar}\right) \\ + O\left(\Delta t^3\right) \approx \prod \hat{W}_i \end{aligned} \tag{7.21}$$

In conclusion, if it is possible to apply a two-body gate to the tensor network representing the state $|\psi\rangle$ preserving its tensor structure and keeping the index dimensions constant, by subsequent iterations the system time evolution can be reproduced. In the following sections we describe multiple strategies available to achieve this goal.

7.2.1 Time-dependent Density Matrix Renormalization Group

The first algorithm that unveiled the fascinating possibility of direct simulation of the real-time evolution of a many-body quantum system is the time-dependent DRMG (t-DMRG) or Time Evolving Block Decimation (TEBD) algorithm [25,302, 303]. They provide the sequence of operations that are necessary to time-propagate a state in MPS form: the former in DMRG language, the latter in the equivalent tensor network formulation. Hereafter, we follow the second one for ease of presentation. With the complete theoretical description and technical tools we have nowadays, this step might appear somehow straightforward; however, its development has been a breakthrough. As depicted in Fig. 7.7a, the contraction of a single gate acting on sites i and $i+1$ is a local operation that involves only three tensors: the gate itself and the two MPS tensors containing the physical indices i and $i+1$. Consequently, it can be performed efficiently, i.e., with an N-independent scaling, in particular

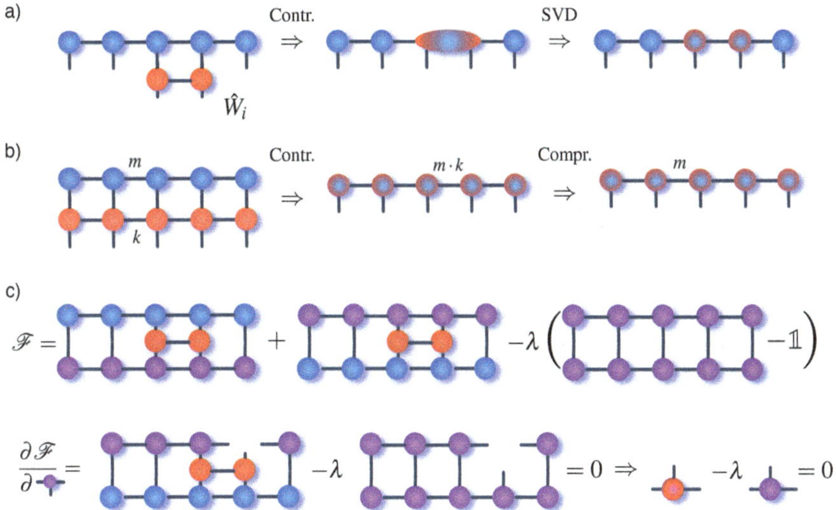

Fig. 7.7 Time-evolution via tensor network: (**a**) The contraction with the MPS (*blue tensors*) and subsequent compression via SVD of a single time evolution gate $\hat{W}_i$ (*red tensors*). (**b**) The contraction with the MPS to be evolved (*blue tensor*) of the MPO representation of the time evolution operator $\hat{U}$ (*red tensors*) followed by an MPS compression via subsequent SVDs. Here the links dimensions are explicitly reported for clarity of presentation. (**c**) Graphical representation of the figure of merit in Eq. (7.22), and of its derivative with respect to a specific tensor $A^{N+j}_{N+j-1,j}$ (in the example $j=4$). The condition $\partial\mathcal{F}/\partial A^{N+j}_{N+j-1,j} = 0$ results in an update rule for $A^{N+j'-1,j}_{N+j'}$

as $O(m^2)$. Finally, the contracted tensor—which contains two physical indexes—can be split again into two rank-3 tensors using an SVD, which is an operator of the order $O(m^3)$. Notice that, after the contraction and the SVD, the auxiliary dimension connecting the two tensors has increased.

It shall be truncated to resume the MPS initial structure, keeping only the m highest singular values. These steps can be iterated, contracting the following gates of the expansion in Eq. (7.21). Eventually, after contracting all gates, a single Δt time evolution has been performed: concatenating all necessary Δt time evolutions, the state at the end of the evolution is obtained in the form of a MPS with a fixed bond dimension m. Once more, the fixed value of the auxiliary bond dimension m introduces an approximation at every algorithm step, which shall be kept under control. It is possible to give bounds on the final error [25, 234, 304], which depends strongly on the time evolutions one simulates.

Finally, we mention that whenever it is convenient to write the time evolution operator in terms of an MPO (for example in case of long-range interactions), the time evolution can be performed contracting the MPO directly to the MPS representing the state as depicted in Fig. 7.7b, and then performing a compression of the whole MPS [305, 306].

As a final remark, the TEBD scheme can also be used to look for ground states of many-body Hamiltonians merely performing an imaginary-time evolution ($t \rightarrow it$) starting from an initial random state. However, despite the simplicity of implementation, for high performing solutions, a DMRG algorithm is typically preferred due to its faster convergence rate. Indeed, the exponential tails of the imaginary time evolutions drastically slow down its convergence.

7.2.2 Fidelity-driven Evolution

Another practical way to evolve a tensor network exists which exploits the computation of the fidelity between the evolved state and another tensor network with constant bond dimension: the idea is to search for the tensor network at fixed dimension which best approximates the evolved state. It can be used for complex tensor networks, and in particular for looped ones [307]. The algorithm is again based on the Trotter decomposition and follows the following steps to apply every gate:

1. Apply the gate to the state to be evolved $|\psi\rangle$ and compute the overlap of the evolved state with another general state described by the same tensor network $|\phi\rangle$, $\mathrm{Re}(\left\langle\phi\middle|\hat{W}_i\psi\right\rangle)$.
2. Minimize the distance between the two tensors with the additional constraint that the new state has to be normalized

$$\min_{\phi} \mathcal{F} = \min_{\phi} \mathrm{Re}(\left\langle\phi\middle|\hat{W}_i\psi\right\rangle) - \lambda(\langle\phi|\phi\rangle - 1). \tag{7.22}$$

The minimization can be achieved, once more, by means of iterative minimization with respect to each tensor $A^{N+J}_{N+j-1,j}$ in the MPS representation of the state $|\phi\rangle$. The gradient $\partial\mathcal{F}/\partial A^{N+j}_{N+j-1,j}$ is depicted in Fig. 7.7c: its extremum can be computed efficiently by contracting the remaining tensor network, resulting in a simple condition for the update of the tensor. If the MPS is gauged with respect to the j-th tensor, the new tensor is proportional to the tensor resulting from the contraction of the term $\left\langle \hat{W}_i \psi \middle| \phi \right\rangle$ apart from the tensor $A^{N+J}_{N+j-1,j}$ itself (see figure Fig. 7.7c).

3. Set $|\phi\rangle = |\psi\rangle$ and iterate the previous step for each tensor until convergence is reached (for small Δt practically few iterations are needed). The resulting $|\phi\rangle$ tensor represents the tensor network approximation at a fixed bound dimension of the time-evolved state.

The algorithm above can be easily adapted to the case where the infinitesimal time-evolution operator $\hat{U}$ is given in an MPO representation.

7.2.3 Time-dependent Variational Principle

An alternative elegant approach to the simulation of time-evolutions via tensor networks is provided by the *Time-dependent variational principle* [308, 309]: the idea is to project the time evolution operator to the manifold tangent to the MPS space with given bond dimension in such a way that the tensor network structure is preserved by construction. Hereafter, we report the main points of the derivation of the algorithm and refer the reader to the original publications for the explicit mathematical derivation [308]. The TDVP algorithm exploits the gauging of the MPS and updates every tensor of the MPS iteratively, solving the time-dependent Schrödinger equation for an infinitesimal time-step Δt, where the generator of the time-evolution is explicitly projected in the tangent manifold of the MPS:

$$\frac{d\left|\psi^{\text{MPS}}\right\rangle}{dt} = -i\hat{P}_{\mathcal{T}}\hat{H}\left|\psi^{\text{MPS}}\right\rangle. \tag{7.23}$$

It can be shown that the projector $\hat{P}_{\mathcal{T}}$ can be expressed in the form depicted in Fig. 7.8a. It then follows that the i-th tensor $A^{N+j}_{N+j-1,j}$ and the vector of singular values $C_{N+J,N+J+1}$—obtained after a SVD of the tensor $A^{N+j}_{N+j-1,j}$ to change gauge (see Fig. 7.8b)—evolve in time according to the equations:

$$A_{\mathbf{k}}(t+\Delta t/2) = \exp(-i\tilde{H}^{\mathbf{j}}_{\mathbf{k}}\Delta t/2)A_{\mathbf{j}}(t) \tag{7.24}$$

$$C_{\mathbf{k}}(t+\Delta t/2) = \exp(+i\tilde{K}^{\mathbf{j}}_{\mathbf{k}}\Delta t/2)C_{\mathbf{j}}(t), \tag{7.25}$$

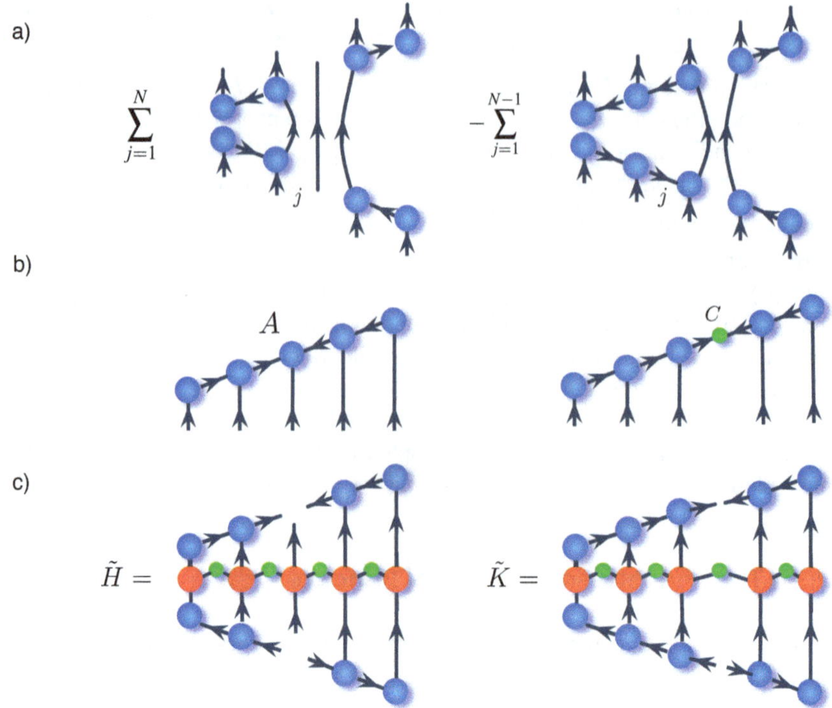

Fig. 7.8 (**a**) Graphical definition of the tangent space projector $\hat{P}_{\mathcal{T}}$. (**b**) Definition of the tensor A (left) and C (right) of Eqs. (7.24) and (7.25). (**c**) Definition of the effective Hamiltonians $\tilde{H}$ (left) and $\tilde{K}$ from the system Hamiltonian in MPO form (*red* and *green tensors*)

where the indexes $\mathbf{k}, \mathbf{j}$ are the indexes resulting by the fusion of the three (two) indexes of the tensor A (C). The effective Hamiltonians $\tilde{H}$ and $\tilde{K}$ are defined graphically in Fig. 7.8c starting from the system Hamiltonian in the MPO representation. In conclusion, the single-site TDVP algorithm can be implemented as follows: starting from a left gauged MPS, the first tensor A_j^{N+j} is evolved according to Eq. (7.24), followed by an SVD of the updated tensor which defines the C tensor to be evolved according to Eq. (7.25). The evolved C tensor is then contracted in the next tensor on the right, and the procedure is iterated throughout the whole MPS. Completing a whole sweep (left to right and backward) correspond to evolving the system wavefunction $|\psi^{\text{MPS}}\rangle$ from t to $t + \Delta t$ with an error of $O(\Delta t^2)$ [308]. This scheme can be straightforwardly implemented also on a general loop-free network [32].

7.3 Measurements

Finally, many interesting physical quantities can be measured efficiently once the entries of the tensor network of choice have been given or optimized to represent the state of interest (e.g., the ground state of a many-body Hamiltonian or the time-evolved state at some final time). We refer the reader to Part IV for an explicit introduction and the physical interpretation of such quantities. Hereafter, we define some of the most interesting quantities and present how it is possible to compute them practically in a tensor network language. We explicitly present the calculation for MPS states, however, a straightforward generalization of such procedures is possible for most tensor network structures.

The first class of quantities that can be directly computed are *local observables*, thus, any expectation value of operators with support only on a single local Hilbert space define over a lattice site,

$$\langle \hat{M}_j \rangle = \left\langle \psi^{\text{MPS}} \middle| \hat{M}_j \middle| \psi^{\text{MPS}} \right\rangle. \tag{7.26}$$

The local observables, as depicted in Fig. 7.9a, can be readily computed similarly to the computation of the norm of the state. Moreover, the computation is reduced

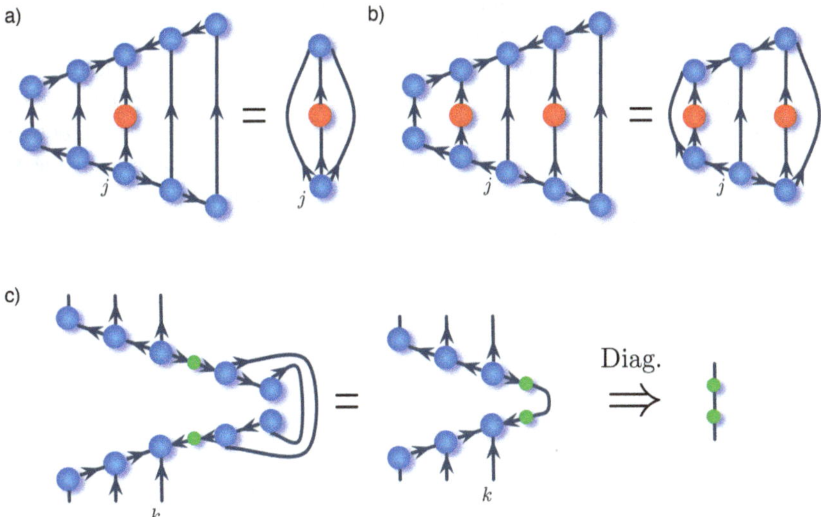

Fig. 7.9 (**a**) Computation of an expectation value of a single-site operator with support on site j on a MPS gauged with respect to the j-th site (**b**) Computation of a two-point correlator on a MPS state properly gauged (see text). (**c**) Computation of the reduced density matrix $\rho_3 = \text{Tr}\{j > 3\rho\}$ from the density matrix of a system of five lattice sites expressed via MPS state $\rho = \left|\psi^{\text{MPS}}\right\rangle\left\langle\psi^{\text{MPS}}\right|$. The *green tensor* contains the singular values λ_1 obtained via the compression of the third and fourth A tensor. The diagonalization of the reduced density matrix can be obtained by means of the unitary operator $U = A_1^6 A_{6,2}^7 A_{3,7}^8$, resulting in the diagonal matrix with diagonal form λ_i^2 (rightmost diagram)

to a system size independent contraction (three tensors) if the MPS is gauged with respect to the j-th site. The second class of efficient measurements is the evaluation of non-local observables such as k-points correlations

$$\langle \hat{M}_{i_1} \hat{M}_{i_2} \dots \hat{M}_{i_k} \rangle = \left\langle \psi^{\text{MPS}} \middle| \hat{M}_{i_1} \hat{M}_{i_2} \dots \hat{M}_{i_k} \middle| \psi^{\text{MPS}} \right\rangle \tag{7.27}$$

which (assuming $i_1 < i_2 < \cdots < i_k$), in a properly gauged MPS (i.e. with respect to a site $i_1 \leq j \leq i_k$) scales as the distance between the lowest and the highest index $i_k - i_1$, see Fig. 7.9b.

Finally, it is possible to have easy access to the reduce density matrix of any system bipartition, containing k and $N - k$ neighbouring lattice sites respectively,

$$\rho_k = \text{Tr} j > k \left| \psi^{\text{MPS}} \right\rangle \left\langle \psi^{\text{MPS}} \right|. \tag{7.28}$$

The trace operation is by definition equivalent to the contraction of the indexes $j > k$. If the MPS is in the correct gauge, it is equivalent to eliminate the contracted tensors. The computation of ρ_3 in a system composed of five lattice sites is depicted in Fig. 7.9c, where for clarity and later use we have explicitly represented the diagonal matrix of the singular values λ_i resulting from the compression of the j-th and $j + 1$-th tensors of the MPS (green rank-two tensor in the figure). While this operation is possible and efficient for any number of sites k, the explicit construction of the reduced density matrix requires an exponential increase memory with k. Thus the explicit computation of ρ_k is still limited to few lattice sites. However, it is possible to compute the first m populations of the reduced density matrix efficiently and thus, an approximation of different entropy measures between the two subsystems bipartitions, as for example all Rényi entropies and the von Neumann entropy. Indeed, the diagonalization of the reduced density matrix can be readily performed applying the remaining MPS tensors themselves which, again due to the gauge condition simplify and result in a diagonal ρ_k (see Fig. 7.9c). The populations of the reduced density matrix are given by

$$p_i = \lambda_i^2. \tag{7.29}$$

As we will see in the subsequent parts of the book, this capability will play a major role in the characterization of equilibrium and out of equilibrium properties of many systems, from the characterization of quantum phase transition to the estimation of the efficiency of quantum computations.

7.4 Problems

1. Exploiting the code developed in the previous exercises, define the object MPS and the basic operations on them: computation of the norm, and evaluation of expectation values of local and nearest neighbor operators.

2. With the tools developed above, write a t-DMRG code for the Ising model in transverse field. Perform the imaginary time evolution starting from a random MPS and compute the resulting ground state energy with those computed using other methods (see exercises in the previous chapter).

8 Tree Tensor Networks

Timo Felser

This chapter is dedicated to an in-depth discussion on the basics of Tree Tensor Networks. Starting in Sect. 8.1, the fundamental ideas of Tree Tensor Network states are presented. After introducing the Tree Tensor Network in a formal definition, we will see that these class of Tensor Network states naturally have a loop-less, hierarchical structure which greatly benefits the underlying algorithms for calculating observables and finding ground states of many-body Hamiltonians. In Sect. 8.2, this idea of Tree Tensor networks is generalised towards the study of higher-dimensional systems. Therein, we will see that the concrete structure of a Tree Tensor Network is crucial for simulating systems in 2D and beyond [189] and how a system can be mapped ideally to suit the topology of a Tree Tensor Network. This mapping is further underlined with a practical step-by-step example for a 4×4-system with nearest-neighbour interactions. Further, we will see that this network itself (without the augmentation introduced in Chap. 9) does not satisfy the area law in higher dimensions, thus eventually fails to faithfully represent physical states for growing system sizes. The last section of this chapter, Sect. 8.2, is dedicated to an in-depth technical description for the numerical implementation of a Tree Tensor Network, featuring discussions on the underlying data structure, important modules and practical note on the High-Performance-Computing aspect of the implementation.

T. Felser (✉)
Tensor AI Solutions GmbH, Pfaffenhofen, DE, Germany
e-mail: timo.felser@tensor-solutions.com

T. Felser, S. Montangero (eds.), *Introduction to Tensor Network Methods*,
Graduate Texts in Physics, https://doi.org/10.1007/978-3-032-17635-6_8

8.1 Tree Tensor Network States

Definition 8.1.1 (Tree Tensor Network) A Tree Tensor Network, abbreviated by TTN, is a Tensor Network following Definition 5.3.1 in which any two tensors are connected by exactly one path.

Definition 8.1.2 (Path) Let $\{\mathcal{T}\}$ be a set of distinct tensors ($\forall i \neq j : \mathcal{T}^{[i]} \neq \mathcal{T}^{[j]}$) and let $\{\nu\}$ be a set of links such that each link ν_i connects the tensors $\mathcal{T}^{[i]}$ and $\mathcal{T}^{[i+1]}$ for $1 \leq i < k$. Then, a *path* is defined as the ordered, alternating list $(\mathcal{T}^{[1]}, \nu_1, \mathcal{T}^{[2]}, \nu_2, ..., \nu_{k-1}, \mathcal{T}^{[k]})$.

Further, we define:

- The *length* of the path as the number of links in the set $\dim \nu = k - 1$.
- $\mathcal{T}^{[1]}$ and $\mathcal{T}^{[k]}$ as the endpoints of the path.
- A cycle as a path with an additional link connecting its endpoints.

Theorem 8.1 *Tree Tensor Network properties Let Ψ be a Tensor Network, then the following statements are equivalent:*

- *Ψ is a Tree Tensor Network.*
- *Ψ contains no cycles, i.e. Ψ is acyclic (loop-free), and the addition of any internal link results in the formation of cycles*
- *Ψ is connected and the removal of any internal link disconnects G.*

Lemma 8.1 *Every Tensor Network without cycles is a Tree Tensor Network.*

8.1.1 Hirarchical Structure of Tree Tensor Networks

As illustrated in Fig. 8.1, each TTN can be arranged in a hierarchical structure with different layers Λ_ℓ when defining a *root link* which will be the topmost-link in the network. In this structure, the first layer Λ_1 consists of two tensors connected via the root link while all other neighbours of the two tensors from the layer Λ_2. Thus, further filling all other layers accordingly leaves us with a well defined hierarchical structure. In this structure, each tensor in layer $\Lambda_{\ell>1}$ has exactly one neighbor in layer $\Lambda_{\ell-1}$ which is defined as its *parent tensor*, and may have several neighbors in layer $\Lambda_{\ell+1}$ which are defined as its *child tensors*. For $\ell = 1$, we define the two tensors to be each other's parents (which might go against traditional biological

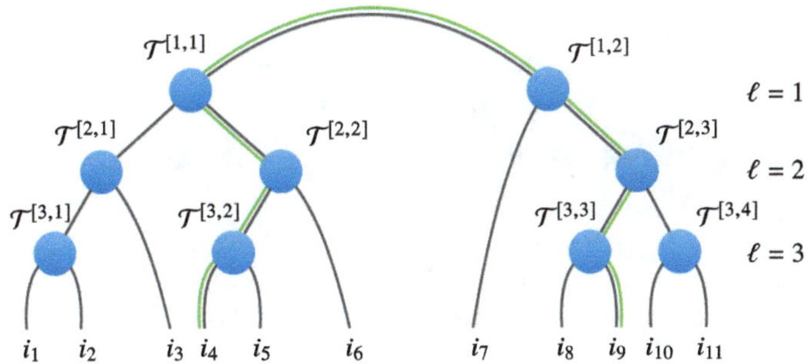

Fig. 8.1 Structure of a general binary Tree Tensor Network (TTN). The green line indicates the only possible path from Tensor $\mathcal{T}^{[3,2]}$ to $\mathcal{T}^{[3,3]}$, or more concretely from site i_4 to site i_9. In a binary TTN, each tensor has three links while in a general TTN, the tensors may have different numbers of links higher than three. (Figure reprinted with permission from [150])

intuition however ensures that every tensor within the network has exactly one parent tensor). Within this structure, the k-th tensor within the ℓ-th layer Λ_ℓ can be enumerated by $\mathcal{T}^{[\ell,k]}$. Further, the *depth* of a hierarchical TTN, is defined as its number of layers $\mathcal{L}$, i.e. the longest path between a tensor and the root link. For sake of compactness, we always choose the root link to be an internal link which minimises the number of layers of the hierarchically-ordered tree.

Given that the computational complexity of the contraction within a network scales with the product of the bond-dimensions of all links attached to the tensors, it is reasonable to minimise the number of links per tensor in an attempt to build an efficiently contractable Tensor Network. Therefore, a particular class of interest are the *binary TTNs* where each tensor can have three neighbours at most, i.e two children in the hierarchical structure. Further, all tensors within this network are of the order-3. Thus, when one tensor does not have two child tensors, it has a physical link attached. Based on the definitions above, we can further categorise a binary TTN into different subclasses:

Perfect TTN —In a perfect binary TTN, each layer ℓ is completely filled with the maximum number of tensors (2^ℓ). Thus, each tensor within the layers $\Lambda_{l<\mathcal{L}}$ has two child tensors while each tensor $\mathcal{T}^{[\mathcal{L},k]}$ within the last layer $\Lambda_\mathcal{L}$ has one internal link upwards connecting it to its parent tensor and two physical links downwards addressing the local Hilbert spaces of the quantum many-body system (see Fig. 8.2a). Consequently, this TTN is only able to capture systems with $N = 2^\mathcal{L}$ sites.

Complete TTN —Analogously to the perfect TTN, each layer within a complete TTN is completely filled, except for the last layer $\Lambda_\mathcal{L}$ which is filled from the left onward, i.e. all tensors within the last layer are as far left as possible (see Fig. 8.2b). Evidently, the *perfect TTN* is a subclass of a *complete TTN*. In contrast to the perfect TTN, however, this TTN is able to describe systems of arbitrary size.

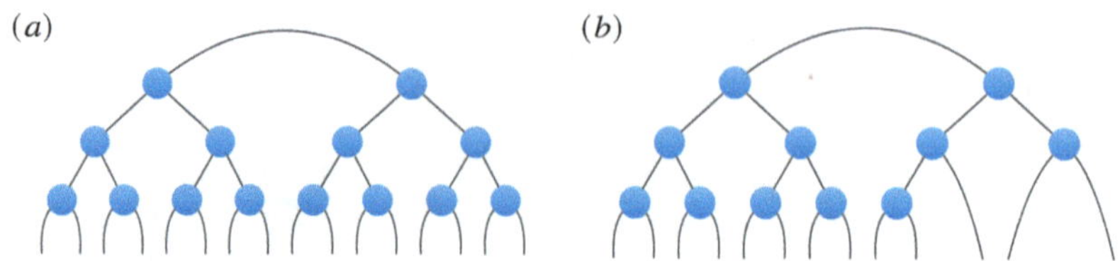

Fig. 8.2 Structure of a perfect TTN (**a**) where each of the $\mathcal{L}$ layers is completely filled with tensors resulting in $N = 2^{\mathcal{L}}$ sites and of a complete TTN (**b**) where the last layer may not be filled entirely. The perfect TTN is a subclass of the complete TTN. (Figure reprinted with permission from [150])

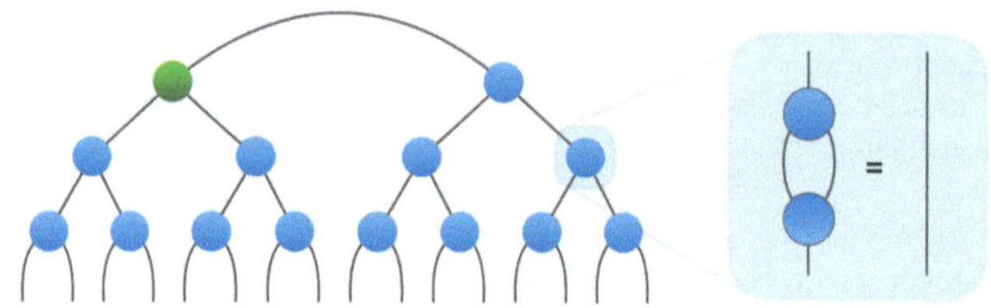

Fig. 8.3 TTN isometrised towards the tensor $\mathcal{T}^{[1,1]}$. All other tensors (blue) are isometries as introduced in Sect. 5.1.2 and obey the isometry condition (right). The tensor $\mathcal{T}^{[1,1]}$ (green) can be an arbitrary tensor. (Figure reprinted with permission from [150])

Matrix Product States —One particular subclass of binary TTNs is the MPS. In the framework of hierarchical trees, the MPS equals a binary TTN where each layer $\Lambda_{l>2}$ consists of one tensor only.

8.1.2 Isometry in a Tree Tensor Network

One crucial property of TTNs is that they can be brought in a certain isometric form by exploiting the gauge freedom of Tensor Networks. In this isometric form, or *unitary gauge*, the TTN mainly consists of isometries following their introduction in Sect. 5.1.2. As illustrated in Fig. 8.3, only one tensor within the network does not obey the isometry condition in the unitary gauge. Thus, this gauge is always defined with respect to a certain tensor $\mathcal{T}^{[l,k]}$ of the network. We call such a TTN *isometrised towards* $\mathcal{T}^{[l,k]}$.

The fact that most of the tensors are isometries and thus vanish when contracted with their complex conjugate over the proper links can be of great benefits in many numerical scenarios. The first benefit on hands is the computation of the norm $\langle \psi | \psi \rangle$ of the TTN state: In this computation, the entire TTN has to be contracted with its complex conjugate. Having the TTN isometrised to any tensor $\mathcal{T}^{[l,k]}$ reduces this computation to the contraction $\sum_{i_1,\dots,i_K} (\mathcal{T}^{[l,k]})_{i_1,\dots,i_K} (\mathcal{T}^{[l,k]})^\dagger_{i_1,\dots,i_K}$ of $\mathcal{T}^{[l,k]}$ with its own complex conjugate over all of its K links. In the next section (Sect. 8.1.3), for instance, we will see that, in the same way, the calculation of observables can be executed far more efficient when taking advantage of a properly isometrised TTN.

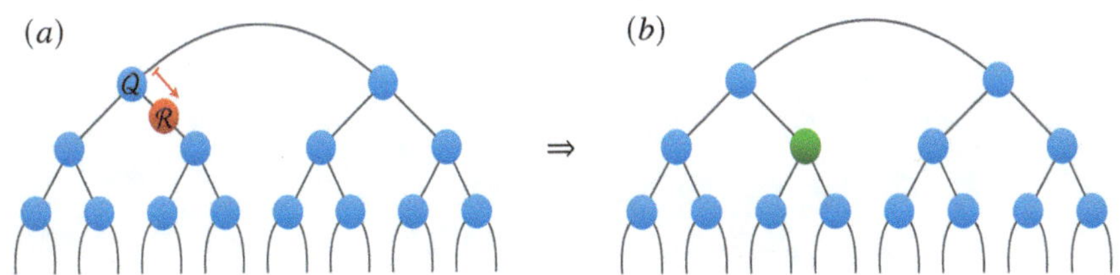

Fig. 8.4 Algorithm to reisometrise a TTN from position tensor $\mathcal{T}^{[1,1]}$ towards tensor $\mathcal{T}^{[2,2]}$: (**a**) The tensor $\mathcal{T}^{[1,1]}$ was decomposed via QR-decomposition into an isometry Q and an order-2 tensor $\mathcal{R}$ (red). Q denotes the new tensor $\tilde{\mathcal{T}}^{[1,1]}$. (**b**) The tensor $\mathcal{R}$ is contracted with the tensor $\mathcal{T}^{[2,2]}$ resulting in a new tensor $\tilde{\mathcal{T}}^{[2,2]}$ which does not fulfil the isometry condition any more (marked in green). Consequently, the TTN is isometrised towards $\mathcal{T}^{[2,2]}$. (Figure reprinted with permission from [150])

Having an isometrised TTN towards a given tensor $\mathcal{T}^{[l,k]}$, we can always change this gauge towards another tensor $\mathcal{T}^{[l',k']}$ in the network, or in other words *reisometrise* the TTN towards $\mathcal{T}^{[l',k']}$. For such a reisometrisation, we have to define the path $\Gamma_{l,k}^{l',k'}$ from the tensor $\mathcal{T}^{[l,k]}$ towards which the network is currently isometrised towards the target tensor $\mathcal{T}^{[l',k']}$. As first step in the procedure, the tensor $\mathcal{T}^{[l,k]}$ becomes decomposed via QR-decomposition into an isometry tensor Q and an order-2 tensor $\mathcal{R}$ along the direction defined by the path. Subsequently, the tensor $\mathcal{T}^{[l,k]} \leftarrow Q$ is replaced by the created isometry while the tensor $\mathcal{R}$ is contracted with the neighbouring tensor along the path $\Gamma_{l,k}^{l',k'}$. This first step already reisometrises the TTN towards the neighbouring tensor of $\mathcal{T}^{[l,k]}$. Consequently, performing this procedure iteratively along the entire path $\Gamma_{l,k}^{l',k'}$, reisometrises the TTN towards the desired tensor $\mathcal{T}^{[l',k']}$. As an example, Fig. 8.4 illustrates the reisometrisation from the tensor $\mathcal{T}^{[1,1]}$ towards its neighbouring tensor $\mathcal{T}^{[2,2]}$.

Since we create an isometry at each step of the reisometrisation, this procedure can be extended for establishing the unitary gauge within an arbitrary TTN by iteratively performing a QR-decomposition through the entire network. In practice, however, an isometrisation is typically created during the initialisation of a TTN as illustrated in Sect. 8.1.4.3. Once created, the executed TTN algorithms are usually designed to manage, keep track and take advantage of the isometrisation.

8.1.3 Calculation of Observables

In what follows, the computation of the expectation values of different observables O are presented for the TTN. Of particular interest in many simulations of quantum many-body systems are the calculations for *(i)* local observables, acting on one site s only, *(ii)* correlation functions, i.e. two-body operators, and *(iii)* general String observables.

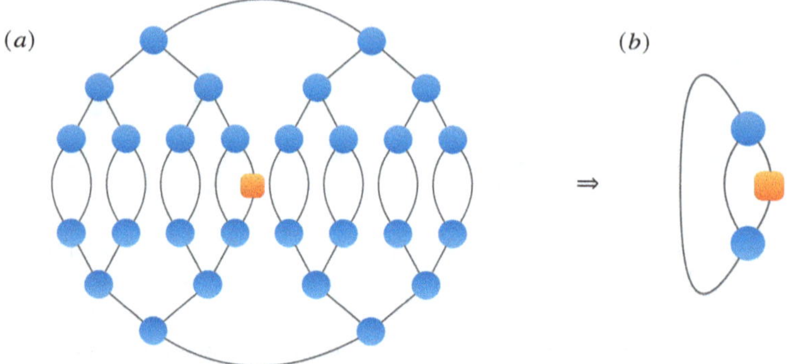

Fig. 8.5 Contraction scheme for the calculation of the expectation value of a local observable (orange) acting on a single site of the TTN (**a**). Isometrising the network towards the tensor directly attached to the operator enables to reduce the computation to a contraction of three tensors since all other tensors are isometries and become an identity when contracted with their corresponding complex conjugate (**b**). (Figure reprinted with permission from [150])

8.1.3.1 Local Observables

Let $O^{[s]}$ be a local observable acting on the site i_s of the TTN state ψ. In order to efficiently compute the expectation value $\langle O^{[s]}\rangle_\psi = \langle\psi|O^{[s]}|\psi\rangle$ the TTN can be isometrised towards the tensor $\mathcal{T}^{[\mathcal{L},\lfloor(s+1)/2\rfloor]}$ which acts on the physical site i_s ($\lfloor\cdot\rfloor$ denotes the floor function). Using this isometry in the TTN, the computation of $\langle O^{[s]}\rangle_\psi$ can be executed by contracting three tensors only with a complexity of $O(d^3m) + O(d^2m)$ where d is the local physical dimension of the system (see Fig. 8.5).

8.1.3.2 Correlators

Let $O^{[s_1,s_2]}$ be a correlator acting on the sites i_s and $i_{s'}$ of the TTN state ψ. In order to compute the expectation value $\langle O^{[s,s']}\rangle_\psi = \langle\psi|O^{[s,s']}|\psi\rangle$, the TTN can be isometrised towards the anchor of the path from the physical site s to s'. Thereby, only the tensors within this path are to be contracted since every other tensor cancels out with its complex conjugate in the global contraction (see Fig. 8.6). This contraction can be performed with a worst-case complexity of $O(m^4\kappa \log N) + O(m^4\kappa \log(N-1)) + O(m^3\kappa)$ where κ denotes the internal dimension of the Tensor Product Operator (TPO) representing the correlator $O^{[s_1,s_2]}$ (for TPO see Sect. 5.3.5) and $\log N$ equals the number of layers within the TTN.

8.1.3.3 String Observables

Let $O^{[s_1,s_2,...,s_\Omega]}$ be a string operator acting on the sites $s_1, s_2, ..., s_\Omega$ of the TTN state ψ. In order to compute the expectation value $\langle O^{[s_1,s_2,...,s_\Omega]}\rangle_\psi = \langle\psi|O^{[s_1,s_2,...,s_\Omega]}|\psi\rangle$, the TTN can be isometrised towards the anchor of the longest path of each combination of two the physical site $s \in \{s_1, s_2, ..., s_\Omega\}$ to $s' \in \{s_1, s_2, ..., s_\Omega\}$. In this scenario, only the tensors within all these different paths are to be contracted since every other tensor cancels out with its complex

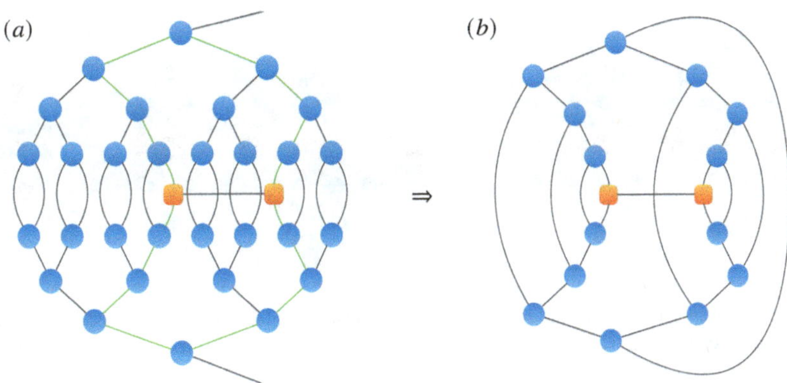

Fig. 8.6 Contraction scheme for the calculation of the expectation value of a correlator (orange) acting on two sites of the TTN (**a**). The open links indicate that the total TTN may be larger than the illustrated subtree. The green line indicates the path combining the two tensors directly attached to the operator. Isometrising the network towards the anchor node of this path enables to reduce the computation to the contraction of all tensors along the path since all other tensors vanish in the contraction due to the isometry condition (**b**). (Figure reprinted with permission from [150])

conjugate in the global contraction. In the most general case, the observable forms an MPO [185, 203]. In this, numerically worst case, non of the tensors can be neglected in the contraction and the complexity of the underlying contraction becomes $O(m^4\kappa^2(\kappa+2)(N-2)) + O(m^2\kappa)$ where κ is the dimension of the TPO representing $O^{[s_1,s_2,\ldots,s_\Omega]}$ and $N-2$ equals the total number of tensors within the TTN.

8.1.4 Ground State Computation

This section describes the optimisation algorithm for a TTN to compute the ground state of an interacting quantum many-body Hamiltonian. As introduced in Sect. 5.4.1, such a ground state search can be done efficiently with Tensor Networks by iteratively optimising a sub-set of its variational parameters minimising the energy $E = \langle\psi|\mathcal{H}|\psi\rangle$. In particular, this optimisation exploits the fact that all single tensors within the network can be treated as independent variables. Consequently, the key idea behind the optimisation of a Tensor Network state $|\psi\rangle$ is to optimise one, or only a limited subset, of the tensors at a time, thus reformulating the global, exponentially large, optimisation problem towards a set of local minimisations in a drastically smaller subspace which is controlled in size by the chosen bond-dimension of the network.

In the following, this optimisation procedure is illustrated in more detail starting from the global optimisation and how it can be reformulated as a local minimisation problem for either single tensors or a subset of locally connected tensors. Afterwards, the local optimisation problem is presented and how we can solve this

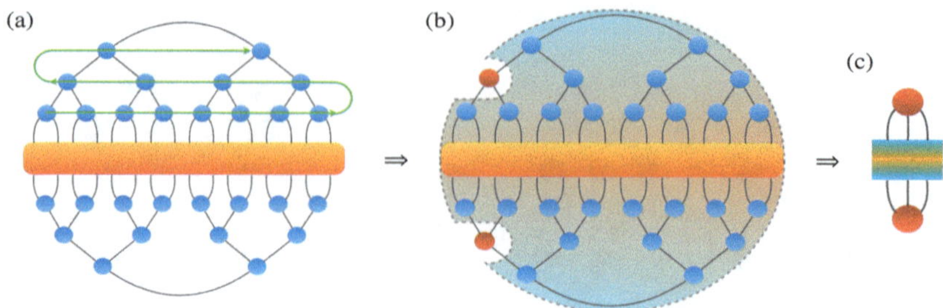

Fig. 8.7 Optimisation algorithm for a Tree Tensor Network to find the ground-state of a quantum many-body system. (**a**) The current energy can be calculated by contracting the Hamiltonian (orange) with the TTN and its complex conjugate. The minimisation of the energy is executed by iteratively optimising the variational parameters of each tensor. Therefore, a sweeping sequence (green) is defined in which we optimise the local tensors. (**b**) To optimise a targeted tensor (red), its environment (shaded area with border in dashed lines) is contracted describing the *effective Hamiltonian* of the local problem. (**c**) The local optimisation problem has been boiled down to the tensor (red), its complex conjungate, and its *effective Hamiltonian* (orange/blue). (Figure reprinted with permission from [150])

minimisation in practical applications before finalising this section with a discussion on the importance of the initialisation of the TTN for an efficient simulation.

8.1.4.1 Solving the Global Optimisation Problem

Let $\mathcal{H} \in \mathscr{H}$ be the Hamiltonian of a quantum many-body system and Ψ a TTN to represent a wave-function of the same Hilbert space. Then, the ground state of $\mathcal{H}$ is the solution to the minimisation problem

$$\min_{\Psi}\{E(\Psi)\} = \min_{\Psi}\{\langle\Psi|\mathcal{H}|\Psi\rangle\} .$$

To find the ground state of $\mathcal{H}$, the variational parameters of the TTN are optimised targeting the minimisation of the energy $E(\Psi)$ of the wave function it represents. Here, the TTN allows performing this optimisation problem by treating each tensor within the network as an independent set of variational parameters. Thus, the global optimization can be executed by *sweeping* through the network (see Fig. 8.7), where we subsequentially perform a local optimisation for each tensor. In order to perform such a local optimisation, the Tensor Network can be differentiated with respect to any single arbitrary tensor $\mathcal{T}^{[\kappa]}$ within the network:

$$\begin{aligned}
\frac{\partial\Psi}{\partial\mathcal{T}^{[\kappa]}} &= \frac{\partial}{\partial\mathcal{T}^{[\kappa]}} \sum_{i_1,i_2,\ldots,i_N} \left(\prod_{k=1}^{K} \mathcal{T}^{[k]}_{\{i\}_k,\{\chi\}_k}\right) \\
&= \sum_{i_1,i_2,\ldots,i_N} \left(\prod_{k=1}^{K} \frac{\partial}{\partial\mathcal{T}^{[k]}} \mathcal{T}^{[k]}_{\{i\}_k,\{\chi\}_k}\right)
\end{aligned}$$

$$= \sum_{i_1,i_2,\dots,i_N} \left(\prod_{k \neq \kappa} \mathcal{T}^{[k]}_{\{i\}_k,\{\chi\}_k} \right) \equiv \Psi^{[\kappa]}_{env}$$

Thus, the derivation of any function linear in $\mathcal{T}^{[\kappa]}$ can be calculated numerically where we call the resulting $\Psi^{[\kappa]}_{env}$ the environment of the tensor $\mathcal{T}^{[\kappa]}$. With this definition, the energy expectation value can be reformulated with respect to a single tensor

$$E = \langle \Psi | \mathcal{H} | \Psi \rangle = \langle \mathcal{T}^{[\kappa]} | \underbrace{\Psi^{[\kappa]}_{env} \mathcal{H} \Psi^{[\kappa]\dagger}_{env}}_{\mathcal{H}^{[\kappa]}_{eff}} | \mathcal{T}^{[\kappa]} \rangle$$

where we further define the *effective Hamiltonian* $\mathcal{H}^{[\kappa]}_{eff}$ for the tensor $\mathcal{T}^{[\kappa]}$. Consequently, the global Lagrangian functional $\mathcal{L}(\Psi, \Psi^\dagger) = \langle \Psi | \mathcal{H} | \Psi \rangle - \epsilon \langle \Psi | \Psi \rangle$ of the optimisation problem can be reformulated into a local Lagrangian functional $\mathcal{L}(\mathcal{T}^{[\kappa]}, \mathcal{T}^{[\kappa]\dagger}) = \langle \mathcal{T}^{[\kappa]} | \mathcal{H}^{[\kappa]}_{eff} - \epsilon \mathcal{N} | \mathcal{T}^{[\kappa]} \rangle$ with respect to $\mathcal{T}^{[\kappa]}$ which is reshaped as a vector $|\mathcal{T}^{[\kappa]}\rangle$. The *effective Hamiltonian* $\mathcal{H}^{[\kappa]}_{eff}$ hereby includes the physical Hamiltonian $\mathcal{H}$ and all tensors in the network surrounding $\mathcal{T}^{[\kappa]}$ while $\mathcal{N}$ denotes a positive operator for the effective square norm constraint. Thus, we can lower the global energy by optimising only the variational parameters within $\mathcal{T}^{[\kappa]}$. In practice, this reformulation is done by contracting $\mathcal{H}$ with the whole environemnt $\Psi^{[\kappa]}_{env}$, i.e. the complete TTN Ψ and its complex conjugate except for the tensors $\mathcal{T}^{[\kappa]}$ and $\mathcal{T}^{[\kappa]\dagger}$. Figure 8.7 illustrates such a contraction of the network towards the local tensor $\mathcal{T}^{[\kappa]}$ with $\kappa = (2, 1)$. Once, the Hamiltonian and the environment of $\mathcal{T}^{[\kappa]}$ is contracted into the effective Hamiltonian, the minimisation problem for $\mathcal{T}^{[\kappa]}$ lives in a space restricted by the chosen maximum bond-dimension m of the network, i.e. $\dim \mathcal{H}^{[\kappa]}_{eff} \leq m^3$. Typically this local optimisation problem can be solved directly by an eigenvalue decomposition $\mathcal{H}^{[\kappa]}_{eff} = E v^\dagger \mathcal{H}^{[\kappa]}_{eff} v$ where we update the tensor $\mathcal{T}^{[\kappa]} \to v_0$ with the eigenvector of $\mathcal{H}^{[\kappa]}_{eff}$ corresponding to the lowest eigenvalue E_0. Thus, optimising the tensor $\mathcal{T}^{[\kappa]}$ within the TTN consists of three different steps:

1. Contract the Hamiltonian $\mathcal{H}$ with the environment $\Psi^{[\kappa]}_{env}$ to the effective Hamiltonian $\mathcal{H}^{[\kappa]}_{eff} = \Psi^{[\kappa]}_{env} \mathcal{H} \Psi^{[\kappa]}_{env}$
2. Compute the eigenvector v_0 with minimum eigenvalue for $\mathcal{H}^{[\kappa]}_{eff}$
3. Update the tensor $\mathcal{T}^{[\kappa]} \to v_0$

Moving on to solve the local optimisation problem iteratively for all tensors within the network gives a high likelihood to converge into the global minimisation problem for the Lagrangian functional $\mathcal{L}(\Psi, \Psi^\dagger)$. In this way, the complexity of the global optimisation can be successfully reduced towards a significantly smaller space bound by the bond-dimension m of Ψ.

The exact *sweeping* sequence of the optimisation, in general, can be random which turns out to be beneficial in complex systems, such as the study of Lattice Gauge systems in higher dimensions (presented in Chap. 11). Another successful sequence is going iteratively from the bottom left to the top right tensor, as indicated in Fig. 8.7a (green). In this way, the information of the system can be propagated upwards through the TTN starting from the tensors directly attached to the Hamiltonian while still keeping a relatively short path to complete one sweep. Iterating this sweeping procedure through the complete network leads to the convergence for the global TTN in terms of energy and selected observables. Equivalently this strategy can be extended to optimising not only one tensor at a time but a small subset of tensors.

8.1.4.2 Local Optimisation

As discussed above, the global minimisation problem can be solved by iteratively solving a local eigenvalue problem for each local tensor $\mathcal{T}^{[\kappa]}$. For this optimisation, i.e. eigenvalue problem, we can exploit well-established algorithms from linear algebra. In particular, the Arnoldi algorithm [310,311] implemented in the ARPACK library [312], a collection of Fortran77 subroutines designed to solve large scale eigenvalue problems, offers an efficient numerical solution. The Arnoldi algorithm solves the local eigenvalue problem for each tensor $\mathcal{T}^{[\kappa]}$ by iteratively diagonalising its corresponding effective Hamiltonian $H_{eff}^{[\kappa]}$. In the practical application, the algorithm only needs to know the action of $H_{eff}^{[\kappa]}$ on $\mathcal{T}^{[\kappa]}$ and thereby provides the lowest eigenpairs of $H_{eff}^{[\kappa]}$ within a predetermined accuracy ϵ (see Sect. 8.2.3 for a hands-on example). The complete numerical complexity for this operation is determined by the dimension of all links of $\mathcal{T}^{[\kappa]}$, upper-bounded by $O(m^4)$. In practical applications, it might be wise to start with a low optimization precision ϵ of the Arnoldi algorithm and increasing it after each sweep such that the solution of the local eigenvalue problems becomes more and more accurate as the TTN comes closer to global convergence.

Interesting strategies are further to extend this single tensor update towards a two-site update, in which two neighbouring tensors $\mathcal{T}^{[\kappa]}$ and $\mathcal{T}^{[\kappa']}$ are optimised at the same time. This can be done analogously to the description above by contracting the two tensors and calculating the effective Hamiltonian, now with the environment $\Psi_{eff}^{[\kappa,\kappa']}$ with respect to the complete space of $\mathcal{T}^{[\kappa]}$ and $\mathcal{T}^{[\kappa']}$. However, since this would lead to a local minimisation problem for a contracted two-tensor space, it is more expensive numerically with a complexity of $O(m^5)$. After the optimisation of this contracted tensor, it is decomposed again into its original form, i.e. two interconnected tensors, keeping the TTN structure alive. This strategy may increase the numerical complexity, but it allows to adapt the bond-link dynamically (including its dimension and for symmetric TTNs its symmetry sectors) and thereby may provide a better or faster convergence to the global minimum.

As a good compromise, the subspace-expansion [170] can be exploited for the local single tensor optimisations. This technique approximates a two-site update by *(i)* first increasing the bond-dimension of the link connecting the two tensors, *(ii)* then iteratively optimising the two local tensors separately before *(iii)* truncating the bond-link back to original dimension. In this way, the local optimisation exploits the best of both worlds: It keeps the favourable algorithmic complexity of $O\left(m^4\right)$ while effectively performing a complete two-site optimisation and thus being able to adapt the bond-link dynamically. For more details on this technique, Ref. [170] offers an in-depth illustration and a comparison between the three strategies.

8.1.4.3 Initialisation

For the ground-state search of a Hamiltonian $\mathcal{H}$, it is beneficial to start from a random guess $|\psi_{init}\rangle$ in order to maximise the probability of an overlap with the target state $|\langle\psi_{\text{init}}|\psi_{\text{target}}\rangle|^2 > 0$. Such a randomly initialised tree can be created with the following bottom-top strategy illustrated in Fig. 8.8.

Starting from the physical indices at the bottom of the TTN, the lowest layer $\Lambda_{l=\mathcal{L}}$ is built up by merging two neighbouring local sites into one internal link introducing a randomly initialized tensor. Subsequently, the next layer $\Lambda_{\mathcal{L}-1}$ can be constructed similarly by taking the resulting internal links from each tensor within the just constructed layer $\Lambda_{\mathcal{L}}$, and again merging two neighbouring internal links by introducing randomly initialized tensors. Following this procedure until $l = 1$, the complete TTN would be initialised with an exponentially increasing bond dimension for each layer $m_{\mathcal{L}-l} = d^l$. Thus, in case we reach the user-defined maximum bond dimension m of the network when constructing each layer from the bottom to the top, we truncate the bond dimension $m_{\mathcal{L}-l} = \min\{d^l, m\}$ accordingly.

Importantly, in the case of present symmetries, this truncation of the maximum allowed bond-dimension has to be done by truncating the single symmetry sectors

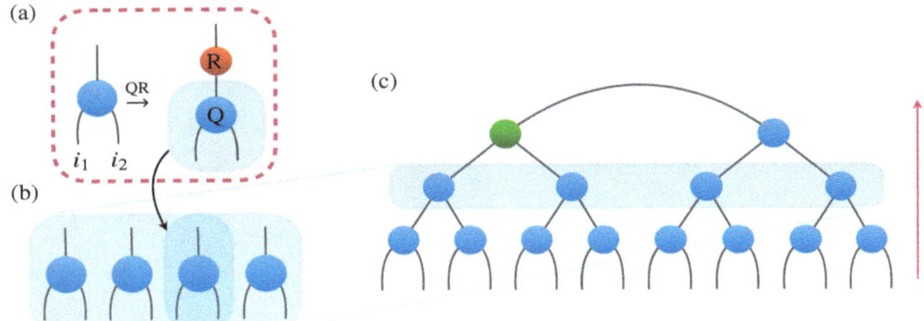

Fig. 8.8 Construction of a Layer for the initialisation of a Tree Tensor Network. (**a**) Given two sites (i_1, i_2) of local state spaces as input, an order-3 tensor is initialised with random elements combining the local spaces (bottom left). Performing a QR-decomposition creates an order-3 isometry Q which is used as the initialised tensor within a TTN layer (**b**). For the next TTN layer, the procedure is repeated, now using the links from the just created tensors of the lower layer. (**c**) The TTN is created layer after layer from the bottom to the top. (`Figure reprinted with permission from [150]`)

within the internal link. Here, we can introduce further randomness in the state by truncating these occupied coupling sectors randomly down until reaching the maximum set bond-dimension for the link in question. Thereby, not only the tensors themselves will be randomly initialised but further the distribution of the coupling symmetry sectors within the complete network. However, to ensure optimal convergence during the optimisation, it is critical to occupy the correct symmetry sectors within the internal links corresponding to the symmetry sectors of the ground state. However, as mentioned before, the local bond-dimension can be dynamically increased during the optimisation algorithm in order to adapt the symmetry sectors within the TTN. Thus, before a symmetrically invariant TTN converges to the ground state, it first converges in the occupied symmetry sectors on each link within the network.

8.1.5 Efficient Handling of the Hamiltonian

As described in Sect. 5.3.4, the complete many-body Hamiltonian $\mathcal{H} = \sum_p \mathcal{H}_p$ typically consists of several separate interactions $\mathcal{H}_p$ which can each be represented as a *Tensor Product Operator* (TPO). Figure 8.9 illustrates an example of such a Tensor Network representation decomposing $\mathcal{H}$ into a set of TPOs. In the TTN analysis, these TPOs act on different physical links of the network while the complete Hamiltonian acts on all open links, i.e. physical links, of the TTN.

In the previous Sect. 8.1.4, it was illustrated that during the optimisation procedure, we repeatedly need to compute the *effective Hamiltonian* $\mathcal{H}_{eff}$ for each tensor $\mathcal{T}^{[l,k]}$ within the TTN. This computation can be performed by contracting all TPOs separately through the network as indicated in Fig. 8.10: Each tensor of all the TPOs $\mathcal{H}_p$ are contracted with the tensor $\mathcal{T}^{[l,k]}$ they act on and its complex conjugate $(\mathcal{T}^{[l,k]})^\dagger$. Thus, we perform a renormalisation procedure of the interactions for each layer in the TTN. This procedure maps the single TPO parts onto the virtual subspace of the internal links of the TTN. In practical applications, we store

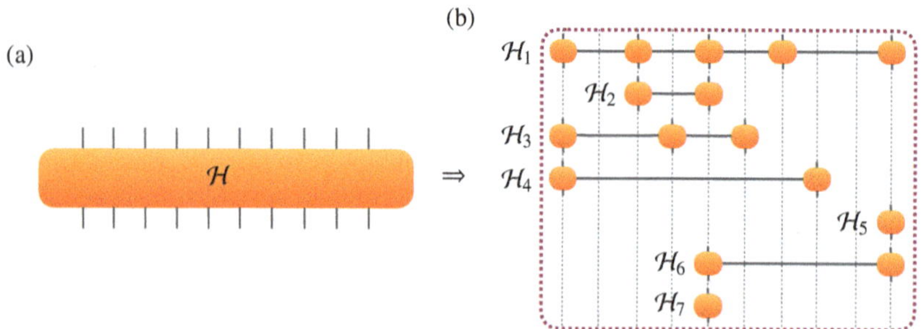

Fig. 8.9 Quantum many-body Hamilton $\mathcal{H} = \sum_p \mathcal{H}_p$ in a Tensor Network framework. The complete Hamiltonian (**a**) is represented as a set of *Tensor Product Operators* (TPOs), one for each interaction $\mathcal{H}_p$ (**b**). (Figure reprinted with permission from [150])

(a)

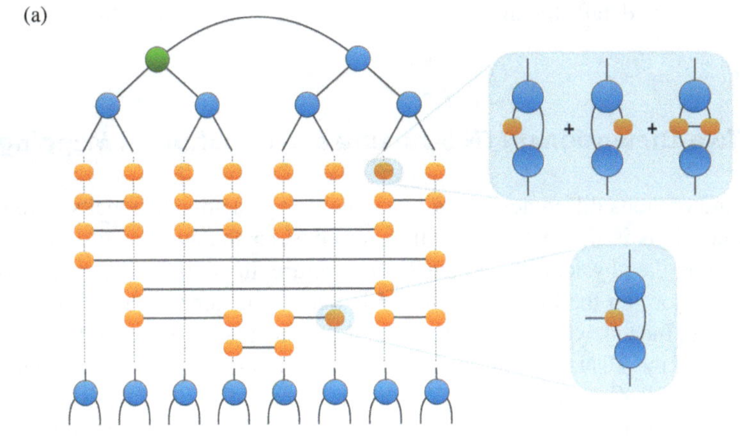

(b)

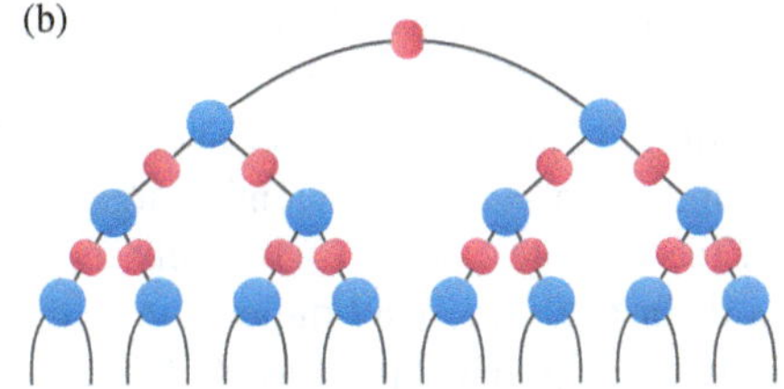

Fig. 8.10 Link-operators of a TTN for efficient computation of effective Hamiltonians $\mathcal{H}_{eff}$. (**a**) Each tensor of the TPO representing the physical interactions $\mathcal{H}_p$ have been contracted with the corresponding tensor in the lowest layer of the TTN and its complex conjugate (see inlet). (**b**) For each link, we store the corresponding coarse-grained operator-tensors as link-operators (purple) to reduce the required computations during the optimisation algorithm. (Figure reprinted with permission from [150])

these renormalised operators at each internal link χ of the TTN as *link-operators*. The explicit storing of the link-operators enables to improve the computation for various TTN algorithms, such as time-evolution or ground state search, by reducing the required computations during the algorithms. For further practical information, Sect. 8.2.3 illustrates a hands-on example including the initialisation of the link-operators and the underlying contraction schemes to calculate the *effective Hamiltonian* $\mathcal{H}_{eff}$ for a targeted TTN tensor $\mathcal{T}^{[l,k]}$.

8.2 Simulating High-Dimensional Systems with TTNs

Once a complete TTN is working for one-dimensional systems, the application to higher dimensions is in theory straightforward. However, there are still non-trivial practical hurdles to overcome in the implementation. The following section describes in more detail this application of a complete TTN to high-dimensional systems.

8.2.1 Two-Dimensional TTN Structure and Hamiltonian Mapping

The first and obvious difference going from analysing one-dimensional systems to higher dimensions is the fact that we have to consider the additional dimensions when attaching the physical sites to the TTN structure. In other words, the question arises of how to map the high-dimensional lattice to the generic one-dimensional TTN structure. Indeed, while for a 1D system the physical site i is applied to the i-th external link of the TTN, each physical site at position $(x_1, x_2, ..., x_D)$ of a general D-dimensional system requires a well-defined mapping to one of the physical links of the TTN. In fact, this mapping is highly critical for the representation power and accuracy of a TTN approach. Thus, for an efficient mapping, we aim to take advantage of the physical principle that the entanglement typically decays with distance in quantum many-body systems. Therefore, the mapped TTN structure shall be arranged in a way that the high entanglement bipartitions correspond to the lower branches of the TTN. For these lower branches $\nu \in \Lambda_l$ with $l \to \log N$, the bond-dimension m_ν is sufficiently large to capture the area law entanglement—or even the complete state—accurately, especially for reasonably small local dimensions d. Instead, the bond-dimension in the higher layers of the TTN $(l \to 1)$ suffers from the required exponentially large bond-dimension.

A straightforward way to implement such a mapping is to map the two-dimensional lattice in a zig-zag pattern towards the physical sites $j \in \{1, \ldots, N\}$ of the TTN. More concretely, each site (x, y) of the two-dimensional lattice with $x, y \in \{1, \ldots, L\}$ becomes mapped to an auxiliary site $j = x + L \cdot (y - 1)$ for the TTN simulation. In this way, the TTN performs a renormalises each "column" in x-direction of the system at its lower layers fist, before the coarse-grained "columns" become merged together in y-direction at the upper layers. Consequently, the resulting TTN topology is not well-suited for capturing correlations in y-directions and the numerical simulations will be biased in x-direction. Thus, this mapping is not ideal for a general TTN approach in two dimensions.

A well-established procedure is to construct the network in a way such that each layer of the TTN alternatingly groups together neighbouring sites in x- or y-direction respectively, as illustrated in Fig. 8.11c. Thereby, each tensor of a layer combines two sites to one coarse-grained virtual site, i.e. internal link, building up the hierarchical tree structure from the bottom to the top. In the case of a $L \times L$ system, this technique effectively coarse-grains the system in local plaquettes and enables to improve the representation power of correlations within these plaquettes in the TTN structure, resulting in a more precise representation of a quantum many-body state of the underlying system in general. However, this procedure by construction is only able to deal with binary-sized dimensions with $L_k = 2^n$.

For this problem, the *generalised Hilbert-curvature* introduced in Ref. [313,314] offers an efficient compromise: Exploiting curvature allows to construct a precise

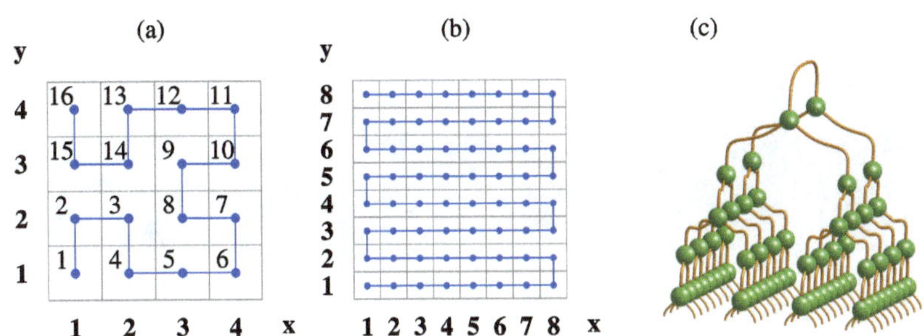

Fig. 8.11 Possible structures of a TTN deployed for two-dimensional systems. The TTN can be aligned along different paths through the system, such as the Hilbert curvature (**a**) or a snake-like pattern (**b**). The chosen path defines the mapping for each local site $(x, y) \rightarrow i \in \{1, ..., N\}$ from the two-dimensional system to the sites i of the TTN giving rise to different structures for the two-dimensional TTN simulation. In (**c**), a binary TTN structure is defined such that its tensors (green) each merge two sites to one bond link (brown). The mapping here groups alternatingly in x- and y-direction from layer to layer starting from the physical sites of the 8×8 lattice. (Figure reprinted with permission from [150])

TTN structure analysing high-dimensional systems while being able to keep the system linear dimensions L_k of arbitrary size. As illustrated in Fig. 8.12, generalised Hilbert-curvature defines the mapping from the high-dimensional system to one-dimensions. Accordingly, each interaction becomes mapped, in general, to an arbitrary one-dimensional long-range interaction. Consequently, the mapped one-dimensional system becomes analysed by the generic TTN structure. In fact, for isotropic systems with nearest-neighbour interactions, the TTN construction described above allows creating a graph together with the physical interactions which is surjectively homomorphic to the resulting graph when mapping the system using the Hilbert curvature. In other words, for binary sized systems, the TTN arising from the x-/y-coarse-graining procedure is equivalent to the TTN using the Hilbert-curvature apart from a structural degeneracy.

A priori, the exact entanglement entropy within a quantum many-body state is usually unknown before analysing the system which makes it challenging to figure out the ideal Tensor Network structure for a given system. However, we know that the entanglement is governed by the interactions within the systems. Moreover, there is even a connection between the coupling strengths g of a given interaction and the amount of introduced entanglement for the ground state: For increasing g the interaction introduces entanglement accordingly with $g \rightarrow 0$ keeping the sited disentangled and $g \rightarrow \infty$ introduces completely entangles the sites. However, in both extremes, disentangled and completely entangled, the introduced entanglement entropy $S = 0$ vanishes. Thus, the entanglement entropy maximises for medium coupling strength g. Thus, knowing the interactions of the system can help to find an efficient high-dimensional TTN structure by optimising the TTN layout such that

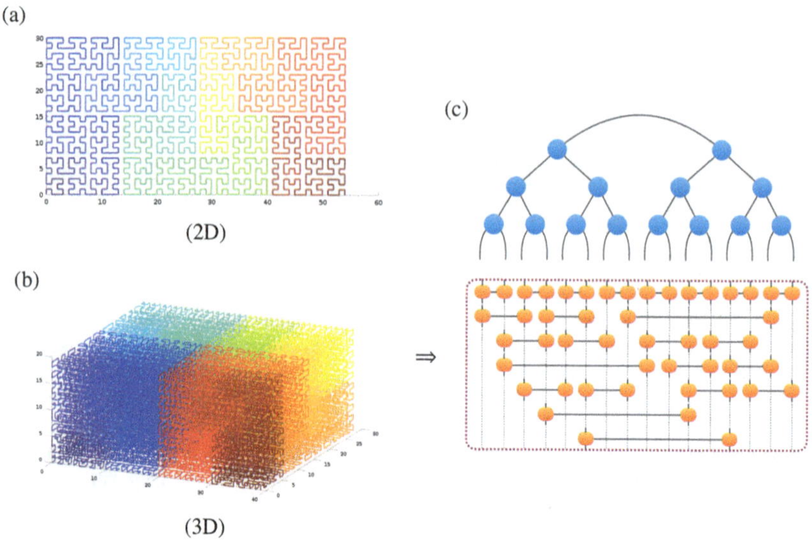

Fig. 8.12 Mapping of a high-dimensional system towards a generic TTN structure. (**a**)+(**b**) The generalised Hilbert-curvature defines the mapping for the local physical sites from the high-dimensional system to a one-dimensional path: (a) 2D mapping $(x, y) \rightarrow i \in \{1, ..., N\}$; (**b**) 3D mapping $(x, y, z) \rightarrow i \in \{1, ..., N\}$. The colour indicates the magnitude of i. (**c**) Mapping all interaction parts $\mathcal{H}_p$ of the high-dimensional Hamiltonian gives rise to a long-range interacting, one-dimensional system (purple-dashed box) in the generic basis of a TTN. Subfigures (**a**)+(**b**) taken from Ref. [313]. (Figure (c) reprinted with permission from [150])

it minimises the average weighted distance between any interacting sites

$$\langle D \rangle = 1/N_p \sum_{p=1}^{N_p} |g_p| D_p$$

where D_p is the path distance of an interaction with corresponding interaction strength g_p and N_p the total number of interactions in the system.

An example illustrating the importance of this mapping can be found by looking at a two-dimensional isotropic model with nearest-neighbour interactions only and periodic boundary conditions. These Hamiltonians of the shape $H = \sum_p \mathcal{H}_p$ with $|g_p| \equiv ||H_p||$ are frequently used to investigate magnetisation, e.g. in the form of the Heisenberg model.

One-Dimensional Systems

Before going to the analysis of the mapping in two dimensions, let us have a look at the one-dimensional case: As illustrated in Fig. 8.13, each interaction is acting on two neighbouring sites of the TTN with the number of interaction $N_{\mathcal{H}} = L - 1$ proportional to the system length L for open boundary conditions ($N_{\mathcal{H}} = L$ for

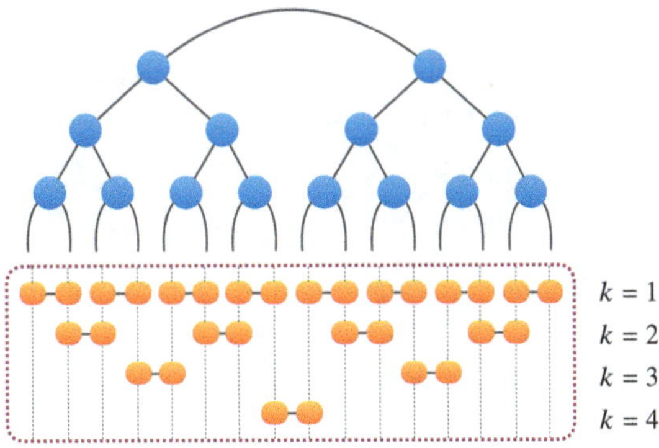

Fig. 8.13 Nearest-neighbour interactions for a one-dimensional TTN analysis. The interactions are ordered with respect to their distance D_p: All interactions in the k-level for $k \in [1, ..., \log L - 1]$ are connected through a path of length $D_p = 2k - 1$ through the network. (Figure reprinted with permission from [150])

periodic boundary conditions). Thus, there are $\frac{L}{2}$ interactions with distance $D_p = 1$, i.e. both physical sites of the interactions are assigned to the physical links of the same tensor. Further, $\frac{L}{4}$ interactions are acting with a path distance $D_p = 3$ on the TTN, i.e. they are connected over a tensor in the second layer, $\frac{L}{8}$ interactions with $D_p = 5$, etc. Following this patter up to the highest layer, the average distance equals

$$\langle D_{1\mathrm{D}} \rangle = \frac{1}{L}\left(\frac{L}{2}\cdot 1 + \frac{L}{4}\cdot 3 + \frac{L}{8}\cdot 5 + ...\right) = \frac{1}{L}\sum_{k=1}^{\log L} \frac{L}{2^k}(2k-1)$$

$$= \sum_{k=1}^{\log L} 2^{-k}(2k-1) \tag{8.1}$$

$$= 3 - \frac{1}{L}(2\log L + 3) \tag{8.2}$$

which converges to $\langle D_{1\mathrm{D}} \rangle \to 3$ for $L \to \infty$. Thus, the average path distance of a nearest-neighbour interaction within a TTN representing a one-dimensional system is finite in the thermodynamical limit.

Two-Dimensional Systems

For the same analysis in two dimensions, the average path distance $\langle D_{2\mathrm{D}} \rangle = \frac{1}{2}(\langle D_x \rangle + \langle D_y \rangle)$ can be separated into the average path distance of all interactions in x- and y-directions respectively. For the sake of simplicity, we may take a binary-sized $L \times L$ system with $L = 2^n$, thus, totalling L^2 interactions in each direction.[1]

[1] L^2 for each direction with periodic boundary conditions and $(L^2 - L)$ for each direction in case of open boundary conditions.

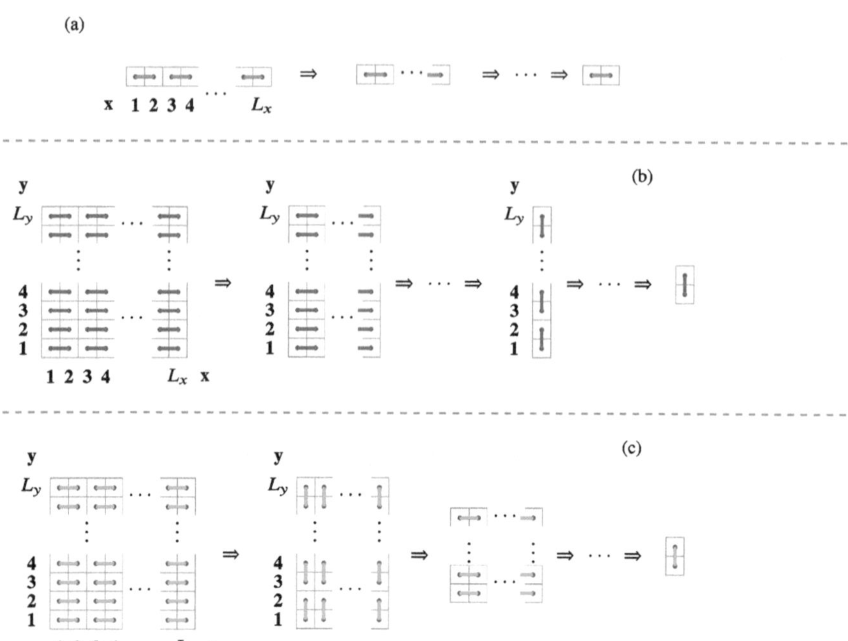

Fig. 8.14 Renormalisation procedures performed by each layer of the TTN for different underlying mappings. (**a**) Coarse-graining procedure of a one-dimensional system with L_x sites: Each tensor in the layer $\Lambda_{\mathcal{L}}$ of the TTN renormalises two neighbouring physical sites into an artificial site. The next higher layers continue merging neighbouring artificial sites until the last two sites are linked by the topmost link in the TTN. (**b**) Coarse-graining procedure of a two-dimensional system when exploiting the *Snake-like pattern* (Fig. 8.11b): The tensors in the layers Λ_l with $l = [\mathcal{L}, ..., \mathcal{L} - \log L_x]$ renormalise the entire first dimension of the system in x-direction, before the layers with $l = [\log L_y, ..., 1]$ coarse-grain the system in y-direction. (**c**) Renormalisation procedure of a two-dimensional system when exploiting the strategies of Fig. 8.11a,c: Each layer Λ_l coarse-grains the system alternatingly along the x- (for even l) and the y-direction (odd l). (Figure reprinted with permission from [150])

Snake Strategy —Applying the Snake strategy to the system, the TTN first coarse grains in x-direction with each interaction in x-direction being connected in the lower half (compare Fig. 8.14b). Thus, we have L coarse-grained subbranches with each one following the one-dimensional case equivalently. Consequently, the computation of $< D_x >$ follows the same structure of Eq. (8.2) except for the sum running to $\frac{1}{2}\log N$ and that there are now L of these sums for each coarse-grained "system column":

$$\langle D_x \rangle = L/N \sum_{k=1}^{\log N/2} \frac{L}{2^k}(2k-1) \overset{(N=L^2)}{=} \sum_{k=1}^{\log L} 2^{-k}(2k-1)$$

For each interaction in y-direction, however, the complete lower half of the TTN with distance $\log N/2$ always has to be gone through in the corresponding paths. In other words, these interactions recombine only in the higher half of the TTN topology offsetting each path distance D_p by $2\frac{1}{2}\log N = 2\log L$ (once going up through the half-tree and once going down). Thus, there are $\frac{L}{2}L$ interactions with distance $D_p = 2\log L + 1$, $\frac{L}{4}L$ with $D_p = 2\log L + 3$, $\frac{L}{8}L$ interactions with $D_p = 2\log L + 5$, etc.:

$$\begin{aligned}\langle D_y \rangle &= \frac{L}{N}\left(\frac{L}{2}\cdot(2\log L + 1) + \frac{L}{4}\cdot(2\log L + 3) + \frac{L}{8}\cdot(2\log L + 5) + \ldots\right)\\ &= \sum_{k=1}^{\log L} 2^{-k}(2\log L + (2k-1)) = 2\log L \underbrace{\sum_{k=1}^{\log L} 2^{-k}}_{=\log L(1-1/L)} + \sum_{k=1}^{\log L} 2^{-k}(2k-1)\end{aligned}$$

Indeed, for the calculation of $\langle D_y \rangle$, this can be seen as equivalently to the one-dimensional case with this offset in D_p and L times the interaction. These lower subbranches completely coarse-grain the system in x-direction reducing the problem to the y-direction at the layer $l = \log L$.

Consequently, the average interaction distance of the complete system with the Snake strategy equals

$$\begin{aligned}\langle D_{\text{Snake}} \rangle &= \frac{\langle D_x \rangle + \langle D_y \rangle}{2} =\\ &= \log L \underbrace{\sum_{k=1}^{\log L} 2^{-k}}_{=(1-1/L)} + \underbrace{\sum_{k=1}^{\log L} 2^{-k}(2k-1)}_{=3-\frac{1}{L}(2\log L+3)} = \log L + 3 - \frac{1}{L}(3\log L + 3)\end{aligned}$$

which diverges logarithmic for $L \to \infty$. Thus, using these naive Snake patters leads to a logarithmic divergence of the average interaction distance and thereby to a with system size increasing incapability of the network to properly capture the entanglement properties of the underlying system.

Hilbert Curvature —The fundamental idea of the Hilbert-curvature is to minimise the increase of the spatial distance when mapping two arbitrary sites from a high-dimension to a one-dimensional system. This is done by iteratively merging plaquettes of the system. The average interaction distance in the TTN can be calculated by analysing the TTN illustrated in Fig. 8.11c which alternatingly groups sites in x- and y-direction at each layer within the TTN (compare Fig. 8.14c). Consequently, when starting coarse-graining in x-direction at the lowest layer, there are immediately $\frac{N}{2}$ interactions acting in x-direction being connected by a tensor in this layer, i.e. with path length $D_p = 1$. The next layer connects $\frac{N}{2}$ interactions in

y-direction with a path distance of $D_p = 3$. Moving further upwards in the network, the $l = \log L - 2$-th layer combines $\frac{N}{4}$ interactions again acting in x-direction with $D_p = 5$, and the $l = \log L - 3$-th layer combines $\frac{N}{4}$ interactions in y-direction with $D_p = 7$. Following this pattern, the average path distance in x-direction reads

$$\langle D_x \rangle = \frac{1}{N}\left(\frac{N}{2} \cdot 1 + \frac{N}{4} \cdot 5 + \frac{N}{8} \cdot 9 + ...\right) = \\ = \sum_{k=1}^{\log N/2} 2^{-k}(4k-3) \overset{(N=L^2)}{=} 5 - \frac{1}{L}(4\log L + 5)$$

while the average path distance in y-direction equals

$$\langle D_y \rangle = \frac{1}{N}\left(\frac{N}{2} \cdot 3 + \frac{N}{4} \cdot 7 + \frac{N}{8} \cdot 11 + ...\right) \\ = \sum_{k=1}^{\log N/2} 2^{-k}(4k-1) \overset{(N=L^2)}{=} 7 - \frac{1}{L}(4\log L + 5)\,.$$

Consequently, the average path distance $\langle D_{2D} \rangle = \frac{1}{2}(\langle D_x \rangle + \langle D_y \rangle) \rightarrow 6 \ (L \rightarrow \infty)$ for this TTN structure converges in the thermodynamical limit making this strategy a solid approach for capturing the entanglement properties of the underlying system via TTN.

8.2.2 Area Law for Tree Tensor Networks

As discussed in Sect. 5.3.3, the von Neumann entanglement entropy for a bipartition of the system is directly linked to the bond-links of a Tensor Network. In this context, the loop-free nature of a TTN enables an efficient way of measuring the entropy captured in this approach of representing a quantum many-body wave function.

In every TTN state $|\psi\rangle$, each internal link v connecting two tensors bipartites the underlying lattice $\mathcal{L}$ into two subsystems $\mathcal{A}^{[v]}$ and $\mathcal{B}^{[v]}$ (see Fig. 8.15). This allows the TTN to be rewritten for each v in the form of the Schmidt decomposition

$$|\psi\rangle = \sum_{\alpha=1}^{m_v} \lambda_\alpha^{[\mathcal{A},\mathcal{B}]} |\psi^{[\mathcal{A}]}\rangle \otimes |\psi^{[\mathcal{B}]}\rangle \tag{8.3}$$

where m_v is the bond-dimension of v and $\lambda_\alpha^{[\mathcal{A},\mathcal{B}]}$ are the Schmidt-values. For each of the bipartitions, the Schmidt rank is upper-bounded by the bond dimension m_v of the corresponding link v. Consequently, the bond-dimension m_v as well limits the amount of bipartite entanglement the TTN is able to capture.

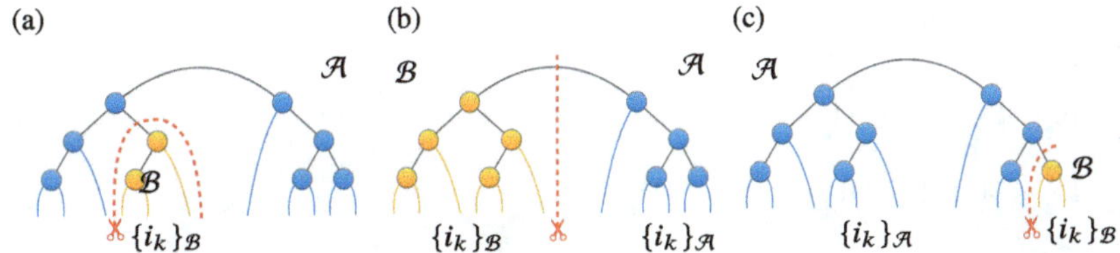

Fig. 8.15 Possible TTN bipartitions with the links $\{i_k\}_{\mathcal{A}}$ (blue) and $\{i_k\}_{\mathcal{B}}$ (yellow) addressing the physical sites within the bipartitions $\mathcal{A}$ and $\mathcal{B}$, respectively. (Figure reprinted with permission from [150])

If we assume the physical state ψ to fulfill the entropic area law, the entanglement entropy $S(\mathcal{A}^{[\nu]})$ of such a bipartition scales with the number of sites γ_ν forming the boundary ∂_ν of $\mathcal{A}^{[\nu]}$:

$$S(\mathcal{A}^{[\nu]}) \sim \gamma_\nu \tag{8.4}$$

For a practical illustration regarding the TTN representation $|\psi\rangle$ of the state ψ, we consider $\mathcal{L}$ to be a two-dimensional system with $N = L \times L$ sites. In this case, the required bond dimension to encode the area law of the physical state ψ scales exponentially with L (see Sect. 5.3.3 for more details on entanglement bounds in Tensor Networks). To be even more accurate, the bond dimension m_ν of each link ν should scale with

$$m_\nu \approx e^{c\gamma_\nu} \tag{8.5}$$

where c is a constant factor. This exponential scaling leads to the fact, that with increasing system size for a two-dimensional system, a TTN representation eventually fails to faithfully represent area law states as it rapidly becomes unfeasible to deal with such an exponentially large network especially with $\gamma_{\nu=1} \sim L$. And going beyond a 2D system to even higher dimensions, this scaling becomes all the more critical for the TTN.

8.2.3 A Practical Example: 4 × 4-System

To give a hands-on illustration, this section will describe the optimisation of a TTN for analysing high-dimensional systems in a step-by-step procedure for a two-dimensional isotropic Hamiltonian on a square 4 × 4-Lattice with open boundary conditions consisting of nearest-neighbour interactions $\mathcal{H}_p$. Therefore, we assume the complete Hamiltonian to be a product of interactions $\mathcal{H} = \sum_p \mathcal{H}_p$ with every interaction $\mathcal{H}_p$ being described as a Tensor Product Operator (TPO) (see Sect. 5.3.5). For sake of simplicity, we will restrict ourselves in this example to Hamiltonians $\mathcal{H}$ containing exclusively *(i)* local terms $\mathcal{H}_p = h^p_{i_p}$ (acting on the site

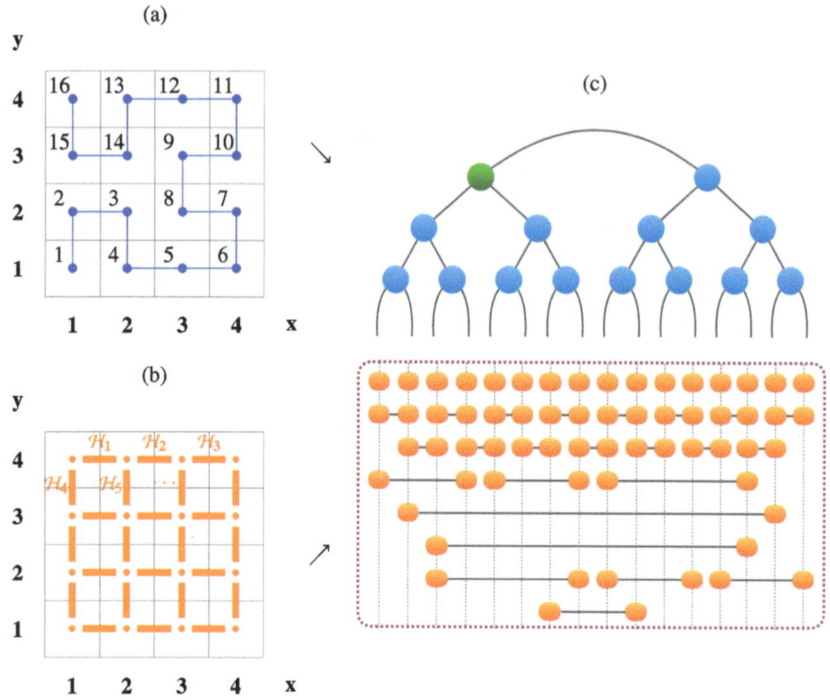

Fig. 8.16 Mapping of a two-dimensional 4×4 system with nearest-neighbour interactions and open boundary conditions to a one-dimensional TTN via Hilbert curvature. (**a**) The Hilbert curvature defines the mapping for the local sites $(x, y) \to i \in \{1, ..., 16\}$ from two into a one-dimensional path. (**b**) All interaction parts $\mathcal{H}_p$ of the two-dimensional Hamiltonian including local terms (orange dots) and nearest-neighbour interactions (orange rectangles), the latter enumerated from $\mathcal{H}_1$ to $\mathcal{H}_{24}$. (**c**) The operators in (**b**) are mapped by the Hilbert curvature in (**a**) towards the TTN. Each interaction $\mathcal{H}_p$, in general, becomes long-range interaction indicated as orange TPOs, making up the mapped one-dimensional model. (`Figure reprinted with permission from [150]`)

i_p) and *(ii)* two-body interaction $\mathcal{H}_p = h^{p,1}_{i_p} h^{p,2}_{i'_p}$ between the physical sites i_p and i'_p. However, the basic idea of the TTN approach can, in general, be applied to more complex Hamiltonians $\mathcal{H}$ and systems with periodic boundary conditions.

This optimisation procedure for the TTN consists of four main parts: *(i)* Mapping of the high-dimensional systems towards the TTN structure, *(ii)* initialising the TTN, *(iii)* initialising the TTN link-operators, and *(iv)* optimising the internal TTN with the mapped Hamiltonian.

Step 1 (Mapping) —Before starting the actual analysis, we map the system as described above in Sect. 8.2.1. For our 4×4-example, Fig. 8.16 illustrates the mapping of the two-dimensional system towards the one-dimensional TTN structure. The Hilbert curvature defines the mapping for the local sites $(x, y) \to$

i in the two-dimensional system to their one-dimensional correspondence $i \in \{1, ..., 16\}$ used for the TTN analysis. Given this mapping, all interaction parts $\mathcal{H}_p$ of the two-dimensional Hamiltonian become transformed accordingly into the one-dimensional basis. Crucially, all nearest-neighbour interactions in the two-dimensional plane are mapped to general long-range interactions in the one-dimensional representation. Thus, the Hamiltonian which in 2D consists of nearest-neighbour interactions only becomes a one-dimensional long-range model.

Step 2 (Initialisation TTN) —Once the system is mapped and the physical sites are assigned to the physical links of the TTN, we start constructing the tree structure. This construction can be done layer-by-layer from the bottom beginning with the physical links up to the top following the instructions in Sect. 8.1.4.3. In this example, we would construct a perfect binary TTN while, in general, for a total number of sites which is not a power of 2, the initialisation of the lowest layer has to be adapted towards a complete TTN since not all physical sites will be attached to a tensor in this lowest layer. After initialising such a random tree which is isometrised towards the tensor $\mathcal{T}^{[1,1]}$, the link operators within the TTN are initialised.

Step 3 (Initialisation Link-Operators) —At this stage, the coarse-grained Hamiltonian parts $\mathcal{H}_p^{[\nu]}$, i.e. link-operators, are calculated for each link ν in the network. This is done in a bottom-top approach following the procedure presented in Sect. 8.1.5. Figure 8.17 illustrates this procedure for our 4×4 example contracting for each layer the corresponding Hamiltonian parts with the proper tree tensor and its complex conjugate. Note, that after this initialisation, the effective Hamiltonian $\mathcal{H}_{eff}^{[1,1]}$ is already present for the tensor $\mathcal{T}^{[1,1]}$ towards which the TTN is isometrised.

Step 4 (Optimisation) —After initialising all link-operators, the algorithm starts sweeping through the network optimising each tensor. The exact sweeping order can, in general, be chosen arbitrarily from the user. When selecting one tensor to optimise, say $\mathcal{T}^{[2,1]}$ as illustrated in Fig. 8.18, the effective Hamiltonian has to be calculated following the description of Sect. 8.1.4.1. Having all link operators initialised and the TTN in an isometric form, this can be done easily by properly contracting the effective Hamiltonian parts $(\mathcal{H}_{eff}^{[l,k]})_p$ with all the tensors along the path from the tensor the TTN is current isometrised towards, to the target tensor. While performing this contraction along the path, it might be wise to isometrise the TTN towards the target tensor along the way by performing a QR-decomposition. Once we isometrised the TTN towards our target tensor, the local optimisation solves the underlying eigenvalue problem as illustrated in Sect. 8.1.4.2. When executed via ARPACK implementation, the optimisation algorithm only requires the knowledge of the action of the effective Hamiltonian $\mathcal{H}_{eff}^{[l,k]}$ on the tensor to optimise $\mathcal{T}^{[l,k]}$. Thus, we iteratively need to perform the contraction of $\mathcal{H}_{eff}^{[l,k]}$ with $\mathcal{T}^{[l,k]}$ which can be done by contracting each parts $(\mathcal{H}_{eff}^{[l,k]})_p$ of the effective Hamiltonian separately. Afterwards, the eigenvalue solver returns the eigenvector with the lowest

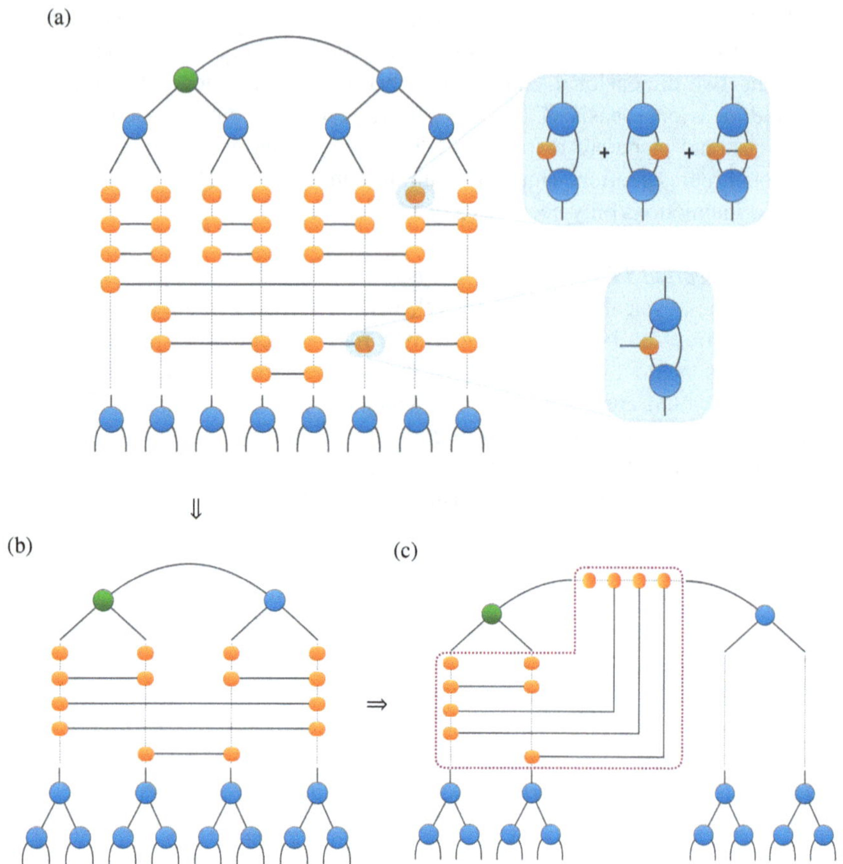

Fig. 8.17 Initialisation of the Hamiltonian parts of Fig. 8.16c within the isometrised TTN as link operators. (**a**) Each tensor of the TPO representing the physical interactions $\mathcal{H}_p$ have been contracted with the corresponding tensor in the lowest layer of the TTN and its complex conjugate (see inlet) making up the TPO tensors of the different link operators. (**b**) Analogously, the link operators of (**a**) are coarse-grained upwards with the next layer. (**c**) The last contraction is done with the tensor $\mathcal{T}[1, 2]$ resulting in all the link operators (in purple box) which constitute the effective Hamiltonian $\mathcal{H}_{eff}^{[1,1]}$ for $\mathcal{T}^{[1,1]}$. (Figure reprinted with permission from [150])

eigenvalue, i.e. energy, which we use to replace the old tensor in the network as a result of the local optimisation. After optimising a single tensor, we move on to the next tensor of the global optimisation sweep and after one sweep, we perform another one until the selected values, such as global energy, expectation values, or entanglement entropy, converge.

Add-On (Towards Numerical Complexity and Periodic Boundary Conditions) — Having presented a 4 × 4 system with open boundary conditions you may ask about the feasibility of simulating periodic boundary conditions as well. In fact,

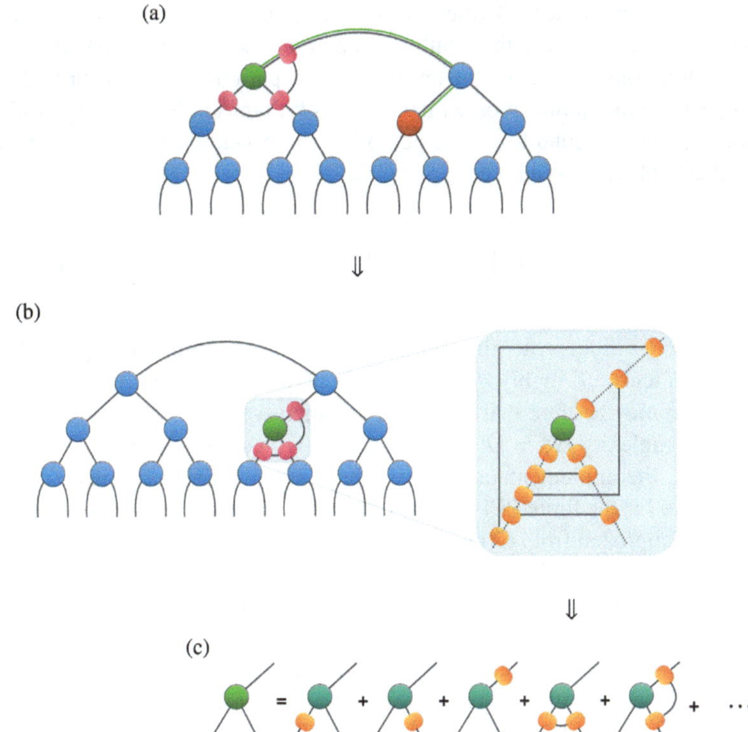

Fig. 8.18 Optimisation of a tensor for the 4 × 4 TTN simulation. (**a**) The TTN is isometrised to $\mathcal{T}^{[1,1]}$ (green) and all link operators are initialised with the effective Hamiltonian $\mathcal{H}_{eff}^{[1,1]}$ (purple) containing all local and interacting TPOs illustrated in Fig. 8.17c. The target tensor $\mathcal{T}^{[2,3]}$ (red) to be optimised is connected to $\mathcal{T}^{[1,1]}$ via the green path. (**b**) Having isometrised the network towards the target tensor (now green) and contracted the link operators accordingly, the effective Hamiltonian $\mathcal{H}_{eff}^{[2,3]}$ (purple) arises for the optimisation consisting of different TPOs (see inlet). (**c**) For the optimisation, the action (turquoise) of the effective Hamiltonian on the target tensor (green) is calculated by summing the contractions of all TPOs in $\mathcal{H}_{eff}^{[2,3]}$ with $\mathcal{T}^{[2,3]}$. (`Figure reprinted with permission from [150]`)

the difference between periodic and open boundary computations in terms of the numerical complexity of the TTN is merely a matter of the number of Hamiltonian parts $N_{\mathcal{H}}$, that is, it impacts sub-leading terms in the overall algorithmic complexity of m^4. Indeed, for open boundary conditions $N_{\mathcal{H}}$ is lower then for periodic once by a margin scaling with system length $N_{\mathcal{H}}^{periodic} - N_{\mathcal{H}}^{open} \sim L$.

The complete optimisation scales maximally with the complexity

$$\zeta_T = O\left(m^4 d^2 (N_{\mathcal{T}} N_{\mathcal{H}} I_{\mathcal{T}} + \sum_k N_{\mathcal{H},\mathcal{T}^{[k(\Gamma)]}})\right)$$

(plus some sub-leading terms) where d is the physical dimension, $N_{\mathcal{T}}$ the number of tensors in the TTN, $N_{\mathcal{H},\mathcal{T}}$ the number of operators acting on the tensor $\mathcal{T}$, $I_{\mathcal{T}}$ the number of iterations optimising the tensor $\mathcal{T}$, Γ the path from one tensor to the next one in the sweeping optimisation with $\mathcal{T}^{[k(\Gamma)]}$ its k-th tensor. Now, going from open to periodic boundary conditions, we get ΔN_H more Hamiltonian parts. Thus, the additional complexity would read

$$\zeta_{periodic} = \zeta_{open} + O\left(m^4 d^2 (N_T \Delta N_{H,T} I_T + \sum_k \Delta N_{H,T^{[k(\Gamma)]}})\right) .$$

While $\Delta N_H \sim L$ holds true, the additional number of Hamiltonian parts $\Delta N_{H,T}$ affecting each tensor T highly depends on the mapping itself. Still, for the sake of simplicity, we may assume that $\Delta N_{H,T} \sim L$ then we get in all leading terms an increase in complexity $\Delta\zeta = O(L)$ which scales with system length L.

In contrast, the implementation of periodic boundary conditions is by construction unfeasible for the finite PEPS and would result in an increasing complexity for a 2D-DMRG approach from $O(m^3)$ to $O(m^5)$.

To summarise, while for alternative methods the implementation of periodic boundary conditions might not be possible or at least require a change in the network structure resulting in a major increase in the scaling with bond-dimension, the aTTN can efficiently cope with the additional complexity of periodic boundary conditions without any fundamental changes where only the numerical complexity of sub-leading terms is affected.

8.3 Technical Implementation

The general structure of the implementation is illustrated as a simplified overview in Fig. 8.19. It shows the hierarchical structure of general Tensor Network algorithms in quantum mechanics. The lowest level is a common linear algebra kernel, such as LAPACK or BLAS, executing the core functionalities on the computational hardware. Using these mathematical libraries, the tensor module extends the linear algebra to the tensor algebra described in Sect. 5.1 including the numerical definition of a tensor type instance and its functions. As described briefly in Sect. 6.2, we can include another level managing the incorporation of symmetries in the code by, loosely speaking, defining the symmetrically invariant tensor as a list of tensors, one for each symmetry subspace. This library includes the definition of symmetric tensors including all operations one may perform to manipulate a symmetrically invariant tensor (see Ref. [170,315] for an in-depth description). For general Tensor Networks, this symmetry layer in the structure is not required. However, it can drastically improve the computation by taking advantage of the well-defined internal structure of symmetrically invariant Tensor Networks on the level of the single tensors within the network. The actual Tensor Network then utilises this library (or the tensor library itself in case of neglecting symmetries) and the therein defined

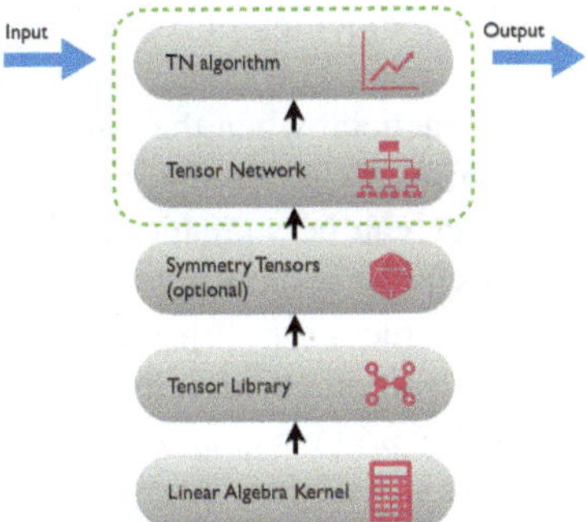

Fig. 8.19 Overview of a typical software structure of a generic Tensor Network code with its different levels ranging from the *Linear Algebra Kernel* performing operations on the hardware up to the main *TN algorithm* interfacing with the user. The main Tensor Network developments performed within this thesis, build on a pre-existing *symmetry tensors* library and consists of the two highest levels (green-dashed box), in particular the implementation of a TTN as *Tensor Network* and a corresponding ground-state optimisation as *TN algorithm*. Figure 8.20 illustrates both of the two levels in more detail. (Figure reprinted with permission from [150])

tensors as the building blocks for the complete network. Here, the exact geometry is defined together with the important functions required, such as calculation of observables, optimisation or time evolution. Finally, on top of everything, the main algorithm for the Tensor Network itself is defined. Later on, the structure of the implementation of such a Tensor Network level is exemplified in more detail for a Tree Tensor Network (TTN). In what follows, each of the nodes for the implementation of the TTN is described in more detail.

8.3.1 Data Structure of a Tree Tensor Network

In general, there are two main ways to implement a Tree Tensor Network for the simulation of quantum many-body systems. On one hand, the TTN can be implemented as a set of nodes where each node consists of a tensor and a list of references, i.e. pointer, to its neighbouring tensors. As an example a binary TTN can be described by the following object *node* written in pseudo-code:

```
1 struct node
2 {
3     tensor T;
4     struct node* left;
5     struct node* right;
6 };
```

In this way of implementation, the topology of the TTN is implicitly stored by the references for the links on each tensor. This concept in general allows very flexible TTN topologies which can be beneficial when the topology shall be adapted to the underlying problem or even on the fly in an optimisation algorithm. The benefit of adapting the network topology can be valuable for its applications in quantum chemistry where the interactions in the system are, in general, not homogeneous but rather determined by molecular structure [246]. Further, a dynamically adaptable topology might turn out the key to automatically finding the most efficient topology for a given problem which is a very promising concept that is not well studied so far and denotes an interesting direction for future research of Tensor Networks.

On the other hand, the TTN can be structured in a static hierarchical order with an array of tensors each being assigned a fixed position within the tree as illustrated in the following pseudo-code.

```
1 struct Tree
2 {
3       tensor[] T;
4       address[] x,y;
5       topology K;
6 };
```

Thus, in this concept, the topology of the TTN is explicitly defined in the data structure. While this static implementation may lack some flexibility in dynamically adapting the topology, it is very well suited for efficient TTN simulations with a well, predefined topology. Fixing the exact topology on a global level allows incorporating the knowledge of the topology in the algorithms manipulating the TTN, such as the optimisation, calculation of observables and computation of entropy, which potentially leads to a more efficient, well-structured implementation and allows for efficient parallelisation techniques on these global algorithms. Consequently, this data structure is well suited for implementing a *perfect TTN*, or a *complete TTN* for which the topology is static and well-defined with each tensor sitting at a predefined position within the tree.

8.3.2 Tree Tensor Network Implementation

The following section is dedicated to the implementation of a TTN software capable of performing ground-state search. In particular, it describes the different modules shown in Fig. 8.20 in more detail and presents some hands-on examples taken in a simplified form from the FORTRAN90 code developed in the course of this thesis [150]. These examples illustrate the most important data structures and most relevant functions for implementing a TTN algorithm for high-performance computations of high-dimensional quantum many-body systems.

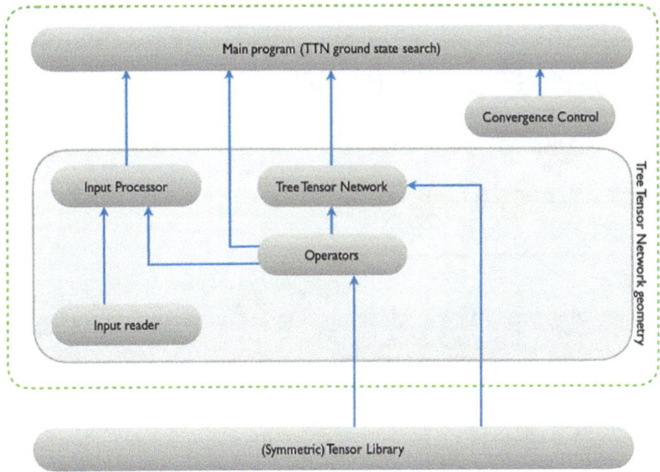

Fig. 8.20 Software structure of a Tree Tensor Network code performing analyses such as ground-state search or time-evolution. Build on a library for (symmetric) tensors including their operations, a TTN geometry can be properly defined by the modules in the shaded box: The *input reader* gets the data from the user which is then pre-processed for the analysis by the *input processor*. The *Operators* module defines all required low-level data types while the TTN itself is defined by *Tree Tensor Network* module. The modules are interfaced with the *main program* controlling the execution of the specific Tensor Network algorithms with the aid of a *Convergence Control* module. (Figure reprinted with permission from [150])

8.3.2.1 Library for (Symmetric) Tensors

As mentioned at the beginning of Sect. 8.3, the actual TTN code utilises a library in which a tensor instance is defined together with all operations one may perform to manipulate a tensor (see Sect. 5.1). This library may optionally feature the ability to deal with symmetries to improve computational time and to target a desired global symmetry sector (see Sect. 6.2 for more details). For the description of the TTN software described in this section, we will assume such a library as given and use it as the building block for the developed code. For an in-depth technical description of such a library, Ref. [170] offers profound literature on the implementation of general tensor libraries with Abelian symmetries and Ref. [315–317] further provide in-depth technicalities for incorporating non-Abelian symmetries onto an existing, ordinary tensor library.

8.3.2.2 Operators

Building on the tensor library, this module is mainly dedicated to the description of operators, and in particular of the Hamiltonian $\mathcal{H} = \sum_p \mathcal{H}_p$ of the underlying system to analyse. The complete operator is stored as an `MPOLayer_type` instance, which stores and keeps track of all single interactions $\mathcal{H}_p$ of the Hamiltonian and on a global level mimics an MPO. Thus, when using this `MPOLayer_type` instance it can be treated like an MPO while it internally

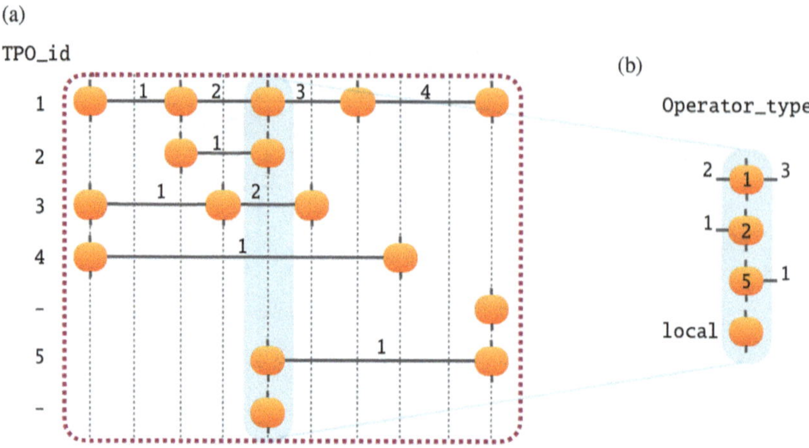

Fig. 8.21 MPOLayer_type instance for storing a Hamiltonian $\mathcal{H} = \sum \mathcal{H}_p$ with different interactions $\mathcal{H}_p$. (**a**) All interactions are represented as TPO. Except for local operations, all TPOs are enumerated (TPO_id). Further, all links are enumerated for each TPO. (**b**) An Operator_type instance models all operations acting on the same site i: All local operations are accumulated into the tensor local while for the interaction parts, the single tensors are stored with the corresponding TPO_id and the link indices. (Figure reprinted with permission from [150])

manages the TPOs it consists of (where each TPO represents an interaction part $\mathcal{H}_p$). However, due to numerical efficiency in the global TTN algorithms, the MPOLayer_type instance is not decomposed directly into a plain list of TPOs, but rather into a list of Operator_type instances which for each site i of the MPO consists of all tensors acting thereon. Thus, effectively, the MPOLayer_type instance describes a layer of operators or, more generally, describes a list of lists of tensors.

Figure 8.21 illustrates the way of storing an Hamiltonian as MPOLayer_type instance. In fact, with the illustration in mind, it becomes evident that there are two-dimensions in the data: The number of TPOs, i.e. interaction parts $\mathcal{H}_p$, and the number of physical sites N of the Hamiltonian. Thus, we order the data first in the direction of the latter dimension in the MPOLayer_type instance and subsequently in the former dimension within the Operator_type. Thus, each Operator_type consists of a list of tensors modelling the overall action of the Hamiltonian on one single site i. Here, we can exploit another numerical benefit: All operator acting locally on a single site can be added together. As example, we can take a Hamiltonian, such as $\mathcal{H} = \sum_i \alpha_i \sigma_x^{[i]} + \beta_i \sigma_z^{[i]} + \sigma_x^{[i]} \sigma_x^{[i+1]}$, for which we can merge the local fields $\mathcal{H}_{i,\text{local}} = \alpha_i \sigma_x^{[i]} + \beta_i \sigma_z^{[i]}$ into one local operator. Such a summation cannot be done locally, however, on the latter interaction part $\mathcal{H}_i = \sigma_x^{[i]} \sigma_x^{[i+1]}$ of the Hamiltonian. Thus, the Operator_type instance only needs to store one local tensor at most modelling the local single-site operators of the Hamiltonian and a list of tensors which together with all other Operator_type instance in an `MPOLayer_type` instance. The two instances might be implemented as follows.

Operator_type

An `Operator_type` instance can be implemented as an extension of a general tensor list. This can be useful, in particular, since a general instance for a tensor list can be exploited on numerous occasions within a TTN algorithm. As shown in the following listing, the tensor list would model all tensors corresponding to a many-body interaction term $\mathcal{H}_p$ while the extension *(i)* adds an optional tensor modelling all local single-site operators of the Hamiltonian and *(ii)* includes an identification array `id` to keep track of the connections of each single interaction term $\mathcal{H}_p$ within the complete Hamiltonian, i.e. the global `MPOLayer_type` instance.

```
1  ! This is a general list of Tensors which can be extended
     for the Operator_type instance:
2  type, public :: TensorList_type
3    private
4    integer :: numtensors = 0
5    type(Tensor_type), dimension(:), allocatable :: tensor
6  end type
7
8  ! This is how an Operator_type instance could look like
     extending the TensorList_type instance above. In
     comments it can be useful to illustrate this instance
     as
9  !                          |
10 !                         -O-
11 !                          |
12 type, extends(TensorList_type), public :: Operator_type
13   private
14   logical :: contains_local
15   type(Tensor_type), allocatable :: local
16   integer, dimension(:,:), allocatable :: id
17 end type
```

The attributes `contains_local` and `local` take care of the possible presence of a single-site term in the Hamiltonian while `id` crucially stores all information about the interaction structure within the Hamiltonian. In particular, this will become a 3×`numtensors` sized array where each triplet entry `id(:,k)=[TPO_id, id_left_link, id_right_link]` contains the all the required information regarding the `k`-th tensor in the list: as indicated in Fig. 8.21, each TPO $\mathcal{H}_p$ becomes a unique identification number `TPO_id` assigned (such as the index p of the interaction $\mathcal{H}_p$ it represents). Further, all tensors of one TPO are ordered with respect to the physical sites they are applied to. In this order, all the internal links of the TPO are enumerated from left to right. Consequently each tensor within the TPO may contain one link to the left and one to the right indicated by `id_left_link` and `id_right_link`, respectively. Additionally allowing these values to be, e.g. `-1`, in case the tensor denotes the endpoint of a TPO, enables `id` to sufficiently capture all relevant information for a typical TTN analysis.

Note that this kind of implementation only allows representing interaction terms $\mathcal{H}_p$ as TPOs where each tensor has two internal links at most. Thus, the modelling of star-shaped interaction terms as illustrated in Fig. 5.7d is not possible. Furthermore, it is worth mentioning that the first index is the running index in FORTRAN while in other languages such as C++ this is not the case. Thus, for the latter scenario, it is numerically more efficient to exchange the indices.

`MPOLayer_type`

The data structure of an `MPOLayer_type` is straight-forwardly a list of these `Operator_type` instances as presented in the following.

```
1  ! This could be an MPOlayer_type instance in FORTRAN
   which might look like this in the code commentation:
2  !
3  !          | | |      |
4  !         -O-O-O- ~ -O-
5  !          | | |      |
6  !
7  ! Designed to be applied to the lower links of a
   TreeLayer_type instance which we will formally
   introduce later on. This could be an example of such an
    application:
8  !          ___|___   _|_   _|_  |  _|_
9  !         (_______) (___) (___) | (___)
10 !          | | | |   | |   | |  |  | |
11 !         -O-O-O-O---O-O---O-O--O--O-O-
12 !          | | | |   | |   | |  |  | |
13 type, public :: MPOLayer_type
14   integer :: NumSites
15   type(Operator_type), dimension(:), allocatable :: site
16 end type
```

8.3.2.3 Tree Tensor Network

This code module describes a general TTN and contains all its relevant functions. In particular, this module includes the object definition for one layer of the TTN (`TreeLayer_type`), for a general path (`path_type`) which can be defined in the underlying tree structure, and ultimately the object definition modelling the TTN itself (`tree_type`). For the ground-state search, important functions to be included in this module for the `tree_type` instance would be in particular the isometrisation of the network towards a targeted tensor including the proper contraction of the effective Hamiltonian $\mathcal{H}_{eff}$ (see Sect. 8.1.4), and the local optimisation of a single tensor.

tree_type

The `tree_type` instance is modelling a Tree Tensor Network with a strict layer orientated topology as illustrated in Sect. 8.1 and following Definition 8.1.1. The complete tree consists of several layers (`TreeLayer_type`, see below) with each layer consisting of several tensors. In general, each tensor can have an arbitrary order and only one link per tensor is connected with the next upper tree layer. A certain tensor within the TTN structure is addressed by two indices, one `layer_index` for the layer and the second index `tensor_index` for the tensor within the specific layer. Thus, a general `tree_type` instance might be implemented as:

```
! This is a tree_type instance in FORTRAN which can be a general TTN structure like this one:
!
!   Layer                     _________________
!                       _____|_____         ___|___
!      1               (___________)       (_______)
!                    ___|__   ____|____     |   __|__
!      2            (______) (_________)    |  (_____)
!                  __|__  | _|_   ___|___   |  _|_  |
!      3          (_____) | (___) (_______) | (___) |
!                  | | |  |  | |   | | | |  |  | |  |
!
type, public :: tree_type
  ! Tree topology attributes
  integer :: NumSites, NumLayers
  type(TreeLayer_type), dimension(:), allocatable :: Layer

  ! Tree isometrisation
  logical :: is_isometrised = .false.
  integer, dimension(2) :: iso_position

  ! Link Operators
  type(MPOLayer_type), dimension(:), allocatable :: Operators
end type

interface tree_type
  module procedure :: tree_construct
end interface
```

where the flag `is_isometrised` together with `iso_position` keeps track of the isometrisation of the TTN. Additionally, the attribute `Operators` can be used to store the link operators as discussed in Sect. 8.1.5.

Crucially, the constructor `tree_construct` of this instance creates a TTN as described in Sect. 8.1.4.3: It initialises a `tree_type` instance for a given number of physical sites N with random tensors where the isometrisation of the TTN can be done on the fly.

`TreeLayer_type`

The `TreeLayer_type` instance describes a tensor layer within a `Tree_type` instance. Therefore, it consists of a list of tensors including the references (`upper_address`) for each of the tensors in this layer to its parent tensor (the corresponding tensor within the next-higher layer to which it is connected). To make this address system consistent, each layer needs to keep track of its own sites which are pointing downwards to child tensors. This track-keeping can be done by ordering and enumerating all the downwards pointing sites of the layer, and then storing the first and the last site indices for each tensor within the layer (`site`). Thus, with the information of `upper_address` from a lower layer and `site` from the next-higher layer, we can reconstruct tensor are connected inter-layers.

```
! This is a TreeLayer_type instance which might look like this:
!
!         ___|___  |  ___|___   _|_   _|_  |  _|_  |
!        (_______) | (_______) (___) (___) | (___) |
!         | | | |  |  | | | |   | |   | |  |  | |  |
!  site:  1 2 3 4  5  6 7 ...
!
type, private :: TreeLayer_type
  ! Tree Layer attributes
  integer :: numtensors = 0
  type(Tensor_type), dimension(:), allocatable :: tensor
  integer :: cutoff

  ! Topology attributes
  integer, dimension(:,:), allocatable :: site
  integer, dimension(:,:), allocatable :: upper_address

  ! Singular values
  type(real_vector), dimension(:), allocatable :: SingVals
end type
```

Additionally, with `sv_present` and `SingVals`, we can keep track of the singular values sitting on each of the link directed upwards in the TTN layer. This might be interesting since *(i)* the singular values are required to calculate the entropy following the description in Sect. 5.3.3, *(ii)* they are beside the global energy a good predictor for convergence of the state representation in a ground state search and *(iii)* can be a good stop criterion for time evolution in which the entanglement may grow beyond the faithful representation power of a TTN with limited bond-dimension m (which is here set as `cutoff`).

path_type

This auxiliary object follows Definition 8.1.2 (Path) and should, among others, help to propagate through the TTN during the optimisation procedure. Each path has an *anchor tensor* which is defined as the highest tensor within the hierarchical network structure. This anchor is stored redundantly by its position within the network `anchor_address` and its position `position_anchor` when traveling along the path. The path itself consists of an `4×length` array describing the `steps` of the path. The quartet `steps(:,k)` contains all necessary information for the `k`-th step of the path: *(i)+(ii)* The address `[layer_index, tensor_index]` of the `k+1`-th tensor as a tuple for the tree layer and the tensor position within the layer, *(iii)* the index for the link of the `k+1`-th tensor which is connected to the `k`-th tensor and accordingly *(iv)* the index for the same link on to the `k`-th tensor. Consequently, the instance can be implemented as shown in the following listing.

```
1  type, public :: path_type
2     ! Path attributes
3     integer :: length, position_anchor
4     integer, dimension(2) :: anchor_address
5
6     ! Single steps of the path
7     integer, dimension(:,:), allocatable :: steps
8  end type
```

tree_optimization_sweep

Having encoded the TTN as a `tree_type` instance, we can implement an optimisation sweep as described in Sect. 8.1.4.1 simply by looping over all tensors within the TTN structure, for each one contracting the effective Hamiltonian towards the targeted tensor, isometrising the network accordingly and optimising this tensor.

```
1   ! This could be an optimisation sweep for a tree_type
      instance:
2   subroutine tree_optimization_sweep( tree , energy ,
      arnoldi_tolerance )
3     class(tree_type), intent(inout) :: tree
4     double complex , intent(out) :: energy
5
6     double precision , intent(in), optional ::
      arnoldi_tolerance
7     integer :: ll , tt
8
9
10    do ll = 1, tree%numlayers
11      do tt = 1, tree%layer(ll)%numtensors
12        !
13        ! Isometrise the network towards the target tensor
      at position [ll,tt]
14        !
15        call tree%isometrise_towards( [ll,tt] )
16
17        !
18        ! Optimise target tensor (one-tensor update)
19        !
20        call tree%optimise_tensor( energy , tolerance =
      arnoldi_tolerance )
21      end if
22    end do
23  end subroutine
```

Note, that as mentioned in Sect. 8.2.3, for the sake of performance, we can implement one function `tree%isometrise_towards()` for performing both, the isometrisation and the contraction of the link operators at once, since they both manipulate tensors along the same path . Thus, in this way, the code will only iterate through the path from the last optimised tensor towards the target tensor once instead of twice. Finally, the function `tree%optimise_tensor()` performs the local optimisation as described in Sect. 8.1.4.2 and basically wraps the ARPACK solver straightforwardly into the code.

In the same manner as shown here, we can further include different optimisation techniques, such as the two-site optimisation, the space expansion technique, or even a gradient descent method, and a different sweeping sequence, for instance optimising in a random sweep order or performing a user-defined sweeping sequence.

8.3.2.4 Input Reader

The input reader module reads in pre-defined files which among others model the Hamiltonian to analyse by defining the interactions and coupling-strengths or set different control parameters for the ground state search, such as the number of maximum sweep iterations or the detailed options of the local optimisation procedure. In particular, from the given interactions $\mathcal{H}_p$ of a generic Hamiltonian $\mathcal{H} = \sum_p \mathcal{H}_p$, the module constructs all interactions in the system as TPOs respecting the chosen boundary conditions. Additionally, we can read single operators which for instance shall be measured as observables during the optimisation procedure.

8.3.2.5 Input Processor

Out of the interactions defined by the input reader module, this module defines the `MPOLayer_type` structure used to model the Hamiltonian. Further, at this level, we can take care of the exact mapping of the operators from the high-dimensional space towards the one-dimensional `tree_type` instance as illustrated in Sect. 8.2. Further, this module is the interface keeping track of the observables to be measured and, in particular, their mapping, thus where they shall be applied in the TTN structure.

8.3.2.6 Convergence Control

This module is independent of the Tensor Network structure itself and designed to track and control the convergence for a Tensor Network simulation. Thus, this module is to be incorporated in the main program and is an interface for managing simulation parameters and for enhanced stopping criteria based on variables (i.e. the current energy) of the simulation.

On one hand, the module defines all the important simulation parameters for a ground-state search, such as tolerance of the local Arnoldi solver, or potentially even parameters for time evolution. The module further tracks the number of iterations, the energy at each iteration as well as the desired numerical precision of the objectives of the final target state.

At the end of each iteration, the current energy has to be provided as an interface for the main program. Based on the number of iterations and the development of the energy, the module can adapt and return important simulation parameters, such as the suggested tolerance in solving the eigenvalue problem. For this particular parameter, it turns out empirically that the estimate

```
max(tol_min, min(tol_max, abs((energy-energy_prev)/energy)/NT)) ,
```

i.e. the relative energy difference after one sweep normalised by the number of tensors `NT` in the network and upper- (lower-)bounded by predefined values `tol_max` (`tol_min`), provides an efficient value for the balance between fast convergence and avoiding to get trapped in local minima.

8.3.2.7 TTN Algorithm

Based on the code structure illustrated above, the algorithm for searching the ground state of a quantum many-body system can be implemented following the description of Sect. 8.1.4 which can be implemented in the following way.

```
1   !{{{ subroutine start_simulation()
2   subroutine start_simulation()
3     type(Link_type), dimension(:), allocatable ::
      physical_links
4     type(MPOLayer_type) :: Hamiltonian
5
6     type(tree_type) :: tree
7     double precision :: current_energy
8     double precision :: tol
9
10    integer :: ii
11    character :: converged
12
13    !
14    ! Reading the Hamiltonian from defined input files
15    !
16    tree = read_physical_system(Hamiltonian, physical_links
      )
17
18    !
19    ! Initialising the Tree Tensor Network
20    !
21    tree = tree_type(Nsites, physical_links, cutoff, .true
      .)
22
23    !
24    ! Initialising the Link Operators
25    !
26    call tree%create_operators(Hamiltonian)
27
28    !
29    ! Calculating energy and observables (optional)
30    !
31    current_energy = tree%calculate_energy()
32    call measure_observables()
33
34    !
      ##########################################################
35    !    Start Iterating
36    !
      ##########################################################
37
38    do ii = 1, control%get_max_number_of_iterations()
39      !
40      ! Start optimization
41      !
42      call control%get_iteration_parameters(
      arnoldi_tolerance = tol )
43
```

```
44       call tree%optimization_sweep( energy = current_energy
     , &
45            arnoldi_tolerance = tol )
46
47       converged = control%check_convergence( current_energy
     )
48
49       select case ( converged )
50          case('f')
51
52          case('a')
53             write(*,*) '
  -------------------------------------------------'
54             write(*,*) 'Converged on absolute deviation'
55             print *, ii
56             exit
57          case('r')
58             write(*,*) '
  -------------------------------------------------'
59             write(*,*) 'Converged on relative deviation'
60             print *, ii
61             exit
62          case('d')
63             write(*,*) '
  -------------------------------------------------'
64             write(*,*) 'Failed converging'
65             exit
66          case default
67             write(*,*) '
  -------------------------------------------------'
68             write(*,*) 'Error in checking convergence'
69             stop
70       end select
71
72       call measure_observables()
73    end do
74 end subroutine
```

The simulation is set up by reading the systems Hamiltonian together with information about its physical sites, initialising the `tree_type` instance based on the physical system, initialising the link operators within the TTN, and optionally calculating observables for the randomly initialised network. Afterwards, the simulation starts by iterating through the number of sweeps we want to perform. As described above, the control module takes care of the hyperparameters for each sweep, such as the Arnoldi tolerance, and tracks the global energy E after each sweep. Based on this tracking, the convergence is checked using the value `converged`. In practice, it is feasible to track the relative value $|\frac{\delta E}{E} < \epsilon_{rel}|$ and the absolute convergence $|\delta E| < \epsilon_{abs}$. Further, at each iteration, we can measure and keep track of the observables of interest allowing us to follow how they develop with convergence. This might be interesting, in particular, to investigate whether the global optimisation has reached a global minimum or is stuck in a local minimum.

8.3.3 Towards High-Performance Computing

The final part of this chapter shall provide a general discussion on the performance of a Tree Tensor Network code which is implemented in the way described in Sect. 8.3.2. In particular, it should illustrate the perspectives concerning the parallelisation capabilities of a TTN simulation performed on High-Performance Computing (HPC) systems.

There are several computations within a TTN algorithm that can be parallelised in different ways. On the lowest level, we can parallelise the tensor contractions, i.e. the LAPACK routines performing the underlying matrix-multiplications. This can be done efficiently using GPUs. However, it is crucial to respect the sizes of the tensors to be contracted, since there can be a detrimental overhead moving the data to the GPU and back for small tensors. Having an educated heuristic, the incorporation of GPUs or even GPU clusters might be an interesting pathway for further development which has not been exploited within the scope of this thesis.

Secondly, the computations performed on symmetrically invariant tensor can be done to a large extent independently for the different symmetry sectors . Since these tensors can be seen numerically as block-diagonal instances, we could parallelise computations, such as tensor decompositions or tensor contractions, similarly to the parallelisation of a matrix decomposition or matrix multiplication, respectively, for a block-diagonal matrix. However, the benefit of this parallelisation highly depends on the distribution of symmetry sectors, i.e. the number and sizes of different blocks in the tensor.

Furthermore, as described in Sect. 8.1.4.2, the local optimisation of a single tensor within the TTN can be parallelised since we need to sum the contraction of several Hamiltonian parts with a single tensor (see Fig. 8.22). Each of these contractions can be performed independently and since they all act on the same space, the computational time should be of the same order for each contraction. In turns out that in the ground state search this parallelisation idea is the most efficient one when using a limited amount of CPUs only, especially since depending on the bond-dimension about 75% of the serial CPU time is spent performing this set of

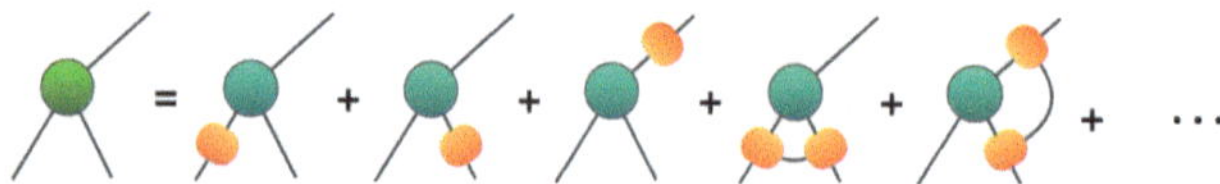

Fig. 8.22 Contraction scheme for a local optimisation. During the optimisation, the action of the effective Hamiltonian $\mathcal{H}_{eff}$ on a local tensor (green) is calculated by summing the contractions of all TPOs in $\mathcal{H}_{eff}$ with $\mathcal{T}$. In practice, these contractions make up for about 75% in a TTN ground search algorithm when performed on one CPU while they can be efficiently performed in parallel. (Figure reprinted with permission from [150])

contractions. However, it clearly has an upper limit of efficiency, namely the number of TPO operators that have to be contracted in the local optimisation which highly depends on the physical system itself. Nevertheless, with this optimisation, we can reach reasonable physical results for systems up to $N = 16 \times 16$ (containing in total 2^{256} degrees of freedom). In this case, we perform the parallelisation with OpenMP and reach typically a ground state with a precision in the energy density of the order $\sim 10^{-5}$, which is sufficiently accurate to obtain valuable insights into the physics behind. However, this parallelisation technique does not scale sufficiently well with system size N. Thus, since the number of tensors to be manipulated in the optimisation scales with the system size N we investigate, it is quite challenging to obtain high-precision results for higher system sizes like $N = 32 \times 32$ or even larger (obviously depending on the system as well).

Another possible form of parallelising a TTN ground-state search would be an MPI parallelisation of the global optimisation technique which could indeed offer the desired scalability with system size N. This parallelisation aims to perform the global optimisation algorithm by optimising all local tensors in parallel. The success of this idea has already been shown in Ref. [318] where it was introduced for an MPS. The main principle therein is to bring the network into the form of the Schmidt-decomposition obtaining the singular values, or Schmidt values, on an internal link. Now, we contract the singular values into both sides of the network keeping the inverse of them on the link. In this way, each side of the network is gauged so that it can perform the local optimisations independently. Applying this idea to a TTN, we can divide the tree into different sub-branches containing a fixed number of tensors. All the sub-branches will be optimised locally on different computational nodes in parallel via MPI. After the optimisation, the information affecting neighbouring branches will be passed to the corresponding nodes.

While at the final stage of this thesis, the latter option has not been implemented, this might be a further major step forward in the development of Tensor Network codes, and thereby the investigations of high-dimensional quantum many-body systems since it can offer an ideal scalability of the code with system size N.

8.4 Problems

1. Exploiting the code developed in the previous exercises, define the object TTN and the basic operations on them: computation of the norm, and evaluation of expectation values of local and nearest neighbour operators.
2. With the tools developed above, write a ground state search algorithm for the Ising model in transverse field. Perform the search for a 1D system starting from a random TTN and compare the resulting ground state energy with those computed using other methods (see exercises in the previous chapter).
3. With the tools developed above, extend the ground state search algorithm towards simulating systems in higher dimensions by implementing different mapping strategies including the *snake strategy* and the *Hilbert curvature*. Perform the ground state search for the 2D Ising model in transverse field and the 2D

Heisenberg model from a random TTN. Compare the performance of the different mapping strategies.

4. Implement a parallelisation strategy for the ground state search algorithm developed in the exercises above.

Augmented Tree Tensor Network

9

Timo Felser

In this chapter, we describe in more detail the augmented Tree Tensor Network (aTTN), a novel Tensor Network geometry for high-dimensional quantum systems. Initially conceived as a means to benchmark upcoming quantum technologies, such as quantum simulators and computers, the aTTN underlyines the impact of Tensor Network methods to experimental groups pioneering quantum technologies [149]. Recent experiments, showcasing the realization of quantum many-body states on diverse platforms like superconducting qubits, ultracold atoms, trapped ions, and Rydberg-atom lattices with unprecedented dimensions, underscore the need for a robust numerical tool for validation and benchmarking [319–322].

As detailed in the preceding chapter (Chap. 5), Tensor Networks have evolved over three decades to simulate quantum many-body systems on classical computers. While the Matrix Product State (MPS) has dominated one-dimensional systems, the quest for Tensor Network algorithms capable of accurately and scalably simulating higher-dimensional systems persists [170, 171]. Existing geometries like Projected Entangled Pair States (PEPS) for two-dimensional systems and Tree Tensor Networks (TTN), as well as MERA, for various dimensionalities, grapple with limitations in representing high-dimensional quantum states' entanglement properties effectively. The augmented Tree Tensor Network (aTTN), however, is capable of capturing intricate entanglement properties in high-dimensional systems while maintaining favorable computational complexity compared to competitors like PEPS and MERA. By merging the strengths of MERA and TTN, the aTTN

T. Felser (✉)
Tensor AI Solutions GmbH, Pfaffenhofen, Germany
e-mail: timo.felser@tensor-solutions.com

T. Felser, S. Montangero (eds.), *Introduction to Tensor Network Methods*,
Graduate Texts in Physics, https://doi.org/10.1007/978-3-032-17635-6_9

empowers simulations of systems both critical and non-critical, pushing the boundaries to unprecedented system sizes.

Subsequently, we delve into the technical details of the aTTN, providing an in-depth exploration of this novel Tensor Network method. This chapter explains how the aTTN addresses the challenge of faithfully capturing the area law of the TTN in high-dimensional systems while retaining its key advantages: (i) low scaling with the bond dimension m in comparison to MERA and PEPS, and (ii) the ability to precisely contract the network. The chapter further details the optimization technique employed for the ground-state search of quantum many-body systems and outlines an efficient procedure for computing observables. Finally, a comprehensive study comparing the aTTN with the TTN on the critical two-dimensional Heisenberg model demonstrates (i) the higher precision of the aTTN in describing two-body correlations, (ii) an analysis of the disentangler positions and their effects on numerical results, and (iii) the more favorable CPU time of the aTTN.

9.1 Technical Description of an aTTN

In this section, we describe the aTTN in more technical detail, providing a comprehensive overview of our novel Tensor Network method. We begin by explaining the structure of the aTTN in more detail and how it compensates the drawbacks of the TTN while maintaining its main advantages: *(i)* the low scaling with the bond dimension m compared to both MERA and PEPS, and *(ii)* the ability to contract the network exactly. Further, we provide a detailed description on the optimisation technique of the aTTN used for the ground-state search of quantum many-body systems and additionally address the procedure on how to efficiently compute observables. Finally, we compare the aTTN in more detail with the TTN in an extensive study on the critical two-dimensional Heisenberg model, demonstrating among others the higher precision of the aTTN in describing two-body correlations $\langle\sigma^{\gamma}_{i,j}\sigma^{\gamma}_{i',j'}\rangle$, an analysis on the engineering of the disentangler positions including their effects on the numerical results, and the more favourable CPU time of the aTTN.

Since the following builds on the presented publication and aims to provide a deeper understanding, this chapter has to be understood in this context only and includes selected parts and material of the original publication without explicit citation.

9.1.1 aTTN Geometry

The augmented Tree Tensor Network (aTTN) represents a wave-function $|\psi\rangle \in \mathcal{H}$ on a lattice $\mathcal{L}$ with the Hilbert space $\mathcal{H}$. In general, the lattice $\mathcal{L}$ can be an D-dimensional lattice containing $N = \prod_{i=1}^{D} L_i$ sites, where each site $j \in \mathcal{L}$ is described by a local Hilbert space $\mathscr{H}_j$ with finite dimension d_j, so that its complete Hilbert space is spanned by $\mathscr{H} = \otimes_j \mathscr{H}_j$.

The geometry of the aTTN is based on a TTN wave-function with an additional layer of disentanglers $\mathcal{D}(u)$ attached to the outgoing links on the bottom of the TTN. Thus the aTTN representation of a pure state $|\psi\rangle \in \otimes_i^N \mathscr{H}_i$ on the lattice $\mathscr{L}$ is given by

$$|\psi_{\mathrm{aTTN}}\rangle = \mathcal{D}^\dagger(u)|\psi_{\mathrm{TTN}}\rangle \tag{9.1}$$

with $|\psi_{\mathrm{TTN}}\rangle$ describing the wave-function parametrised by the internal TTN. Therein, the appended layer $\mathcal{D}(u)$ contains N_D disentanglers $\{u_k\}$, which all act independently of each other on different sites of the lattice $\mathscr{L}$. Each of the disentanglers u_k is a unitary when fusing its first two and its last two indices respectively, thus obeying the isometry condition

$$\sum_{k_3,k_4} (u_k)^{k_1,k_2}_{k_3,k_4} (u_k^\dagger)^{k_3,k_4}_{k_1',k_2'} = \delta_{k_1,k_1'}\delta_{k_2,k_2'}. \tag{9.2}$$

Hence, one disentangler u_k performs a unitary transformation on two physical sites $(i_1^{[k]}, i_2^{[k]})$ towards the attached TTN. This local transformation aims to decouple—or *disentangle*—relevant degrees of freedom in the quantum many-body state which consequently disappear for the TTN. Thus, the complete layer $\mathcal{D}(u)$ maps a pure state ψ of the lattice $\mathscr{L}$ to another pure state ψ_{aux} within the same Hilbert space $\mathscr{H}$ by applying all of its disentanglers u_k:

$$\mathcal{D}(u) : \mathscr{H} \rightarrow \mathscr{H} \tag{9.3}$$

$$\mathcal{D}(u)|\psi\rangle = u_1 u_2 \ldots u_K |\psi\rangle = |\psi_{\mathrm{aux}}\rangle \tag{9.4}$$

Note, that the different disentanglers u_k commute with each other, as they all act on different spaces $\mathscr{H}_{k_1} \otimes \mathscr{H}_{k_2}$. In this manner, $\mathcal{D}(u)$ can be as well seen as a unitary mapping for a given physical Hamiltonian $\mathcal{H} \in \mathscr{H}$ to an auxiliary Hamiltonian $\mathcal{H}_{\mathrm{aux}} = \mathcal{D}(u)\mathcal{H}\mathcal{D}^\dagger(u)$ towards the TTN within the aTTN. This preconditioning of the Hamiltonian $\mathcal{H}$ for the internal TTN can be performed in a way, such that it introduces an area law for higher-dimensional systems of the complete network while keeping the complexity for the optimisation at $O\left(m^4\right)$. Thus the aTTN avoids the weakness of a TTN—being the lack of an area law—and still maintains its main advantages, namely *(i)* the reasonably low scaling with bond dimension m compared to both MERA and PEPS, and *(ii)* the ability to contract the network exactly, which in general is not guaranteed for a PEPS (see also Table 9.1 for comparison). Let us point out as well, that the aTTN is not restricted to a certain dimensionality of the underlying system, but can be applied for a general D-dimensional system.

In Fig. 9.1 we give an illustrative example of an aTTN for a two-dimensional 8×8 system with its disentangler layer $\mathcal{D}(u)$ consisting of 6 different disentanglers u_k

Table 9.1 Comparison of the most prominent Tensor Networks discussed in the main text: Numerical complexity as a function of the bond-dimension m, obeying the entanglement area law, the typical bond-dimension to be used in current high performance simulations and the calculation of expectation values, i.e. the exact contractability

Tensor network	Complexity	Area law in 2D	Typical bond dimensions	Exact contractable
MPS/DMRG	$\mathcal{O}\{m^3\}$	No (Only in 1D)	> 10.000	Yes ($\mathcal{O}\{m^3\}$)
TTN	$\mathcal{O}\{m^4\}$	No (Only in 1D)	$\approx 1.000 - 2.000$	Yes ($\mathcal{O}\{m^4\}$)
PEPS	$\mathcal{O}\{m^{10}\}$	Yes	~ 10	No ($\mathcal{O}\{m^L\}$)
MERA	$\mathcal{O}\{m^8\}$ (1D)	Yes	~ 10	Yes ($\mathcal{O}\{m^8\}$)
	$\mathcal{O}\{m^{16}\}$ (2D)			
aTTN	$\mathcal{O}\{m^4\}$	Yes*	≈ 500	Yes ($\mathcal{O}\{m^4\}$)

*The fullfilment of the area law becomes a geometrical enginnering problem

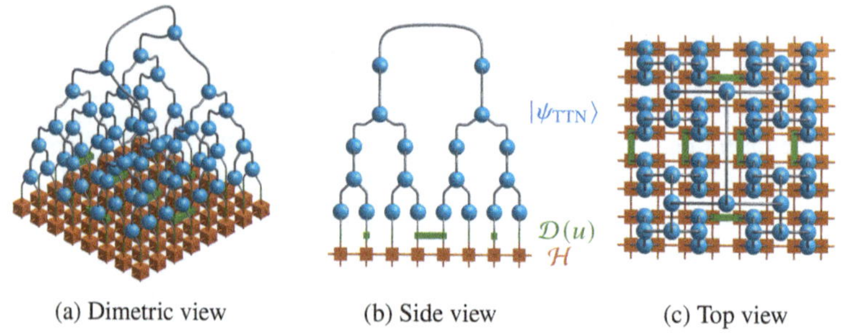

(a) Dimetric view (b) Side view (c) Top view

Fig. 9.1 Example of an aTTN for a 2D system with $N = 8 \times 8$ sites in different views (**a**)–(**c**). The aTTN contains an internal TTN ψ_{TTN} (illustrated with tensors in *blue* and links in *gray*) with a layer of disentanglers $\mathcal{D}(u)$ (*green*) attached to the outgoing links of the TTN. For illustration, we indicated a Hamiltonian $\mathcal{H}$ (*orange*) acting on the physical sites of the aTTN. (**c**) The disentangler layer $\mathcal{D}(u)$ supports the most critical links of the internal TTN with a number of disentanglers K_ν connecting the bipartitions introduced by the link ν. To encode the area law, K_ν scales with the boundary length γ_ν. (Figure reprinted with permission from [150])

(green). As shown therein, not every physical site j is addressed by a disentangler, which—as we will explain later on—is key for a better numerical complexity. Thus the positioning of the disentanglers u_k for a general aTTN is critical in order to *(i)* keep an optimal numerical complexity for the optimisation and *(ii)* efficiently encode an area law in the Tensor Network.

Resuming, the total number of parameters for an aTTN state scales with $\mathcal{O}(Nm^3 + N_D d^4)$ where m is the bond dimension of the TTN within, and N_D is the number of disentanglers u_k in the disentangler layer $\mathcal{D}(u)$.

9.1.2 Area Law in aTTN

For the description of the area-law captured by the aTTN, we illustrate the case of a two-dimensional square lattice $\mathcal{L}$ with $N = L \times L$ sites, where furthermore $L = 2^n$. The fundamental idea anyhow holds true for an D-dimensional lattice structure with arbitrary dimensions L_i. Furthermore, we here assume a binary TTN, which is arranged so that the tensors within the tree alternatingly in x- and y-direction coarse-grain neighboring sites going from layer to layer, as it is the case for the internal TTN in Fig. 9.1 (the internal TTN is illustrated by its tensors in blue and its links in gray). The topmost link of such a tree bipartites the whole system $\mathcal{L}$ into the two equally $(L \times L/2)$-sized subsystems $\mathcal{A}$ and $\mathcal{B}$. Going from the topmost link downwards in the TTN, each link within a layer further divides a smaller $l_x \times l_y$-dimensional sublattice.

When we now consider an area-law state ψ, the bipartition entanglement scales with the boundary ∂_ν of the divided subsystem, as stated in Eq. (8.4). In Table 9.2 this area is presented depending on the boundary conditions of $\mathcal{L}$. Therein, we introduced the notation $\gamma_{\nu[k]}$ for the boundary of a subsystem $\mathcal{A}^{[\nu]}$ emerging from the partitions introduced by a link ν within the k-th layer of the TTN (with subsystem size $N/2^k$). As mentioned above in Eq. (8.5), the TTN itself would require an in γ_ν exponentially large bond dimension m_ν in order to capture the area-law entanglement for the state ψ. In order to prevent this scaling, we now place K_ν disentanglers on the bottom of the TTN connecting the subsystems $\mathcal{A}^{[\nu]}$ and $\mathcal{B}^{[\nu]}$ given by the bipartition introduced by the link ν. More precise, we position the disentanglers u_k such that one of its physical sites $i_1^{[k]}$ or $i_2^{[k]}$ belongs to the subsystem $\mathcal{A}^{[\nu]}$, while the other one corresponds to $\mathcal{B}^{[\nu]}$. Each one of the disentanglers can maximally asses information in a d^2-dimensional space belonging to two local Hilbert spaces. Thus it can reduce the entanglement for the TTN up to the order of d^2. Consequently, all the K_ν disentanglers together can support the TTN link ν by disentangling information in the order of

$$m_{\nu,\mathrm{aux}} \approx (d^2)^{K_\nu} \tag{9.5}$$

When we now scale the number of disentanglers K_ν according to the boundary γ_l of the bipartition introduced by the link ν, the information captured within the complete aTTN for the cut through the network introduced by ν scales with

$$m_{\nu,\mathrm{eff}} \approx m_{\mathrm{aux}} m_\nu = d^{2K_\nu + \xi_\nu} \tag{9.6}$$

Table 9.2 Size of boundaries of different bipartitions in a 2D system $\mathcal{L}$ for different boundary conditions of $\mathcal{L}$

	$\gamma_{\nu[1]}$		$\gamma_{\nu[2]}$		$\gamma_{\nu[3]}$		$\gamma_{\nu[4]}$	...
Open BC	L		L		$\frac{5}{4}L$ or $\frac{3}{4}L$			
Cylindrical BC	L		$\frac{3}{2}L$		$\frac{3}{2}L$ or L			
Toroidal BC	$2L$	$\longrightarrow$	$2L$	$\xrightarrow{\cdot 3/4}$	$\frac{3}{2}L$	$\xrightarrow{\cdot 2/3}$	$\frac{3}{2}L$	$\xrightarrow{\cdot 3/4}$

Thus, the proper positioning of the disentanglers results in an exponential scaling of the effective bond dimension where we defined the parameter $\xi_n u \equiv \log_d m_\nu$ reflecting the contribution of the TTN bond dimension m_ν. In order to encode the area law for the two-dimensional aTTN state according to Eq. (8.5), the number of disentanglers K_ν connecting the different bipartitions introduced by the links ν has to be directly proportional to the boundaries γ_ν and thereby for $\nu = 1$ to the length L of the system.

In practice we know, that ξ_ν in Eq. (9.6) is sufficiently large for the lower branches of the tree in order to capture the area law entanglement—or even the complete state—accurately, especially for reasonably small local dimensions d. Thus we waive disentanglers within the smaller plaquettes of the TTN. Following the same idea, we focus on supporting the links higher up in the TTN with disentanglers $K_\nu \sim \gamma_\nu$, as the contribution of ξ_ν compared to the required exponentially large bond dimension becomes negligibly small. Equivalently, while going to lower layers, the contribution of ξ_ν weights increasingly stronger and thus we do not need to strictly enforce $K_\nu \sim \gamma_\nu$ in order to capture area law within the complete network.

9.1.3 Connection to Other Tensor Networks

In the prior sections, we described the connection of the aTTN to the well-established TTN geometry in detail. In what follows, we will elaborate on the similarity of the aTTN to the MERA and the connections to a PEPS.

MERA—The aTTN can be seen as a particular subclass of a MERA, where we sacrifice the disentanglers at higher renormalisation levels, and thereby the scale-invariance, for a higher numerical efficiency. In particular, the scale-invariance is not strictly necessary to ensure the area law. Indeed, in the aTTN we shift the problem of encoding the area law to a non-trivial positioning problem of the disentanglers at the bottom of the aTTN which we can solve with several approaches as discussed later on. Thus, the aTTN does not necessarily have the predefined, straight-forward scale-invariance structure of the aTTN but, with a clever positioning of the disentanglers, it is indeed able to encode area law with a leading complexity of $O\left(m^4\right)$ instead of $O\left(m^{16}\right)$ in case of the MERA.

PEPS—It has already been shown that the MERA can directly be mapped to a PEPS [323]. Following this procedure, and keeping in mind that the aTTN is a subclass of the MERA, we in principle can map the aTTN to a PEPS with in general unisotropic bond dimensions. In fact, the bond-dimensions of the resulting PEPS would highly depend on the positioning of the disentanglers and the internal TTN bond-dimensions indicating that this interplay between disentanglers at the bottom and the internal TTN is the crucial key to the success of the aTTN. We mention, however, that performing this mapping numerically might be not beneficial from a computational point of view.

9.2 Optimisation of the aTTN

For the optimisation of the aTTN, we assume the complete Hamiltonian to be a product of interactions $\mathcal{H} = \sum_p \mathcal{H}_p$. Thus every interaction $\mathcal{H}_p$ can be described as a Tensor Product Operator (TPO). For sake of simplicity, we will restrict ourselves to Hamiltonians $\mathcal{H}$ containing exclusively *(i)* local terms $\mathcal{H}_p = h^p_{i_p}$ (acting on the site i_p) and *(ii)* two-body interaction $\mathcal{H}_p = h^{p,1}_{i_p} h^{p,2}_{i'_p}$ between the physical sites i_p and i'_p. Note, that the basic idea of the aTTN approach can, in general, be applied to more complex Hamiltonians $\mathcal{H}$, but depending on the Tensor Network representation of $\mathcal{H}$ the computational complexity in secondary terms may increase. Furthermore the numerical implementation becomes more challenging as well when going beyond two-site interactions.

Given the Hamiltonian $\mathcal{H}$ we optimise the variational parameters of the aTTN wavefunction ψ in order to find the ground state of the system by minimising the energy

$$E = \langle \psi | \mathcal{H} | \psi \rangle \, . \tag{9.7}$$

This optimisation procedure for the aTTN consists of three different parts: *(i)* The optimisation of the disentanglers u_k, *(ii)* the mapping of the Hamiltonian $\mathcal{H} \xrightarrow{\mathcal{D}(u)} \mathcal{H}_{\text{aux}}$, and *(iii)* the optimisation of the internal TTN with the auxiliary Hamiltonian $\mathcal{H}_{\text{aux}}$.

Below we will describe each part of the optimisation in more detail, followed by some general remarks towards the practical application of the optimisation.

9.2.1 Disentangler Layer $\mathcal{D}(u)$

We optimise the layer $\mathcal{D}(u)$ of an aTTN by optimising all of the disentanglers forming $\mathcal{D}(u)$ one-by-one individually. In Fig. 9.2 we illustrate the procedure for one disentangler u_k to optimise. The fundamental idea hereby origins from the general MERA optimisation [147] with some minor adaptions for the aTTN geometry. Thus for the disentangler u_k the energy $E = \langle \Psi | H | \Psi \rangle$ to minimise depends bilinearly on u_k and $u_k^\dagger$,

$$E(u_k) = \underbrace{\text{tr}\left\{ \sum_p u_k M_p u_k^\dagger N_p \right\}}_{\equiv E_k(u_k)} + c_k \, , \tag{9.8}$$

where M_p and N_p are two sets of matrices corresponding to different contractions of the environment of u_k and $u_k^\dagger$ with the Hamiltonian part $\mathcal{H}_p$. Note that p only runs over the Hamiltonian parts $\mathcal{H}_p$ which act on one or both sites of the disentangler

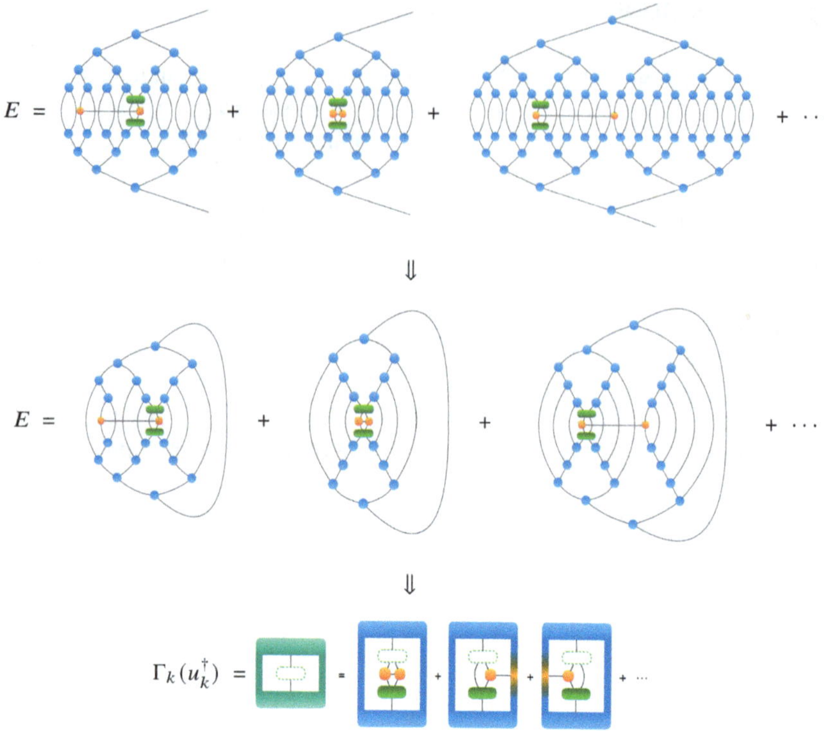

Fig. 9.2 Optimisation procedure for one disentangler u_k (*green*) of the aTTN. [top] Energy expectation value $E(u_k)$ of the aTTN. Each Hamiltonian part $\mathcal{H}_p$ (orange) is contracted with the complete aTTN and adding up to $E(u_k)$. For the optimisation of u_k only the parts $\mathcal{H}_p$ attached to u_k have to be considered. [middle] Contraction for $E(u_k)$ when isometrising the internal TTN towards the anchor node of the causal cone [147]. All tensors outside annihilate to identities. [bottom] Environment Γ_k for u_k including all the relevant contributions $\mathcal{H}_p$. By construction $E = \mathrm{tr}\{\{u_k\Gamma_k\}\} + c_k$ holds true. By decomposing $\Gamma_k = U\sigma V^\dagger$, we optimise the disentangler u_k with $u_k \equiv -VU^\dagger$. Figure reprinted with permission from [150]

to optimise. All the other Hamiltonian parts $\mathcal{H}_{p'}$ contribute to the overall energy independently of the disentangler u_k and thus are all included in the constant c_k for the Energy $E(u_k)$. Consequently, for the optimisation of u_k, we minimise the first part $E_k(u_k)$ of Eq. (9.8) only and neglect the constant contributions in c_k. The sum over the different Hamiltonian parts p thereby is illustrated in Fig. 9.2 in terms of the Tensor Network representation. We point out, that in Eq. (9.8), the disentangler u_k and its complex conjugate $u_k^\dagger$ are reshaped as unitary matrices by fusing their legs respectively.

In order to contract the complete network towards the matrices M_p and N_p, we can exploit the isometry of the internal TTN. As introduced for the MERA structure [147], the positions of the disentangler u_k and the interaction part $\mathcal{H}_p$ for the contraction determine a *causal cone* of tensors. We hereby define the *anchor* of a causal cone as the topmost tensor (with respect to the hierarchical TTN structure), which is still included in the causal cone. Thus, isometrising the internal TTN towards the anchor node of the causal cone for u_k and $\mathcal{H}_p$ results in all tensors outside of the causal cone vanishing to identities due to their enforced isometry. Consequently, the complete contraction of the network can be reduced to the tensors within the causal cone only.

Anyhow, after contracting, the environment resulting in the matrices M_p and N_p, there is no generic algorithm that solves the bilinear problem of Eq. (9.8) while additionally fulfilling the isometry constraint in Eq. (9.2). Thus, we solve this optimisation problem iteratively by linearising the function for the Energy $E_k(u_k)$ with respect to the disentangler u_k following the idea in Ref. [147]. Thereby, u_k and $u_k^\dagger$ are temporarily considered as independent tensors within one iteration step. This allows to optimise the Energy $E_k(u_k)$ with respect to u_k while keeping its complex conjugate $u_k^\dagger$ constant. For this linearised problem, the energy functional now reads

$$E_k(u_k) = \mathrm{tr}\left\{u_k \Gamma_k(u_k^\dagger)\right\}\,, \tag{9.9}$$

$$\text{with} \quad \Gamma_k(u_k^\dagger) = \sum_p M_p u_k^\dagger N_p \tag{9.10}$$

where Γ_k is a matrix obtained by the contraction of the complete environment around the disentangler u_k, as it is illustrated in Fig. 9.2. Thus, each iteration step, we optimise $E_k(u_k)$ with respect to u_k, while including $u_k^\dagger$ in the environment Γ_k. Thus, assuming Γ_k to be independent of u_k, the energy $E_k(u_k)$ is minimised by setting $u_k = -VU^\dagger$ where U and V are obtained via singular value decomposition of $\Gamma_k = U\sigma V^\dagger$. Thus by construction u_k' obeys the isometry constraint of Eq. (9.2) as both U and V are unitary matrices. The minimised energy for this iteration step becomes

$$\tilde{E}_{k,min} = \mathrm{tr}\left\{u_k \Gamma_k\right\} = \mathrm{tr}\big\{-VU^\dagger U\sigma V^\dagger\big\} = -\sum_i \sigma_i \tag{9.11}$$

Note, that the total energy is $E(u_k) = E_{k,min} + c_k$, as we neglected the constant c_k for the minimisation.

Concluding one iteration step, the complete procedure will be repeated with the new disentangler u_k', until the singular values σ_i of the SVD convergence which equals a convergence in the minimisation of the Energy to E_{min}.

Summarising the optimisation of the disentangler u_k:

(i) Contract the environment matrices M_p and N_p for all Hamiltonian parts $\mathcal{H}_p$ addressed by the disentangler u_k
(ii) Compute the environment Γ_k of the u_k for one iteration
(iii) Decompose $\Gamma_k = U\sigma V^\dagger$ via SVD
(iv) Update the disentangler $u_k \rightarrow u_k' = -VU^\dagger$
(v) Restart from (ii) until all σ_i converge.

Once the disentangler u_k is optimised we move on the next disentangler, until all the disentanglers within the layer $\mathcal{D}(u)$ have been optimised. Due to the disentanglers being positioned in a way, that they don't share interaction parts $\mathcal{H}_p$ of the physical Hamiltonian $\mathcal{H}$, the optimisation of each disentangler u_k is completely independent of all the other disentanglers $u_{k'}$. Thus it is sufficient to optimise each disentangler just once to obtain the optimised disentangler layer $\mathcal{D}(u)$—and furthermore, all optimisations for each disentangler can be fully parallalised.

The complete optimisation of one disentangler can be done with the complexity $O(m^4 d^2 + m^3 d^4 + d^6)$, where the contractions for part (i) of the summary above scale with $O(m^4 d^2 + m^3 d^4)$, for (ii) with $O(d^6)$, and the decomposition in (iii) as well with $O(d^6)$. Note that the most expensive part (i) can be done once for the complete optimisation of one disentangler u_k and is not required to be recomputed during the iterative procedure.

9.2.2 Hamiltonian Mapping

After each optimisation of the disentangler layer $\mathcal{D}(u)$, the physical Hamiltonian $\mathcal{H}$ mapped by the newly optimised disentangler layer $\mathcal{D}(u)$ for the subsequent TTN optimisation. For the aTTN, we consider the systems Hamiltonian $\mathcal{H} \in \mathscr{H}$ to be a product of interactions $\mathcal{H} = \sum_p \mathcal{H}_p$. As mentioned above, the aTTN introduces a layer of disentanglers $\mathcal{D}(u)$ which maps $\mathcal{H}$ to an auxiliary Hamiltonian $\mathcal{H}_{\text{aux}} \equiv \mathcal{D}(u)\mathcal{H}\mathcal{D}^\dagger(u)$ as preconditioning for the TTN. The energy expectation value

$$\langle \psi_{\text{aTTN}} | \mathcal{H} | \psi_{\text{aTTN}} \rangle = \langle \psi_{\text{TTN}} | \underbrace{\mathcal{D}(u)\mathcal{H}\mathcal{D}^\dagger(u)}_{\equiv \mathcal{H}_{\text{aux}}} | \psi_{\text{TTN}} \rangle \tag{9.12}$$

equals the expectation value of the mapped Hamiltonian for the internal TTN wave-function. This transformation of the complete Hamiltonian $\mathcal{H} \overset{\mathcal{D}(u)}{\longrightarrow} \mathcal{H}_{\text{aux}}$ is done by contracting each interaction part $\mathcal{H}_p$ separately with $\mathcal{D}(u)$.

In what follows, we illustrate this mapping for a general operator $\mathcal{T}$, which we describe as a Tensor Product Operator (TPO)

$$(\mathcal{T})^{\{i_j\}}_{\{i'_j\}} = \sum_{\{\gamma_\nu\}} \prod_j (\mathrm{t}^{[j]})^{\{\gamma_{\nu'}\}}_{i_j, i'_j} \tag{9.13}$$

following the definition in Ref. [170]. Thereby, the j-th tensor $\mathrm{t}^{[j]}$ is acting locally on the site i_j and is connected by the links $\{\gamma_{\nu'}\}$ to other tensors within the TPO. Thus, the complete TPO acts on the physical sites $\{i_j\}$. In this TPO formalism, we can describe for instance a local observable $\mathcal{T}_i$, an interaction part $\mathcal{H}_p$ of the Hamiltonian $\mathcal{H}$, a string observable $\mathcal{T}_{\{i_j\}}$ or more general structures like an MPO.

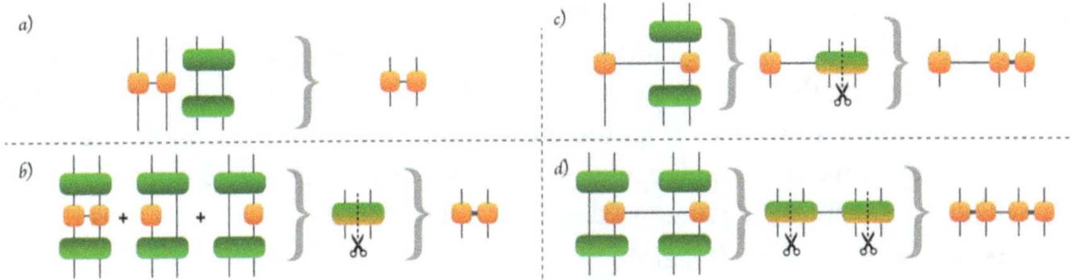

Fig. 9.3 (**a**) Trivial mapping of an operator that is not attached to a disentangler. (**b**) Mapping of several Hamiltonian parts $\mathcal{H}_p$ acting exclusively on the subspace addressed by the disentangler u_k. After contracting each operator $\mathcal{H}_p$ individually, the resulting tensors act on the same space and can be summed up. The obtained tensor is decomposed concluding the mapping. (**c**) Mapping of an operator with one site addressed by the disentangler u_k and one site free of disentanglers. After contracting the operator with u_k and $u_k^\dagger$, the contracted tensor is decomposed resulting in the mapped TPO. (**d**) Mapping of an operator with both site addressed by different disentangler u_k and u'_k. After contracting both tensors, the contracted tensors are decomposed resulting in the mapped TPO. Figure reprinted with permission from [150]

Consequently, the mapping $\mathcal{T} \overset{\mathcal{D}(u)}{\longrightarrow} \mathcal{T}_{aux}$ is done by contracting the TPO $\mathcal{T}$ with the proper disentanglers of layer $\mathcal{D}(u)$. This mapping can transform the operator $\mathcal{T}$ according to four different cases depicted in Fig. 9.3.

(*a*) *Trivial mapping*—Let there be no disentangler u_k in $\mathcal{D}(u)$ applied to neither one of the physical sites $\{i_j\}$ of the operator $\mathcal{T}$ ($\nexists u_k : i_k \in \{i_j\} \vee i_{k'} \in \{i_j\}$). Therefore, $\mathcal{T}$ is unaffected by the disentangler layer $\mathcal{D}(u)$ and the mapping effectively becomes an identity resulting in $\mathcal{T}_{\text{aux}} = \mathcal{T}$.

(*b*) *Mapping to two-site TPO*—Let the disentangler u_k be applied to all the sites $\{i_j\}$ on which the operator $\mathcal{T}$ is acting ($\forall j : i_j \in \{k_1, k_2\}$). In this case, $\mathcal{T} \equiv \mathcal{T}_i$ is either a local operator acting on the single site $i_1 \in \{k_1, k_2\}$ only or a two-site operator $\mathcal{T} \equiv \mathcal{T}_{k_1,k_2}$ acting on the exact same sites $\{k_1, k_2\}$ addressed by the disentangler u_k.
For the first step of the mapping, we contract the TPO with u_k and $u_k^\dagger$ to

$$(\mathcal{T}_{\text{contr}})_{k'_1,k'_2}^{k_1,k_2} = \sum_{i_1,i_2,i'_1,i'_2} (u_k)_{i_1,i_2}^{k_1,k_2} (\mathcal{T})_{\{i'_j\}}^{\{i_j\}} (u_k^\dagger)_{k'_1,k'_2}^{i'_1,i'_2} . \tag{9.14}$$

Afterwards, we decompose the resulting tensor $\mathcal{T}_{\text{contr}}$ via singular value decomposition obtaining two tensors—one for each site addressed by u_k—forming the mapped TPO $\mathcal{T}_{\text{aux}}$. Note, that if the original TPO was a two-site operator with an internal bond dimension κ, this mapping may lead to a different bond dimension for $\kappa' \neq \kappa$ within the mapped operator $\mathcal{T}_{\text{aux}}$. Anyhow, this new bond dimension is restricted by $\kappa' \leq d^2$.
Let us mention as well, that in case of mapping the Hamiltonian $\mathcal{H}$, we can have several interaction parts $\mathcal{H}_p$ which are all exclusively acting within the subspace

$\mathcal{H}_{k_1} \otimes \mathcal{H}_{k_1}$ addressed by the disentangler u_k. In Fig. 9.3 we illustrate such an example with one two-site interaction and two local Hamiltonian parts, all to be contracted with the same disentangler u_k. Here it is worth to point out, that after the contraction of Eq. (9.14) all the interaction parts $(\mathcal{H}_p)_{\text{contr}}$ can be added up, as now all of them act as one tensor on the same subspace. The summed tensor will subsequently be decomposed resulting in the mapped two-site TPO for the auxiliary Hamiltonian $\mathcal{H}_{aux}$.

(c) *Mapping with one disentangler*—Let there be only one disentangler u_k addressing at least one but not all of the sites $\{i_j\}$ $((\exists i_k : i_k \in i_j) \wedge (\exists i_m \in i_j : i_m \notin \{i_k\}))$. For the mapping we here proceed in a similar way as in the prior scenario. First, we contract the operator $\mathcal{T}$ over the connected links with the diesentangler u_k and its c.c $u_k^\dagger$:

$$(\mathcal{T}_{\text{contr}})_{k_1,k_2,\{i'_m\}}^{k_1,k_2,\{i_m\}} = \sum_{i_1,i_2,i'_1,i'_2} (u_k)_{i_1,i_2}^{k_1,k_2} (\mathcal{T})_{\{i'_j\}}^{\{i_j\}} (u_k^\dagger)_{k'_1,k'_2}^{i'_1,i'_2} . \tag{9.15}$$

After the contraction, we decompose the tensor $\mathcal{T}_{\text{contr}}$ in order to keep a TPO structure where each tensor within corresponds to one physical site. Thereby, we obtain the mapped TPO which now contains either the same number of tensors (when u_k is contracted with two tensors of the TPO) or one additional tensor (u_k addresses only one tensor of $\mathcal{T}$). Note, that here again the bond dimension for $\kappa' \neq \kappa$ within the TPO might increase but still is restricted by $\kappa' \leq d^2$ for H_p being a two-site TPO or to $\kappa' \leq \kappa d^2$ for some cases with larger TPO sizes respectively.

(d) *Mapping with two or more disentanglers*—Let there be K disentanglers $\{u_k\}$ be applied for the sites $\{i_j\}$. In this most general case, the operator $\mathcal{T}$ becomes mapped highly non-trivial. Here, the contraction with the disentangler layer is done by contracting the operator $\mathcal{T}$ with all the K disentanglers attached:

$$(\mathcal{T}_{\text{contr}})_{\{i_m\}}^{\{i_m\}} = \sum u_{k_1} u_{k_2} \dots u_{k_K} \mathcal{T}_{\{i'_j\}}^{\{i_j\}} u_{k_1}^\dagger \dots u_{k_2}^\dagger u_{k_1}^\dagger . \tag{9.16}$$

For the sake of compactness, we dropped the indices for the disentanglers. Furthermore, we introduced the set of indices $\{i_m\}$ containing all the indices i_k of all K disentanglers $\{u_k\}$ and additionally all the physical legs i_j of the original operator $\mathcal{T}$ over which we did not contract. Note as well that the order of application of the disentanglers u_k is not important, as all u_k act on different subspaces of the complete Hilbert space $\mathcal{H}$ and thus commute with each other. Here again the resulting operator $\mathcal{T}_{\text{contr}}$ is decomposed back into a TPO structure with its tensors acting on one single site of the TTN. For a general TPO which may even contain loops in its structure, like e.g. a PEPO [204–206], this decomposition is a highly non-trivial task. Furthermore, for loopless TPOs, this mapping might again result in an increased internal bond dimension within the TPO.

Considering the Hamiltonian $\mathcal{H}_{\text{aux}}$ is used to optimise the TTN, an arbitrary loopless interaction $\mathcal{H}_p$ which becomes mapped to a general MPO might in the worst case still increase the complexity of the overall optimisation to a with system size exponential scaling $O(m^4 d^L)$. But if we restrict ourselves to all Hamiltonian parts $\mathcal{H}_p = h_{p,1} h_{p,2}$ being nearest-neighbor interactions only, the additional cost is restricted to a worst-case prefactor of d^4. In order to go even further and prevent this prefactor, we can constraint the aTTN in its construction to prevent an interaction part $\mathcal{H}_p$ being addressed by two—or more—disentanglers. Thus, for nearest-neighbor interactions, we strictly forbid to attach two disentanglers u_k and u'_k onto neighboring sites, as can be seen in Fig. 9.1. This not only decreases the second-order scaling of the complete algorithm but additionally simplifies the numerical implementation and introduces a higher potential of parallelisation, as now all the disentanglers can be treated completely independent within the optimisation.

9.2.2.1 Tree Tensor Network

The optimisation of the TTN follows the prescriptions of Chap. 8. Each pair of connected tensors within the TTN is optimised via the subspace-expansion technique which approximates a two-site update [170]. This technique allows to keep the favourable scaling of $O\left(m^4\right)$ compared to the higher numerical complexity of $O\left(m^6\right)$ in case of the direct two-site optimisation (see Sect. 8.1.4.2). The local optimisations of each single tensor optimisation are performed via Arnoldi algorithm as discussed in Sect. 8.1.4.2 which solves the local eigenvalue problem by iteratively diagonalising the corresponding effective Hamiltonian H_{eff}, returning the lowest eigenvalues of H_{eff} within a predetermined accuracy ϵ.

We optimise the global TTN state by performing these local optimisations including the space-expansion technique step-by-step from the bottom of the TTN to its top. The space expansion, in particular, is always performed for the local target tensor together with its *parent* tensor as space expansion partner.

9.2.2.2 General Remarks

In practice, we start the complete optimisation of the aTTN with a randomly initialised TTN. We first iterate a few times threw the TTN with the physical Hamiltonian $\mathcal{H}$ disregarding the disentangler layer $\mathcal{D}(u)$. In this part, we use a relatively large space expansion and a low precision for solving the local eigenvalue problems for each tensor. Thereby, we aim to efficiently *(i)* adapt the randomly initialised symmetry sectors within the TTN for a qualitative description of the ground state and *(ii)* use this qualitative TTN wavefunction for an advanced the initialisation of the aTTN. Consequently, we start with the aTTN consisting of the resulting TTN with a set of identities $\{u_k\} \equiv \{\mathbb{1}\}$ attached at the sites we aim to disentangle forming the initial disentangler layer $\mathcal{D}(u)$. We subsequently optimise the aTTN by optimising the disentaglers once, mapping $\mathcal{H}$ to $\mathcal{H}_{\text{aux}}$ and performing n steps of optimising the TTN, where we usually choose $n \in \{1, 2, 3\}$.

Furthermore, in the results obtained for this paper, we improved the TTN optimisation by parallelising the contractions over different Hamiltonian parts $\mathcal{H}_p$ within the optimisation procedure. Additionally, we want to mention, that the computational time for optimisation of the disentanglers and the mapping of the Hamiltonian was negligible compared to the TTN optimisation in our practical applications. As of last we point out, that in the aTTN approach there is still more room for improvement by parallelisation, as the TTN optimisation can be further parallelised in a similar way as shown for parallel-DMRG [318] as well as the optimisation of the disentanglers.

9.3 Computation of Observables

9.3.1 Local Observables

Let $O^{[s]}$ be a local observable acting on the site s of the aTTN state ψ. In order to compute the expectation value $\langle O^{[s]} \rangle_\psi = \langle \psi | O^{[s]} | \psi \rangle$ we distinguish the two different cases:

(i) A disentangler u_k is placed at site s—thus connecting it with the site $\tilde{s}$, or
(ii) there is no disentangler attached to the physical site s.

In the latter case, the one-site operator $O^{[s]}$ is simply passed to the TTN and calculated by the standard computation of a local observable in TTN, as $O^{[s]}$ is unaffected by the disentangler layer $\mathcal{D}(u)$ and the isometry property $\mathcal{D}(u)\mathcal{D}^\dagger(u) = \mathbb{1}$ holds true. Using the proper isometry in the TTN, this calculation can be done by contracting three tensors only [170].

In case *(i)* however, the local operator becomes mapped to a two-site operator by the disentangler layer $\mathcal{D}(u)$, as it is contracted with u_k and $u_k^\dagger$ to

$$(O_{aux}^{[s]})_{k_1',k_2'}^{k_1,k_2} = \sum_{s,s',\tilde{s}} (u_k)_{s,\tilde{s}}^{k_1,k_2} (O^{[s]})_{s'}^{s} (u_k^\dagger)_{k_1',k_2'}^{s',\tilde{s}} . \tag{9.17}$$

Consequently, in this case, the computation of $\langle O^{[s]} \rangle_\psi$ effectively becomes a calculation of the expectation value of the correlator $\langle (O_{aux}^{[s]})_{k_1',k_2'}^{k_1,k_2} \rangle_\psi$ for the TTN. The worst case complexity of the underlying contraction in the TTN is $O(m^4 \kappa \log N) + O(m^4 \kappa \log N - 1) + O(m^3 \kappa)$.

9.3.2 Correlators

Let $O^{[s_1,s_2]}$ be a correlator acting on the sites s and s' of the aTTN state ψ. In order to compute the expectation value $\langle O^{[s,s']} \rangle_\psi = \langle \psi | O^{[s,s']} | \psi \rangle$ we distinguish the following four different cases for the mapping $\mathcal{D}(u)$—some of which are similar or even identical to the mapping described in Sect. 9.2.2:

(a) *Trivial mapping*—Let there be no disentangler u_k in the layer $\mathcal{D}(u)$ attached to neither one of the physical sites s_1 or s_2 of the correlator. Therefore, the correlator $O^{[s]}$ can—as described for the local observable—simply passed threw to the TTN and calculated by the standard computation of a correlator in TTN, as here again $O^{[s_1,s_2]}$ is unaffected by the disentangler layer $\mathcal{D}(u)$ and the mapping effectively becomes an identity. This subsequent contraction of the TTN can be done with the worst-case complexity of $O(m^4 \log N) + O(m^4 \log N - 1) + O(m^3)$.

(b) *Mapping to two-site TPO*—Let the disentangler u_k connect both sites s_1 and s_2 on which the correlator $O^{[s_1,s_2]}$ acts. In this case, the correlator becomes mapped to another two-site operator by the disentangler layer $\mathcal{D}(u)$, as it is contracted with u_k and $u_k^\dagger$ to

$$(O_{aux}^{[s_1,s_2]})_{k_1',k_2'}^{k_1,k_2} = \sum_{s_1,s_2,s_1',s_2'} (u_k)_{s_1,s_2}^{k_1,k_2} (O^{[s_1,s_2]})_{s_1',s_2'}^{s_1,s_2} (u_k^\dagger)_{k_1',k_2'}^{s_1',s_2'} . \quad (9.18)$$

Therefore, the computation of $\langle O^{[s_1,s_2]}\rangle_\psi$ becomes a calculation of the expectation value of the correlator $\langle O_{aux}^{[s_1,s_2]}\rangle_\psi$ for the TTN. The worst-case complexity of the underlying contraction in the TTN is $O(m^4\kappa \log N) + O(m^4\kappa \log N - 1) + O(m^3\kappa)$ where κ describes the internal bond dimension of the TPO after the mapping.

(c) *Mapping to a three-site TPO*—Let there be only one disentangler u_k addressing at either site s_1 or s_2, and connecting it with the site $s_3 \notin \{s_1, s_2\}$. Here the layer $\mathcal{D}(u)$ maps the correlator to a three-site operator, as it is contracted to

$$(O_{aux}^{[s_1,s_2]})_{k_1',k_2',s}^{k_1,k_2,s} = \sum_{s_c,\tilde{s},s_c'} (u_k)_{s_c,\tilde{s}}^{k_1,k_2} (O^{[s_c,s]})_{s_c',s'}^{s_c,s} (u_k^\dagger)_{k_1',k_2'}^{s_c',\tilde{s}} . \quad (9.19)$$

Note that we here changed the notation of $O^{[s_1,s_2]}$ to $O^{[s_c,s]}$ without loss of generality where s_c indicates the link of the correlator attached to the disentangler u_k and thus is to be contracted over while s indicates the remaining link for the resulting three-site operator. After the mapping the computation equals a calculation of the expectation value $\langle O_{aux}^{[s_1,s_2]}\rangle_\psi$ for the TTN—describing a string observable which in this case scales with a worst-case complexity of $O(m^4\kappa \log N) + O(m^4\kappa \log N - 1) + O(m^3\kappa)$.

(d) *Mapping to a four-site TPO*—Let the disentanglers u_{k_1} and u_{k_2} be placed at the sites s_1 or s_2, and connecting them with the sites $s_3 \notin \{s_1, s_2\}$ and $s_4 \notin \{s_1, s_2\}$ respectively. In this most complex case, the correlator becomes mapped to a four-site operator, the contraction reads

$$(O_{aux}^{[s_1,s_2]})_{k_1',k_2',k_3',k_4'}^{k_1,k_2,k_3,k_4} = \sum_{s_1,s_2,k',k''} (u_{k_1})_{k',s_1}^{k_1,k_2} (u_{k_2})_{k'',s_2}^{k_3,k_4} (O^{[s_1,s_2]})_{s_1',s_2'}^{s_1,s_2} (u_{k_2}^\dagger)_{k_3',k_4'}^{k'',s_2'} (u_{k_1}^\dagger)_{k_1',k_2'}^{k',s_1'} .$$

After this mapping the computation again equals a calculation of the expectation value of a string observable for the TTN, which in this case exhibits a worst-case complexity of $O(m^4\kappa^2 \log N) + O(m^4\kappa^2 \log N - 1) + O(m^3\kappa)$.

9.4 Numerical Studies

In order to benchmark the aTTN algorithm, we analyze the antiferromagnetic two dimensional Heisenberg model

$$\mathcal{H} = \sum_{i,j=1}^{L} \sum_{\gamma \in \{x,y,z\}} \sigma_{i,j}^{\gamma}\sigma_{i+1,j}^{\gamma} + \sigma_{i,j}^{\gamma}\sigma_{i,j+1}^{\gamma},$$

with $\sigma_{i,j}^{\gamma}$ ($\gamma \in \{x, y, z\}$) the Pauli matrices acting on the site (i, j). We consider a $L \times L$ lattice with system sizes $L = \{8, 16, 32\}$ and set the boundary terms with $\sigma_{i,N+1}^{x} = \sigma_{i,1}^{x}$ and $\sigma_{N+1,j}^{x} = \sigma_{1,j}^{x}$ as periodic boundary conditions. The ground state of this Heisenberg Hamiltonian is characterised by power-law decaying correlations and has been established by several numerical approaches as challenging benchmark.

9.4.1 Energy Density Calculation

In Fig. 9.4 we compare the energy density obtained by the TTN with the ones obtained via aTTN with increasing bond-dimension m. Evidently, the aTTN is more accurate than the TTN when compared at same bond-dimension. However, due to a constant prefactor c in the numerical complexity, the CPU time $t \approx O\left(cm^4\right)$ for the aTTN is higher at same bond-dimension. Therefore, we compute the relative error obtained by the different ansätze and the best known results, obtained via Quantum Monte Carlo [324]. As expected, when keeping the bond dimension equal for both approaches, the aTTN obtains a higher precision for the energy density compared to the TTN but requires more CPU time for the complete optimisation. However, when we look at constant CPU times, the aTTN eventually still outperforms the TTN in terms of accuracy in the energy density for sufficiently large bond dimension m. Thus, it is evidently the more efficient Tensor Network for these two-dimensional simulations. In Fig. 9.4 c and d we compare the CPU time of each calculation against the accuracy as a figure of merit for the efficiency of the ansatz, i.e. which of the Tensor Networks provide a lower relative error at constant CPU time. We see, that for low CPU time (consequently, low bond dimensions) it is more efficient to scale up the bond-dimension of the TTN, while eventually the aTTN becomes the more efficient ansatz when increasing the simulation time, i.e. the bond-dimension. We mention that this prefactor c depends on the number of Hamiltonian parts $\mathcal{H}_p$ addressed by the disentanglers of the aTTN.

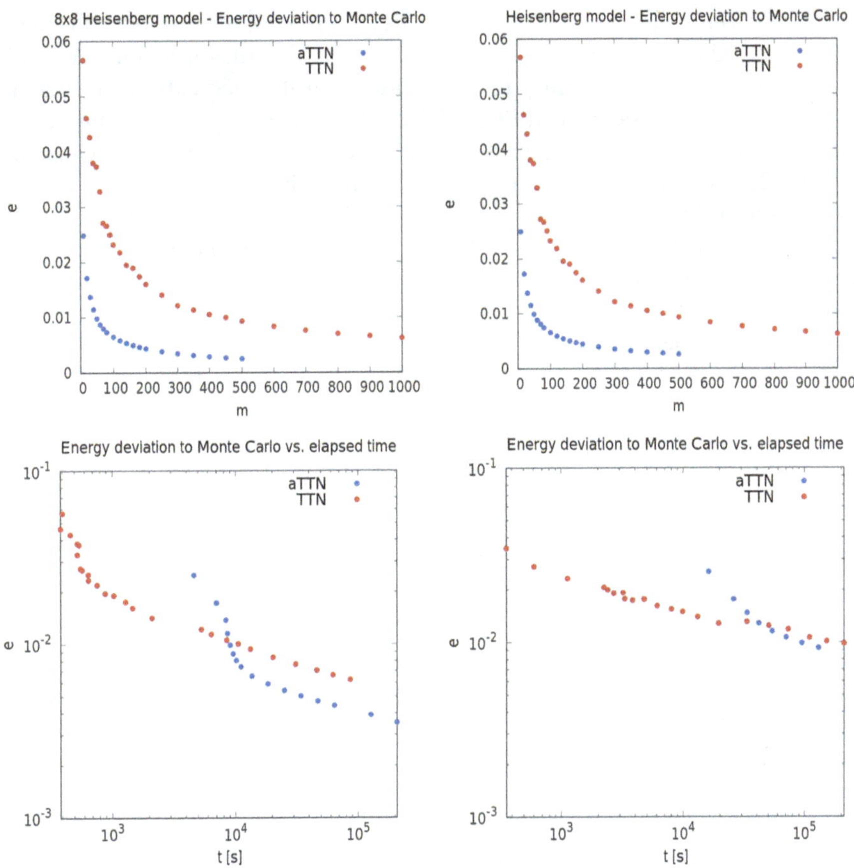

Fig. 9.4 Relative error δE of the 2D Heisenberg ground-state energy compared with the best available estimates obtained via Quantum Monte Carlo [324] as a function of (**a**) the bond dimension m at $L = 8$, (**b**) the bond dimension m at $L = 16$, (**c**) the CPU time $t[s]$ at $L = 8$ and (**d**) CPU time $t[s]$ at $L = 16$. The TTN results in *red*, aTTN in *blue*. `Figure reprinted with permission from [150]`

When we extrapolate with the bond dimension, we further confirm that the aTTN is more accurate compared to the TTN for both system sizes with a deviation of $\epsilon_8 = 2.22 \cdot 10^{-5}$ and $\epsilon_{16} = 7.11 \cdot 10^{-4}$ in the energy density for both sizes $L = 8$ and $L = 16$ (with respect to Monte Carlo). The TTN, in contrast, performs reasonably well in the 8×8 case but struggles in precision when going to $L = 16$. Interestingly, the aTTN for $L = 16$ obtains an even more precise ground state energy density compared to alternative variational ansätze at lower finite system size of $L = 10$, such as Neural Network states [325] ($\epsilon_{10} \approx 7.5 \cdot 10^{-4}$), Entangled Pair States [326] ($\epsilon_{10} = 2.5 \cdot 10^{-3}$) or PEPS [39] ($\epsilon_{10} = 1.5 \cdot 10^{-3}$). It turns out that for this model a very competitive variational approach is the 2D-DMRG [188], which outperforms the alternative methods for finite size with open or cylindrical boundary conditions up to the system size $L = 10$ ($\epsilon_{10} \approx 2 \cdot 10^{-5}$), but struggles with periodic boundary conditions and with increasing system sizes $L \gtrsim 12$. We further mention that, PEPS is very efficient with its ability to work directly in the thermodynamic limit in describing infinite systems as iPEPS [327, 328]. We as well point out that further more complex Neural Network states, such as the Neural Autoregressive Quantum States [329], obtain comparable results for $L = 10$ with $\epsilon_{10} \approx 3.5 \times 10^{-5}$.

Furthermore, we extended our analysis to reach the system size of $N = 32 \times 32$ for which, to the best of our knowledge, no public result of alternative methods is yet available as a comparison. In consistency with the finite size scaling analysis of Ref. [324], we obtain an energy density $\epsilon_{32} = -0.669(1)$ as extrapolated value $m \to \infty$. We mention that there exist techniques combining Tensor Networks with Variational Monte Carlo, shown with the PEPS++ [330, 331] providing a high potential with strong results at system sizes up to $N = 32 \times 32$. However due to the underlying PEPS geometry in this case, the approach still suffers from an unappealing numerical scaling of $O\left(m^{10}\right)$ and requires an tremendous computational effort for obtaining these results.

9.4.2 Correlations

Next to the energy density, we compared the correlations captured within both approaches, the TTN and the aTTN. In particular, we computed the two-body correlations $\langle \sigma^z_{i,j} \sigma^z_{i',j'} \rangle$. In Fig. 9.5 we show the correlations for the 8×8 size (a) and 16×16 size (b) for different bond dimensions. It is evident that the computed correlations of the ansätze are growing and getting more accurate with an increasing number of variational parameters, i.e. with increasing bond dimension m. Indeed, when comparing both approaches at equal bond dimension, the obtained

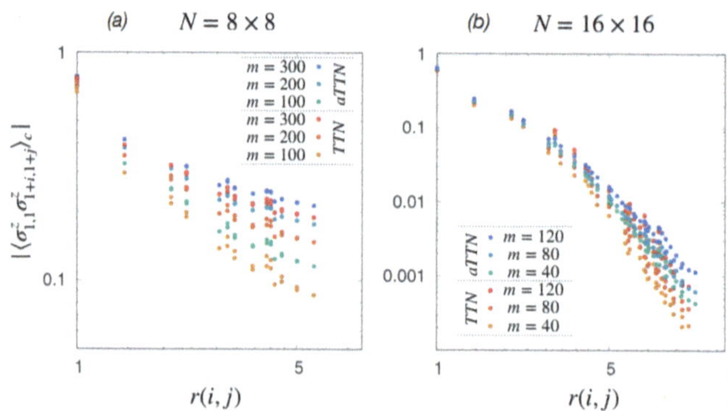

Fig. 9.5 Connected correlation function in the TTN (in shades of *red* for different bond dimensions) and the aTTN (in shades of *blue*) representation of the ground state of the $L = 8$ (**a**) and $L = 16$ (**b**) Heisenberg Hamiltonian plotted against the distance $r(i, j) \equiv [i^2 + j^2]^{1/2}$, where $i \leq j \in \{0, L/2\} \times \{0, L/2\}$. Figure reprinted with permission from [150]

correlations of the aTTN simulations are clearly more accurate, demonstrating the higher precision of the aTTN in describing two-body correlations. Interestingly, for lower distances $r(i, j) < 5$ in the $L = 16$ case, some of the correlations computed via TTN are more precise then the aTTN correlations at equal bond-dimension, while the TTN correlations clearly decrease faster for higher distances $r(i, j) < 5$. At the most far distant measurement at $r(8, 8) = 8\sqrt{2}$, we observe that the TTN correlations for $m_{\text{TTN}} = 120$ even end up at the same precision as the aTTN correlations for lower bond dimension $m_{\text{aTTN}} = 80$ and for $m_{\text{TTN}} = 80$ clearly below the aTTN correlations for $m_{\text{aTTN}} = 40$.

We point out that the here performed aTTN simulations (as well as the TTN simulations) exploit a $U(1)$ symmetry. However, for this model, we could further drastically improve the performance of the aTTN by incorporating the present $SU(2)$ symmetry in the simulation framework [170, 316, 332]. In particular, this would *(i)* drastically improve the connected correlations since we enforce $\langle \hat{\sigma}_j^\alpha \rangle = 0$, as well as $\langle \hat{\sigma}_i^\alpha \hat{\sigma}_j^\alpha \rangle_c$ to be equivalent independent on $\alpha \in \{x, y, z\}$, and *(ii)* reduce the effective bond dimension and thereby the computational time since for non-abelian symmetry we may only work within the symmetry multiplet spaces. Both advantages are expected to lead to dramatically increase the accuracy in the estimated energy.

9.4.3 Disentangler Engineering

Additionally, we performed an analysis on the positioning of the disentanglers in the aTTN based on the Heisenberg model. In particular, we perform the simulation with three different geometries (see Fig. 9.6) investigating the influence of the aTTN geometry on the overall precision for $L = 8, 16$:

(a) In this strategy, we start placing as many disentanglers as possible to reinforce the top-most link (its bipartition indicated with red-dotted lines in Fig. 9.6a), then place as many disentangler as possible to reinforce the second highest links (indicated in blue), and so on. Following this strategy for the 8×8 system, we position all $N_D = K_1 = 8$ disentanglers supporting the highest TTN layer with no place left for disentanglers supporting the second layer.

(b) Starting from strategy (a), we now sacrifice disentanglers reinforce the top-most link in order to place some disentanglers for the second highest tree layer (its bipartition indicated with blue-dotted lines in Fig. 9.6b). In this way we are able to increase the density of disentanglers applied. For the 8×8 system, we position $K_1 = 6$ disentanglers supporting the highest TTN layer and $K_2 = 4$ disentanglers for the second highest TTN layer, totalling $N_D = 10$ disentanglers.

(c) In the last strategy, we change the structure of the internal TTN: Now, instead of coarse-graining alternatingly in $x-$ and $y-$direction, now we coarse-grain the complete y-direction first, before we go on to coarse-grain in x-direction when

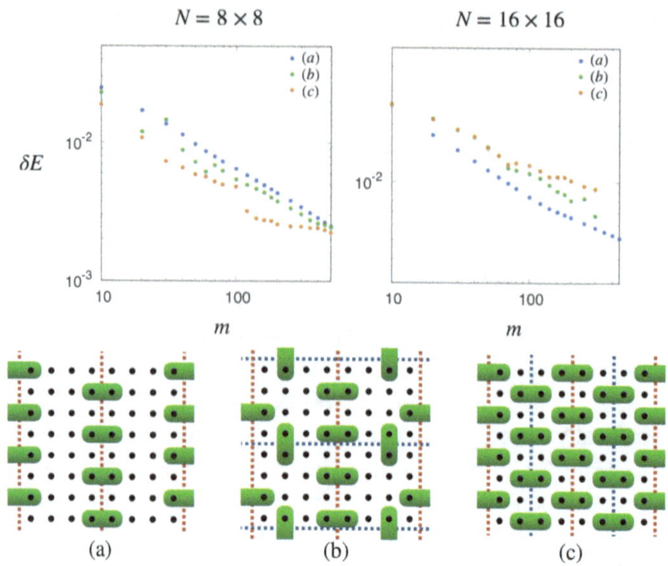

Fig. 9.6 (top) Relative error ϵ of the 2D Heisenberg ground-state energy for different geometrical configurations (exemplified for $L = 8$ at the bottom) of the aTTN compared with the best available estimates obtained via Quantum Monte Carlo [324] as a function of the bond dimension m at $L = 8$. (bottom) The green shapes represent the disentanglers, dashed lines indicate the bipartitions introduced by the top-most (*red*) and the second highest (*blue*) links within the tree. `Figure reprinted with permission from [150]`

initialising the tree from the bottom to the top. Thus, as indicated by the dashed lines in Fig. 9.6c, the topmost links bipartite the system in x-direction. In this way, we can place the maximum amount of disentanglers to reinforce the two top-most links, namely $N_D = 16$ with $K_1 = K_2 = 8$. Thereby, in the $L = 8$ case, the lowest 3 layers will coarse-grain each column of the system with each of the lowest branches grouping the 8 sites in y-direction together while for $L = 16$, each column of 16 sites will be joined in the lowest 4 layers.

In Fig. 9.6 we present the deviation in the energy density for the different geometries with respect to Quantum Monte Carlo [324]. In the 8×8 analysis, we see that all of the geometries give very similar results along the complete range of m. Surprisingly, reshaping the entire tree structure, i.e. geometry (c), offers the best results out of the three configurations, while the initial strategy (a) which was used in the previous analysis comes out worst. When moving to the $L = 16$ case, however, geometry (c) clearly fails to keep up with the other two and strategy (a) comes out ahead. Indeed this can be traced back to the same problem of 2D-DMRG for $L \gtrsim 12$: Keeping L reasonably small (e.g. $L = 8$), the lower branches of the internal TTN are able to efficiently capture the necessary information to faithfully represent the ground state. However, moving to higher system sizes, the internal TTN struggles to keep up efficiently joining the 16 "column" sites together. Therefore, comparable to the 2D-DMRG, the strategy (c) should offer the best results for an aTTN simulation for limited second dimension L_y and might be a good compromise for non-squared systems ($L_x > L_y$) with periodic boundary conditions. On the other hand, we find that the most efficient strategy for general $L \times L$ systems indeed is strategy (a). However, we point out that finding the best possible geometry is a highly non-trivial engineering task and depends on many factors, such as the system size, the interactions within the system and the boundary conditions. The optimisation of the Tensor Network structure towards the particular problem might be an interesting problem to solve in future work.

9.5 Problems

1. Exploiting the code developed in the previous exercises, define the object aTTN (augmenting the TTN code developed in the previous exercises) and implement the basic operations on them: computation of the norm, and evaluation of expectation values of local and nearest neighbor operators.
2. With the tools developed above, extend the ground state search algorithm for the TTN in higher dimensions towards the aTTN.
3. With the tools developed above, perform the ground state search for the 2D Ising model in transverse field and the 2D Heisenberg model from a random TTN. Compare the performance of different disentangler layers and of the ordinary TTN.

Part IV

Applications

Quantum Phase Transitions 10

Simone Montangero

The understanding of the physics of many-body quantum systems at equilibrium encompasses a plethora of processes at the heart of our understanding of nature. Indeed, the properties of materials, of chemical molecules, and lattice gauge theories for high energy physics belong to this class, just to name a few. In this chapter, we present the necessary basis to enter into this fascinating field, mostly from the perspective of a condensed matter physicist, thus focusing on the properties of quantum matter. However, hereafter we do not present a complete introduction to these fascinating topics for which extensive literature exists, see, e.g., [137, 141, 264, 333, 334]: We introduce a useful collection of concepts to start working in the field, in particular having in mind to address this class of problems with TN methods. We first present the basics of phase transitions, quickly reviewing Landau theory of ferromagnetism and the statistical quantum-classical correspondence among system of different dimensionality. Finally, we present local and non-local (entanglement) measures that has been used to characterize such phenomena that can be easily addressed via TN methods.

10.1 Phase Transitions

Everybody is familiar with *phase transitions*, that is, drastic change of matter appearance and properties as a function of temperature: water freezes at zero and boils at hundred Celsius degrees. Less familiar might be other kinds of phase transitions, which occur in magnetic materials [15]. Indeed, experimentally one can verify that below a precise temperature some material sharply exhibits non-zero spontaneous magnetization, even in absence of external magnetic fields. Moreover,

S. Montangero (✉)
Università di Padova, Padova, IT, Italy
e-mail: simone.montangero@unipd.it

T. Felser, S. Montangero (eds.), *Introduction to Tensor Network Methods*,
Graduate Texts in Physics, https://doi.org/10.1007/978-3-032-17635-6_10

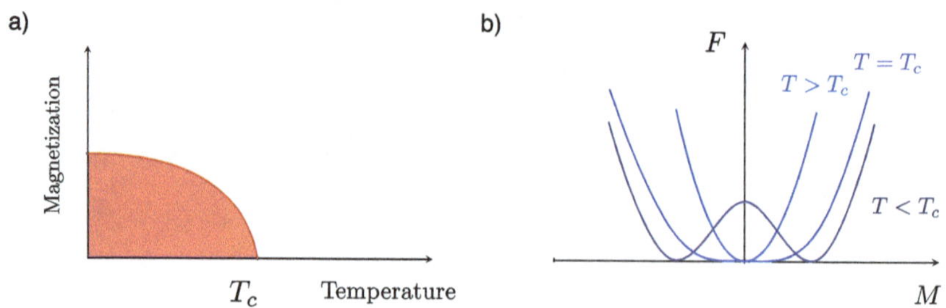

Fig. 10.1 (**a**) Sketch of the magnetization as a function of the temperature undergoing a phase transition at critical temperature T_c. (**b**) Landau theory of ferromagnetism, the free energy as a function of magnetization for different temperatures: at $T = T_c$ the stable minima at $M = 0$ becomes unstable, and the free energy becomes a quartic function with two degenerate minima reflecting the $\mathbb{Z}_2$ symmetry of the problem

the magnetization as a function of the temperature is non-differentiable at the *critical temperature* T_c, where the magnetization appears (see Fig. 10.1a). The discontinuity of the derivative at the *critical point* signals that something drastic is occurring in the system: the system's microscopic components are rearranging themselves, and a novel collective phenomenon arises, which necessarily calls for correlations spread across the system (all magnetic spins align along the same direction) and, as we will see, the spontaneous breaking of a symmetry occurs.

The first attempt to model such phenomena from a phenomenological and macroscopical point of view has been the Landau theory of ferromagnetism [335, 336]. The idea is to describe the main aspects of the phenomena using an equation of state which respects the symmetry of the problem and relate the two most important thermodynamical quantities describing the system: the free energy F (that shall be minimized to have the system at equilibrium) and the magnetization M, the independent variable of the system. Assuming a power expansion of the function $F(M)$ we obtain

$$F(M) = F_0 + F_1 M + F_2 M^2 + F_3 M^3 + F_4 M^4, \tag{10.1}$$

where higher orders are neglected assuming that they are not needed to explain the main features we are interested in. This is a completely arbitrary assumption justified only a posteriori by the success of the theory. We assume that there is no preferred direction, that is, the spins interact with each other with some specific Hamiltonian that cannot distinguish among "up" and "down" direction: the system displays a $\mathbb{Z}_2$ symmetry (invariant under parity operator). Thus, the odd terms in Eq. (10.1) shall vanish, $F_1 = F_3 = 0$. Finally, we assume that $F_4 > 0$ and the quadratic coefficient depends on the temperature (which shall appear somewhere) in the simplest possible way, i.e. $F_2(T) = F_2(T - T_c)$.

The ground state is by definition the state with a magnetization M that minimizes the free energy: the condition

$$\frac{\partial F}{\partial M} = 2M\left[F_2(T - T_c) + 2F_4M^2\right] = 0 \tag{10.2}$$

results in the solutions, sketched in Fig. 10.1a,

$$\begin{cases} M = 0 & T > T_c \\ M = \pm\left[\frac{F_2(T-T_c)}{2F_4}\right]^{1/2} & T < T_c \end{cases}, \tag{10.3}$$

which reproduces the typical experimental results. Despite the tremendous success of this theory to describe the main features of such complex phenomena with an intuitive picture, there are of course limitations to its predictive power. Indeed, the quantitative values of T_c and the other constants shall be fitted from experiments and, more importantly, the square-root scaling law predicted by Eq. (10.3) (the first critical exponent we encounter, see Sect. 10.3.1) is typically not matching the measured one. Indeed, the prediction of Eq. (10.3) corresponds to the mean-field one, and more sophisticated approaches are necessary to extract the correct scaling values. Among others, TN methods are nowadays routinely applied to attack the problem of computing the different phases of quantum matter and their properties. However, before proceeding with the quantum counterpart of phase transition, it is worth spending some time having a closer look at the potential $F(M)$ as sketched in Fig. 10.1b. Imagine an experiment where the temperature of the system is lowered slowly, starting from a disordered state (no magnetization, the spins are randomly aligned and $T > T_c$): the system lays at the bottom of the single well and $M = 0$. Suddenly, at the critical temperature T_c the minima at $M = 0$ becomes instable while two other solutions appear for which $M \neq 0$. However, the system still sits in the unstable minima at $M = 0$ until a small perturbation (introduced via the thermal activity of the material), unbalances it and force the system to relax to one of the two new minima, either the positive or the negative one. In conclusion, thermal fluctuations break the symmetry of the problem and allow for the appearance of the spontaneous nonzero magnetization. Finally, notice that once the system has broken the symmetry, it will relax to the new stable minima: this means that once the system underwent the phase transition, it will acquire some rigidity, that is, a resistance to a new change or to restore the original symmetry. For example, crystals (which break translational invariance) do not bend easily, and spontaneous magnetization lasts for an extended period of time (it costs energy to climb up the potential $F(M)$). Similarly, new excitations are present around the new minima, e.g. lattice waves in crystals, spin waves in ferromagnets etc.. Last but not least, as the symmetry breaking is due to thermal fluctuations which acts randomly along all the systems, it might occur that regions far apart break it differently: in such a case, different ordered domains appear, with defects at their interface such as domain walls in magnets, crystals oriented in different directions, etc. We will see in the next chapter how a simple and powerful theory, the Kibble-Zurek mechanism, allows us to describe such phenomena and predict the final density of defect using an elegant argument.

We shall now spend some time to recall that quantum and classical systems are deeply connected form the statistical mechanic's point of view. Thus, also quantum many-body systems display behaviors reflecting what has been introduced here, that is *quantum phase transitions*.

10.2 Quantum-Classical Statistical Correspondence

The deep connection between the classical and quantum worlds in terms of statistical mechanics springs from the fact that the partition function of a D-dimensional quantum system Z_q is equivalent to that of a (D+1)-dimensional classical one Z [337, 338]. As all thermodynamical quantities can be derived from the partition function (for example, the free energy $F = -k_B T \log Z$), phase transitions also occur in quantum many-body systems. The equivalence mentioned above becomes apparent once the *transfer matrix* and the path integral methods to compute Z and Z_q respectively are introduced.

The transfer matrix approach exploits the fact that the classical partition function for N sites can be written as

$$Z_N = \sum_{\alpha_i,\dots,\alpha_N} e^{-\beta_T H(\alpha_i,\dots,\alpha_N)}, \tag{10.4}$$

where the sum runs over all accessible classical states and $\beta_T = 1/k_B T$, and we specialize it to one dimension with periodic boundary conditions, i.e. $\alpha_1 = \alpha_{N+1}$. Assuming nearest neighbour interactions and translational invariance, that is $H = \sum_j H(\alpha_j, \alpha_{j+1})$, the partition function can be rewritten as

$$Z_N = \sum_{\alpha_1} \cdots \sum_{\alpha_N} T_c(\alpha_1, \alpha_2) \dots T_c(\alpha_{N-1}, \alpha_N) = \mathrm{Tr}\{T_c^N\} \tag{10.5}$$

where the classical transfer matrix $T_c(\alpha_1, \alpha_2) = e^{-\beta_T H(\alpha_j, \alpha_{j+1})}$ is independent of j and has a (typically small) dimension such that it can be easily diagonalized to obtain its eigenvalues μ_i (we assume $\mu_1 > \mu_2 > \dots$). This condition results in a straightforward computation of the partition function at the thermodynamical limit as μ_1^N plus exponentially small corrections. In conclusion, for example, the free energy per site is

$$\lim_{N\to\infty} \frac{F}{N} = -\beta_T \log \mu_1. \tag{10.6}$$

The partition function of a quantum system described by the Hamiltonian operator $\hat{H}_q$, is by definition,

$$Z_q = \mathrm{Tr}\left\{e^{-\beta_q \hat{H}_q}\right\} = \mathrm{Tr}\left\{e^{-\delta\tau \hat{H}_q} e^{-\delta\tau \hat{H}_q} \dots e^{-\delta\tau \hat{H}_q}\right\}, \qquad (10.7)$$

where we split the imaginary time in n intervals such that $\beta_q/n = \delta\tau \ll 1$. Indeed, the expression above is equivalent to the solution of time-dependent Schrödinger equation in imaginary time. In the spirit of Feymann's idea that that the net transition amplitude between two states of the system can be calculated by summing amplitudes for all possible paths between them, we can insert identities (sums over a complete set of basis states, here for convenience labeled as $\mathbb{1} = \sum_{\gamma_j} |\gamma_j\rangle\langle\gamma_j|$) obtaining

$$Z_q = \sum_{\gamma_1} \cdots \sum_{\gamma_n} [e^{-\delta\tau \hat{H}_q}]_{\gamma_1,\gamma_2} \dots [e^{-\delta\tau \hat{H}_q}]_{\gamma_{n-1},\gamma_n} = \mathrm{Tr}\{T_q^n\}. \qquad (10.8)$$

It shall be evident that Eqs. (10.5) and (10.8) are formally equivalent, with the only difference that the number of multiplied matrices in Eq. (10.8) is not given by the number of elements in the system, but by the number of slices in the imaginary time evolution. This is indeed true in general: the partition function of a quantum Hamiltonian of D-dimensional is equivalent to that of a classical system with $D+1$ dimensions provided that $\beta_T H = \delta\tau \hat{H}_q$ [337]. The difference between the dimensions comes from the fact that a quantum Hamiltonian is an operator, while for a classical system is a scalar. Thus, for example, the classical Hamiltonian of a chain of spins in an Ising chain is $H = -J \sum_{i=1}^{N} s_i s_{i+1}$ where the classical variable $s_i = -1, 1$ and the transfer matrix is

$$T_c = e^{\tilde{J}} \begin{pmatrix} 1 & e^{-\tilde{2}J} \\ e^{-\tilde{2}J} & 1 \end{pmatrix} = e^{\tilde{J}} (\mathbb{1} + e^{-\tilde{2}J} \sigma^x),$$

where $\tilde{J} = -J/kT$. For small enough $\delta\tau$, we can write the quantum transfer matrix as $T_q \simeq \mathbb{1} + H_q \delta\tau$, and thus the quantum and the classical partition function are equivalent provided that $n = N$, and $H_q \delta\tau = e^{-\tilde{2}J} \sigma^x$ (neglecting the rescaling factor $e^{\tilde{J}}$). Thus, the quantum partition function of a single spin (dimension $D = 0$) is equivalent to that of a chain of classical spins ($D = 1$). Notice, however, that the time dimension is infinite only in the limit $\beta_q \to \infty, T \to 0$. Given that a D-dimensional quantum system is statistically equivalent to a classical one in $D+1$-dimensions and that classical systems undergoes phase transitions for different temperatures; also quantum systems at $T = 0$ shall undergo phase transitions. However, the driving field will not be the system temperature (it is always zero) but some system's parameter individuated by the relation $\beta_T H = \delta\tau \hat{H}_q$.

10.3 Quantum Phase Transition

Quantum phase transitions, and in particular second-order quantum phase transition on which we will focus on here, describe the drastic changes of the ground state properties of some many-body quantum systems as a function of an external parameter. The role of temperature fluctuations is now played by quantum fluctuations, i.e., the uncertainty introduced by the Heisenberg principle, that is, by the presence of noncommuting terms in the Hamiltonians. As we have seen in the previous section, this shall correspond to the physics of classical phase transitions, where we expect symmetry breaking to occur, the divergence of correlations and in general discontinuous behaviors in the derivatives of the system properties. The characterization of these phenomena has been pursued studying different quantities and their scaling relations with the distance from the critical point (as for Eq. (10.3)), in time and as a function of the system size: critical system can be classified by means of the exponents of such scalings, allowing one to introduce *universality classes* which group all systems with the same exponents [137, 141, 333]. Universality classes are formed by systems characterized by the same symmetries, and the low-energy physics of such systems does not depend on the details of the Hamiltonian: the critical properties are dictated only by the symmetries of the Hamiltonian [141, 333]. Also the systems entanglement properties have been used to classify critical systems, providing one of the first strong and unexpected bridges between quantum information theory and strongly correlated systems [140, 339].

We are then interested in the study the ground state properties of a many-body Hamiltonian of the form

$$H(g) = H_0 + gH_1, \tag{10.9}$$

where the system is invariant under a symmetry group, and the competition between the two noncommuting terms H_i determines the system properties. Then, it shall be possible to find an *order parameter*, a local operator whose expectation value on the system ground state $\langle \hat{O}_i \rangle$ (that plays the role of the magnetization in the Landau theory) signals the occurrence of the quantum phase transition at some particular value of the parameter g_c: a change from the disordered phase where $\langle \hat{O}_i \rangle = 0$ to the ordered one where the symmetry is broken and $\langle \hat{O}_i \rangle \neq 0$. The point where this abrupt change occurs is the critical point g_c. Finding the right order parameter to identify a transition is a task which can be straightforward as for the case of the Ising model in the transverse field, or highly nontrivial for more complex theories, requiring experience and skills.

Hereafter, we introduce some of the most common ways used to characterize critical phenomena, with particular attention to those that are readily accessible using TN methods introduced in the previous chapters.

10.3.1 Critical Exponents

As we have seen before, a quantum phase transition is signaled by the expectation value of the order parameter that becomes non-zero in the ordered phase. In particular, the functional dependence in the ordered phase is given by

$$\langle \hat{O}_i \rangle \sim (g - g_c)^{-\beta}, \tag{10.10}$$

where β is the first critical exponent we introduce, and is referred as *spontaneous magnetization*. This quantity is easily accessible by TN simulations, being the expectation value of a local observable. The relation given in Eq. (10.10) is exact only at the thermodynamical limit, where the non-analyticity of the function appears.

Moreover, for every finite N, the symmetry breaking could not occur. Indeed, for every finite N, the symmetry-broken ground states have a non-zero coupling term between them, which restore a symmetric superposition of them as the true ground state. In other words, the ground states are not exactly degenerate for finite N; thus the symmetric superposition is still energetically favorable, even though the gap closes with N, and especially when describing such phenomena via numerical methods the results might be unstable. A common way to overcome this problem is to artificially break the symmetry, introducing a small field (that eventually will go to zero) to stabilize the computation. However, this procedure introduces another parameter which shall be carefully taken care of, and it does not solve the problem in the limit where the field goes to zero. As we will see later on, the solution to this problem lies in the exploitation of the calculation of the *structure factor*.

As already mentioned, the correlations $C_{i,j}$ as given in Eq. (4.9) diverge at the critical point, meaning that if fitted as an exponential function of the distance between the sites $\ell = i - j$, $\exp(\ell/\xi)$, the correlation length ξ diverges. In other words, the correlation $C_{i,j}$ has another functional dependence with the distance: a power law decay $C(\ell) \propto \ell^{-\eta}$, where η is another critical exponent characterizing the system, the *anomalous dimension*. More generally, away from the critical point, the correlation scales as [141]

$$C(\ell) \propto \ell^{-\eta} \exp(-\ell/\xi). \tag{10.11}$$

The previous relation is practically very important as in numerical investigations, apart from some trivial cases, the location of the critical point is always approximate and thus the correlation length always finite. Employing numerical methods it is possible to compute the correlation functions $C(\ell)$ and fit the correlation length using the above relation.

Numerical simulations typically work at finite system sizes N (or effectively introduce a cut-off), and this aspect shall also be taken into account when studying critical systems to extract relevant quantities such as the critical exponent β. The solution to these problems comes from our theoretical understanding of critical system and from the self-similarity we expect at all scales, which allow us to introduce scaling relations. In particular, by means of finite-size scaling [141,340] it is possible to acquire two critical exponents at once: the spontaneous magnetization β and the *correlation length divergence exponent* ν, which describes how the correlation length changes as a function of the distance from the critical point

$$\xi \sim (g - g_c)^{-\nu}. \tag{10.12}$$

In accordance to renormalization group analysis, the order parameter obeys a precise scaling with the system size L and the parametric distance from the critical point

$$\langle \hat{O}_i \rangle(g) \simeq L^{-\beta/\nu} f\left((g - g_c) \cdot L^{1/\nu}\right), \tag{10.13}$$

where f is a non-universal function, depending on the microscopic details of the model. Given that the relation Eq. (10.13) is valid, after computing the order parameter for different system sizes $\langle \hat{O}_i \rangle(g, L)$, one can first tune the unknown $\gamma_1 = \beta/\nu$ until all curves $\langle \hat{O}_i \rangle(g, L) \cdot L^{\gamma_1}$ cross in a single point, which individuate the critical point g_c (the only point invariant for different L, according to Eq. (10.13)). Then, rescaling the curves $\langle \hat{O}_i \rangle(g_c + (g - g_c)L^{\gamma_2}, L) \cdot L^{\gamma_1}$ by means of $\gamma_2 = -1/\nu$ until all collapse onto one another. The two values γ_1, γ_2 uniquely defines the critical exponents ν, β [340].

Another important critical exponent is the *dynamical-scaling* exponent z which gives the scaling of the imaginary-time correlations with the distance from the critical point of an operator evaluated at different imaginary-time in $C(\tau) = \langle \hat{O}(\tau)\hat{O}(0) \rangle \sim \exp(-\tau/\xi_\tau)$,

$$\xi_\tau = \xi^z \sim (g - g_c)^{-\nu z}. \tag{10.14}$$

Going back to the quantum-classical correspondence, if the system can be mapped to a homogeneous quantum system in $D + 1$ dimensions, then the time-direction of the Feynman integral in Eq. (10.8) is driven by the same Hamiltonian than in the space directions. That is, the correlations in space and time are equivalent, and $z = 1$. However, this is not always the case and in general $z \neq 1$. We also mention that using the analytical continuation, one can also infer on the real-time correlations of the system [337].

Finally, an important relation between correlations in imaginary time and the gap of the system is worth mentioning: indeed writing the evolved operator in the Heisenberg picture $\hat{O}(\tau) = e^{H\tau}\hat{O}(\tau)e^{-H\tau}$, it is easy to show that the correlations are given by

$$C(\tau) = \sum_j e^{-(E_j - E_0)\tau} |\langle 0|\hat{O}(\tau)|j\rangle|^2; \tag{10.15}$$

where we inserted a set of complete of system's eigenfunctions $|j\rangle$ and E_j are the corresponding eigenenergies. It shall then be clear that a finite principal gap $\Delta = E_1 - E_0$ implies that for (long) imaginary time the correlations decay exponentially, i.e. $C(\tau) \sim e^{-\Delta\tau}$. Thus, on the contrary, if the correlations decay algebraically, the gap shall be vanishing, as it occurs at the quantum critical point. The expression above together with Eq. (10.14) allows relating the scaling of the gap around critical points with the critical exponents

$$\Delta \sim (g - g_c)^{\nu z}. \tag{10.16}$$

10.3.1.1 Structure Factor

As mentioned in the previous section, whenever one works on quantum systems with finite-size, spontaneous symmetry-broken phase does not occur. It is possible to overcome this problem by exploiting an order parameter definition insensitive to the symmetry breaking. For example, one can adopt the *square root of the structure factor density*, precisely

$$\mathbf{S}(k, g, L) = \sqrt{\frac{1}{N^2}\sum_{i,j}^{N} e^{ik(i-j)}\langle\psi(g, L)|\hat{O}_i \otimes \hat{O}_j|\psi(g, L)\rangle}\ , \tag{10.17}$$

where $|\psi(g, N)\rangle$ is the many-body ground state calculated at g and size N. For example, if applied to study a standard ferromagnetic transition, instead of computing the order parameter based on $\hat{O}_i = \sigma_i^x$, which will be always zero at finite sizes, one can evaluate the structure factor at $k = 0$: it can be shown to coincide to the standard ferromagnetic order parameter $\bar{m} = N^{-1}\sum_j^N \langle O_j\rangle$ exactly at the thermodynamical limit. Indeed, given that by definition $\langle\hat{O}_i\hat{O}_j\rangle = \langle\hat{O}_i\rangle\langle\hat{O}_j\rangle + C(\ell)$, then the correlation function $C(\ell)$ becomes irrelevant towards quantity in Eq. (10.17). In fact, we have that either $C(\ell)$ decays to zero either exponentially (gap different from zero) or algebraically (critical point). In both cases, we have

$$\left|\sum_{i,j}^{N}\frac{C(\ell)}{N^2}\right| \le \frac{1}{N}\int_0^\infty |C(\ell)|\,dx \to 0. \tag{10.18}$$

In conclusion, we have that

$$\mathbf{S}(0, g, N) = \sqrt{\sum_{i,j}^{N}\frac{\langle\hat{O}_i\rangle\langle\hat{O}_j\rangle}{N^2}} \simeq \sqrt{\bar{m}^2} = \bar{m}, \tag{10.19}$$

which shows that $\mathbf{S}(0, g, N)$ is equal to the ferromagnetic order parameter without suffering from finite-size symmetry breaking issues, since it is based on two-point correlation measurements and not on local observations.

10.4 Entanglement Measures

As we have seen in Chap. 5, the von Neumann entropy of the singular values of a system bipartition plays a fundamental role in determining the efficiency of TN methods. Hereafter, we review another important relation between entanglement measures and the physics of the system of interest. The von Neumann entropy is only one of the possible measures of entanglement that can be used to characterize many-body quantum systems [140]: One of the first attempts to connect entanglement and physical properties of the system, in particular in systems undergoing a quantum phase transition, has been made using the concurrence, a commonly used entanglement measure of two qubits [339,341]. Given the reduced density matrix of two spins one-half embedded in a larger system $\rho_{i,j}$, it is possible to calculate the concurrence C diagonalizing the matrix $R = \sqrt{\rho_{i,j}}\tilde{\rho}_{i,j}\sqrt{\rho_{i,j}}$, and computing

$$C = \max(\lambda_1 - \lambda_2 - \lambda_3 - \lambda_4, 0), \tag{10.20}$$

where $\lambda_1 > \cdots > \lambda_4$ are the ordered squared root of the eigenvalues of R and $\tilde{\rho}_{i,j} = (\sigma^y \otimes \sigma^y)\rho^*_{i,j}(\sigma^y \otimes \sigma^y)$. It has been shown that the derivative with respect to the parameter g of the concurrence between the two spins diverges exactly at the critical point, signaling the critical behavior and thus working as a non-local order parameter. A posteriori, this beautiful result can be easily interpreted, as the elements of the reduced density matrix are given by the two-point correlations in the systems [140], and thus, their divergence at the critical point is reflected in the concurrence. This observation has been one of the first witnesses of the deep relation between quantum information science and condensed matter theory and has triggered enormous research activities [140]. In particular, it is nowadays widespread the use of the von Neumann entropy of a system bipartition to characterize critical systems. Given a system bipartition in two subsystems A and B, the von Neumann entropy is defined as

$$S = -\mathrm{Tr}\{\rho_A \log \rho_A\} = -\mathrm{Tr}\{\rho_B \log \rho_B\} = -\sum_i p_i \log(p_i), \tag{10.21}$$

where ρ_A, ρ_B are the reduced density matrices of the subsystems and p_i their eigenvalues. As can be recalled from Eqs. (7.28) and (7.29), this quantity—and its scaling with the system size for bipartitions of one-dimensional quantum systems—can be readily be computed using MPS. In the next section, we review some important results which relate the von Neumann entropy scaling and physical properties of the system.

10.4.1 Central Charge

The main feature of the scaling of the von Neumann entropy of a one-dimensional system is that it can be related to the central charge c of the correspondent conformal field theory [140, 229, 334]. Conformal field theory gives a classification of one-dimensional critical quantum systems based on symmetry considerations, independently from the detailed Hamiltonian of the system. The central charge—defined by the conformal anomaly of the Virasoro algebra—plays a central role being one of the unique elements which identify the conformal theory [334] and can be used as the critical exponents to classify the system of interest in a universality class.

For a discrete system with open boundary conditions divided in two parts of size ℓ and $N - \ell$ respectively with a single boundary, the scaling of the von Neumann entropy of the reduced density matrices reads

$$S = \frac{c}{6} \log \left(\frac{N}{\pi} \cdot \sin \frac{\pi \ell}{N} \right) + c' . \tag{10.22}$$

where c' is a non-universal constant [140, 229]. For periodic boundaries, or more generally in the case of a bipartition of the system with two boundaries, the scaling above is doubled, taking into account the two boundaries which contribute to the correlations between the two system's partitions. More generally, the von Neumann entropy can also be used to measure the correlation length of the system ξ, as slightly away from criticality, it scales for $\ell \to \infty$ as

$$S \sim \frac{c}{6} \log (\xi) . \tag{10.23}$$

An extension to the finite temperature case is also available, where the effective temperature $\beta_T = 1/(k_B T)$ introduces a cutoff to the correlation length and thus plays the role of an effective system size (even at the thermodynamical limit), resulting in a scaling of the form [140]:

$$S = \frac{c}{6} \log \left(\frac{\beta_T}{\pi} \cdot \sin \frac{\pi \ell}{\beta_T} \right) + c' . \tag{10.24}$$

It is worth mentioning that the above results are still valid in the case of disordered systems, provided that the proportionality factor—the central charge—is renormalized in a system-dependent way [140]: for example, for a spin one-half Ising model in transverse field with uncorrelated disorder $\tilde{c} = c \ln 2$ [342]. However, the renormalization depends also on the correlations of the disordered terms, and eventually, correlated noise might results again in the clean case scenario [343–346].

Finally, the relations above can be extended to Rényi entropies,

$$S_q = \frac{1}{1-q} \log(\mathrm{Tr}\{\rho_A^q\}), \tag{10.25}$$

where the von Neumann entropy is the special case $q \to 1$. Indeed, in this case we have that Eq. (10.22) shall be corrected by the factor

$$S_q = \frac{c}{6}\left(1+\frac{1}{q}\right)\log\left(\frac{N}{\pi}\cdot\sin\frac{\pi\ell}{N}\right)+c'\,. \tag{10.26}$$

10.4.2 Topological Entanglement Entropy

We conclude this very concise overview of the use of von Neumann entropy to extract relevant information on the physics of correlated quantum system introducing briefly the concept of *topological entanglement entropy* [38, 140, 334, 347, 348]. Topological systems do not break any symmetry; thus they cannot be characterized by a local order parameter. The ground state is generally not unique, and on a two-dimensional surface, the degeneracy grows exponentially with the number of holes in the surface (genus). Local excitations on top of the degenerate ground states are protected by large energy gaps, making the topological state robust against local perturbations [334]. Finally, topological phases cannot be distinguished by local observations, and thus a global characteristic shall be used such as the entanglement between system bipartitions [347, 348]. In particular, it has been shown that the Von Neumann entropy can be used to characterize the topological order of the system: it obeys an area law, i.e., scaling with the boundary of the size bipartition L

$$S = \alpha \cdot L - \gamma \tag{10.27}$$

where α is a nonuniversal constant but $\gamma = \log D \geqslant 0$, the total quantum dimension of the theory, characterizes the topological content of the state. For example, in the case of discrete gauge theories, D is simply the number of elements in the gauge group, while for a Laughlin state in a fractional quantum Hall system with filling factor $\nu = 1/r$ where r is an odd integer, $D = \sqrt{r}$. Despite the challenges still present in applying TN methods to two dimensional systems, in the last years, they have been used to extract the topological entropy of different systems, see, e.g., [180, 349–352].

10.5 Problems

1. Compute the mean field critical exponents of the one-dimensional Ising model in transverse field via tensor networks (i.e., using $m = 1$) and confirm the analytical predictions.
2. Compute the ferromagnetic order parameter of the one-dimensional Ising model in transverse field both directly and using the structure factor. Compare the results.
3. Perform a finite size scaling to compute the critical exponents of the critical Ising model in transverse field (compute the order parameter for different size and values of the transverse field) both with RG and DMRG and compare the results.

Hamiltonian Lattice Gauge Theories 11

Giovanni Cataldi

This chapter explores the application of Tensor Network (TN) algorithms to Lattice Gauge Theories (LGTs), a versatile framework for investigating quantum field theories with gauge symmetries on discretized spacetime lattices [354–358]. LGTs form the foundation for understanding fundamental interactions in nature, governed by gauge invariance—the principle that physical laws remain unchanged under local symmetry transformations. Symmetry groups such as U(1) in Quantum Electrodynamics (QED) and SU(3) in Quantum Chromodynamics (QCD) define the dynamics of fundamental forces in the Standard Model.

LGTs offer a non-perturbative approach to studying quantum field theories, especially in regimes where traditional methods fail [359–365]. By discretizing spacetime, LGTs enable numerical simulations that unveil compelling phenomena such as confinement [366–371], chiral symmetry breaking [372–375], the hadronic spectrum, and the dynamics of quarks and gluons [376]. Moreover, LGTs have applications beyond high-energy physics, providing valuable insights into low-energy phenomena, such as high-temperature superconductivity [377–379] and spin liquids [380, 381].

These theories pose unique challenges compared to other quantum many-body (QMB) systems due to the strongly interacting nature of models like QED and QCD. Analytical methods often fall short, and numerical approaches face significant difficulties. Traditional Monte Carlo (MC) simulations of LGTs in the Lagrangian formulation [382–393], while being extremely powerful, are limited by the infamous sign-problem [365, 394, 395], which hampers their effectiveness in real-time dynamics and finite-density regimes. In contrast, the Hamiltonian formulation of LGTs is particularly well-suited for TN approaches and quantum simulations, as it allows direct access to these challenging regimes [282, 396–400].

G. Cataldi (✉)
Max Planck Institute of Quantum Optics, Garching, DE, Germany
e-mail: Giovanni.cataldi@mpq.mpg.de

T. Felser, S. Montangero (eds.), *Introduction to Tensor Network Methods*,
Graduate Texts in Physics, https://doi.org/10.1007/978-3-032-17635-6_11

This chapter provides a comprehensive introduction to Hamiltonian LGTs, beginning with a review of the lattice discretization of spacetime in Sect. 11.1. The discussion includes staggered fermions in Sect. 11.1.3 and gauge fields in Sect. 11.1.4, essential components for modeling matter and interactions in these theories. The original contributions of this chapter start with the introduction of the dressed-site formalism in Sect. 11.2, a gauge-invariant framework designed to address the complexities of LGTs using TN algorithms and quantum hardware. This approach is particularly advantageous in higher dimensions, where the simulation of LGTs becomes increasingly challenging. Two specific implementations of the dressed-site formalism are explored: the U(1) (Abelian) LGT in Sect. 11.3 and the SU(2) (non-Abelian) Yang Mills LGT in Sect. 11.4. These implementations have been successfully applied to produce key numerical results in various studies [282, 396–403], furthering our understanding of LGTs and their role in describing the fundamental forces of nature.

11.1 Lattice Discretization

Let us consider continuous (D+1) spacetime domain ($\mathbb{R}^D \times \mathbb{R}$) identified by the (D+1)-dimensional point $r = (t, \mathbf{r})$, where $\mathbf{r} = (\mathbf{r}_1, \dots \mathbf{r}_D)$. When performing spatial discretization, the time variable t remains continuous, while the spatial domain $\mathbb{R}^D$ is replaced by a hyper-cubic D-dimensional lattice Λ, whose sites are equally spaced by lattice constant a_0 (see Fig. 11.1). Lattice sites are then labelled by coordinates $\mathbf{n} = (\mathbf{n}_1, \dots, \mathbf{n}_D)$, where $r_k = a_0 n_k$ and $n_k \in \{1 \dots N_k\}$, with N_k denoting the number of lattice sites along the kth direction $\forall k = 1, \dots, D$. Then, if $N = \prod_{k=1}^{D} N_k$ is the total number of lattice sites, with the corresponding lattice volume given by $V = a_0^D N$. Additionally, we define the unit lattice vectors $\pm \boldsymbol{\mu}_k$ pointing along the $\pm k^{\text{th}}$ direction, assuming $\pm \boldsymbol{\mu}_k = \pm \boldsymbol{\mu}_k$. We use the convention where summations over lattice vectors include only positive directions. Then, the lattice link connecting two neighboring sites $\mathbf{n}$ and $\mathbf{n} + \boldsymbol{\mu}_{kk}$ along the direction $+\boldsymbol{\mu}_k$, is denoted by the couple $(\mathbf{n}, \boldsymbol{\mu}_{kk})$ or equivalently with $(\mathbf{n} + \boldsymbol{\mu}_k, -\boldsymbol{\mu}_k)$. For future purposes, we refer to an *even* (*odd*) site, if $(-1)^{\mathbf{n}} = (-1)^{\sum_k n_k}$ is even (odd).

11.1.1 Preliminaries

Lattice discretization of spatial dimensions affects all the mathematical objects: for a general scalar field $f(\mathbf{r})$ whose absolute value is *square-integrable*

$$\int_{\mathbb{R}^D} dr^D |f(\mathbf{r})|^2 = L < \infty , \tag{11.1}$$

we have to consider the following replacements:

$$\int_{\mathbb{R}^D} dr^D \longrightarrow a_0^D \sum_{\mathbf{n}}, \qquad f(\mathbf{r}) \longrightarrow \hat{f}_{\mathbf{n}}, \qquad \partial_k f(\mathbf{r}) \longrightarrow \frac{\hat{f}_{\mathbf{n}+\mu_k} - \hat{f}_{\mathbf{n}-\mu_k}}{2a_0}. \tag{11.2}$$

Such a discretization prescription introduces an *ultraviolet cut-off*, in energy and momentum, which depends on the inverse of the lattice spacing a_0. Namely, for each spatial direction k, momentum is anisotropic and assumes N_k discrete values $p_k(i) \in \left[-\frac{\pi}{a_0}, \frac{\pi}{a_0}\right]$, where $i \in \{1 \dots N_k\}$. The maximal momentum along each direction is then $p_{\max} = \pi/a_0$, and corresponds to the maximal energy $E_{\max} = \sqrt{\mathbf{p}_{\max}^2 + m_0^2}$, where $\mathbf{p}_{\max} = (\pi/a_0, \pi/a_0, \pi/a_0)$. Correspondingly, assuming tha lattice Λ with periodic boundary conditions, the continuous Fourier transform is adapted as follows:

$$f(\mathbf{r}) = \int_{\mathbb{R}^D} \frac{d\mathbf{p}}{(2\pi)^D} \tilde{f}(\mathbf{p}) e^{i\mathbf{r}\cdot\mathbf{p}} \quad \longrightarrow \quad f(\mathbf{n}) = \left(\frac{a_0}{2\pi}\right)^D \sum_{\mathbf{p}}^{BZ} \tilde{f}_{a_0}(\mathbf{p}) e^{ia_0\mathbf{n}\cdot\mathbf{p}}, \tag{11.3}$$

where BZ stands for the 1st Brillouin zone $\left[-\frac{\pi}{a_0}, \frac{\pi}{a_0}\right]^D$ and $\tilde{f}_{a_0}(\mathbf{p}_{\max}) = \tilde{f}_{a_0}(-\mathbf{p}_{\max})$. We can then express $\tilde{f}_{a_0}(\mathbf{p})$ in (finite) Fourier-series

$$\tilde{f}_{a_0}(\mathbf{p}) = a_0^D \sum_{\mathbf{n}} \hat{f}_{\mathbf{n}} e^{-ia_0\mathbf{n}\cdot\mathbf{p}}. \tag{11.4}$$

If we set $\hat{f}_{\mathbf{n}} = (2\pi)^{-D}$, then (11.4) reduces to the Fourier series representation of the δ function in the 1st BZ, namely

$$\delta_p(\mathbf{p}) = \left(\frac{a_0}{2\pi}\right)^D \sum_{\mathbf{p}}^{BZ} e^{-ia_0\mathbf{n}\cdot\mathbf{p}}, \tag{11.5}$$

where p stands for *periodic*. Note that, unlike the continuum limit, where $a_0 \to 0$, here $\delta_p(\mathbf{p})$ has "non-vanishing" support in $\mathbf{p} = 0$ (modulo $2m\pi$ for $m \in \mathbb{N}$). Similarly, the Dirac δ function $\delta(\mathbf{r} - \mathbf{r}')$ becomes a Kronecker $\delta_{\mathbf{n}\mathbf{n}'}$ (multiplied by a factor a_0^{-D})

$$\delta_{\mathbf{n}\mathbf{n}'} = \left(\frac{a_0}{2\pi}\right)^D \sum_{\mathbf{p}}^{BZ} e^{ia_0\mathbf{n}\cdot\mathbf{p}}. \tag{11.6}$$

11.1.2 Continuum Limit and Lorentz Invariance

The main physical consequence of the ultraviolet cutoff is the following: the discretized theory does not describe physics at scales (momenta) smaller (larger) than the lattice spacing a_0 (π/a_0), i.e., all the short-distance (high-energy) phenomena [354].

Another effect of lattice discretization is the loss of the Lorentz invariance (the symmetry of special relativity), as the lattice spacing a_0 introduces a preferred reference frame. In the Lagrangian formalism, where both space and time are discretized, the spacetime lattice does not support continuous Lorentz transformations, but the theory retains a discrete version of spacetime symmetry, such as discrete rotations or translations. On the contrary, the spatial discretization performed in Hamiltonian LGTs breaks the symmetry between space and time, which is a crucial part of Lorentz invariance. In these terms, the Hamiltonian formalism is said to explicitly break Lorentz invariance.

Restoring the continuum theory and the Lorentz invariance requires to take the continuum limit, where $a_0 \to 0$, $N \to \infty$, while keeping $a_0 \cdot N$ fixed [405]. In this regime, maximum momentum $\mathbf{p}_{\max}$ and energy $E_{\max}$ become infinite and the lattice approximates the continuous space. The continuum limit is typically associated with the process of *renormalization*, where the theory is adjusted to correctly reproduce physical observables (which should be insensitive to the discretization) as the lattice spacing goes to zero. Unfortunately, the continuum limit location is non trivial and requires a lot of attention in understanding the relation between the lattice spacing and the Hamiltonian parameters. A comprehensive dissertation about the continuum limit would require a dedicated effort out of the purposes of this chapter. Nonetheless, we will partially discuss it in Sects. 11.1.3 and 11.1.4 by using simple dimensional analysis. For a deeper discussion, see [355, 366, 383, 405–408].

11.1.3 Discretization of Fermions Fields

Discretization of more complex objects such as fermion fields, i.e. spinors, requires more attention and encounters a set of problems whose solutions are not univocal nor perfect. Throughout this section, we will cover these challenges by focusing on one of the possible solutions: *staggered fermions*. For a more general and complete discussion about fermion discretization, see [366, 405, 409].

Let us start from the continuum Dirac equation for a massive free fermion. Assuming the spacetime coordinate $r = (t, \mathbf{r}) = (t, \mathbf{r})$ and the metric $\eta^{\mu\nu} = \text{diag}(+1, -1, -1, -1)$, we have:

$$(i\hbar\gamma^\mu\partial_\mu - m_0c)\psi(r) = \left(i\frac{\hbar}{c}\gamma^0\partial_t - i\hbar\gamma^k\partial_k - m_0c\right)\psi(r) = 0\,, \tag{11.7}$$

where c is the speed of light, $\hbar$ is the Planck constant, m_0 is the mass parameter, and γ^μ are the 4×4 Dirac matrices satisfying Clifford's algebra

$$\{\gamma^\mu, \gamma^\nu\} = 2\eta^{\mu\nu} . \tag{11.8}$$

The field $\psi(r)$ is a 4-dimensional spinor, whose Hermitian conjugate is $\overline{\psi}(r) = \psi^\dagger(r)\gamma^0$. The corresponding quantized version satisfies the canonical equal-time commutation relations

$$\left\{\hat{\psi}_\alpha(t, \mathbf{r}), \hat{\psi}^\dagger_\beta(t, \mathbf{r}')\right\} = i\hbar\delta_{\alpha\beta}\delta^3(\mathbf{r} - \mathbf{r}') , \tag{11.9}$$

where the α and β indices label spinor components. In the Hamiltonian form, Eq. (11.7) reads:

$$-i\hbar\partial_t\hat{\psi}(t, \mathbf{r}) = \left(ic\hbar\gamma^0\gamma^k\partial_k + m_0c^2\gamma^0\right)\hat{\psi}(t, \mathbf{r}) . \tag{11.10}$$

Omitting the time dependence, the Dirac Hamiltonian in this form becomes

$$\hat{H}_{\text{Dirac}} = \int d\mathbf{r}\hat{H}(\mathbf{r}) = \int d\mathbf{r}\left[\hat{\psi}^\dagger(\mathbf{r})(ic\hbar\gamma^0\gamma^k\partial_k + m_0c^2\gamma^0)\hat{\psi}(\mathbf{r})\right], \tag{11.11}$$

When discretizing the space on a D-dimensional spatial lattice Λ, we replace $\hat{\psi}(t, \mathbf{r}) \rightarrow \hat{\psi}_\mathbf{n}(t)$, where $\mathbf{r} = a_0\mathbf{n}$. Correspondingly, the canonical equal-time commutation relations become:

$$\left\{\hat{\psi}_{\mathbf{n},\alpha}(t), \hat{\psi}^\dagger_{\mathbf{n}',\beta}(t)\right\} = i\hbar\delta_{\alpha\beta}\delta_{\mathbf{n},\mathbf{n}'} . \tag{11.12}$$

Unless required, from now on, we will omit the temporal dependence, by simply focusing on $\hat{\psi}_\mathbf{n}$. Similarly, spatial derivatives $\partial_k\hat{\psi}(\mathbf{r})$ are replaced by (symmetric) finite differences:

$$\partial_k\hat{\psi}(\mathbf{r}) \longrightarrow \frac{1}{2a_0}\left[\hat{\psi}_{\mathbf{n}+\boldsymbol{\mu}_k} - \hat{\psi}_{\mathbf{n}-\boldsymbol{\mu}_k}\right]. \tag{11.13}$$

By recalling Eq. (11.2), the lattice Dirac Hamiltonian reads

$$\hat{H}^{\text{latt}}_{\text{Dirac}} = a_0^D \sum_{\mathbf{n},k} \hat{\psi}^\dagger_\mathbf{n}\left[ic\hbar\frac{\gamma^0\gamma^k}{2a_0}(\hat{\psi}_{\mathbf{n}+\boldsymbol{\mu}_k} - \hat{\psi}_{\mathbf{n}-\boldsymbol{\mu}_k})\right] + a_0^D m_0c^2 \sum_\mathbf{n} \hat{\psi}^\dagger_\mathbf{n}\gamma^0\hat{\psi}_\mathbf{n} . \tag{11.14}$$

We then redefine $\hat{\psi}_\mathbf{n} \rightarrow a_0^{-D/2}\hat{\psi}_\mathbf{n}$ and obtain

$$\hat{H}^{\text{latt}}_{\text{Dirac}} = \sum_{\mathbf{n},k} \hat{\psi}^\dagger_\mathbf{n}\left[ic\hbar\frac{\gamma^0\gamma^k}{2a_0}(\hat{\psi}_{\mathbf{n}+\boldsymbol{\mu}_k} - \hat{\psi}_{\mathbf{n}-\boldsymbol{\mu}_k})\right] + m_0c^2 \sum_\mathbf{n} \hat{\psi}^\dagger_\mathbf{n}\gamma^0\hat{\psi}_\mathbf{n} . \tag{11.15}$$

This choice makes the lattice fermion fields adimensional, even if their hidden dependence on the lattice spacing becomes important when considering the continuum limit location.

11.1.3.1 Fermion Doubling Problem

An artifact of the lattice discretization of fermions is the appearance of extra solutions of the Dirac equation, known as *doublers* [355]. In a D-dimensional space, these doublers lead to the appearance of $2^D - 1$ extra fermion species with respect to the continuum theory: namely, each spatial direction k contributes a factor of 2 to the number of solutions (one at $p_k = 0$ and one at $p_k = \pi/a_0$).

In order to formally visualizes these extra solutions, let us transform the lattice spinor field to the momentum space using the previously defined Fourier transform:

$$\hat{\psi}_{\mathbf{n}} = \frac{1}{V}\sum_{\mathbf{p}} e^{i\mathbf{p}\cdot\mathbf{n}a_0}\hat{\psi}(\mathbf{p}), \tag{11.16}$$

where $\mathbf{p}$ is the lattice momentum, while V is the lattice volume. Substituing Eq. (11.16) into the discretized derivative, we have:

$$\begin{aligned}\frac{\hat{\psi}_{\mathbf{n}+\mu_k} - \hat{\psi}_{\mathbf{n}-\mu_k}}{2a_0} &= \frac{1}{V}\sum_{\mathbf{p}} e^{i\mathbf{p}\cdot\mathbf{n}a_0}\frac{e^{ip_k a_0} - e^{-ip_k a_0}}{2a_0}\hat{\psi}(\mathbf{p}) \\ &= \frac{1}{V}\sum_{\mathbf{p}} e^{i\mathbf{p}\cdot\mathbf{n}a_0}\frac{i\sin(p_k a_0)}{a}\hat{\psi}(\mathbf{p}).\end{aligned} \tag{11.17}$$

Then, the lattice Dirac Hamiltonian in the momentum space becomes

$$\hat{H}(\mathbf{p}) = \hat{\psi}(\mathbf{p})^\dagger\left[-\sum_{k=1}^{3}\gamma^0\gamma^k\frac{c\hbar}{a_0}\sin(p_k a_0) + m_0c^2\gamma^0\right]\hat{\psi}(\mathbf{p}). \tag{11.18}$$

From the eigenvalue equation $H(\mathbf{p})\hat{\psi}(\mathbf{p}) = E(\mathbf{p})\hat{\psi}(\mathbf{p})$, we have

$$\left[-\sum_{k=1}^{3}\gamma^0\gamma^k\frac{c\hbar}{a_0}\sin(p_k a_0) + m_0c^2\gamma^0\right]\hat{\psi}(\mathbf{p}) = E(\mathbf{p})\hat{\psi}(\mathbf{p}) \tag{11.19}$$

Squaring both sides of the eigenvalue equation, we have:

$$\begin{aligned}\left[\gamma^0\left[-\sum_{k=1}^{3}\frac{c\hbar\gamma^k}{a_0}\sin(p_k a_0) + m_0c^2\right]\right]^2 &= E^2(\mathbf{p})\hat{\psi}(\mathbf{p}) \\ \left[\sum_{k=1}^{3}\left(\frac{c\hbar}{a_0}\sin(p_k a_0)\right)^2 + m_0^2c^4\right]\hat{\psi}(\mathbf{p}) &\underset{*}{=}\end{aligned} \tag{11.20}$$

where, in the second passage, we canceled all the mixed terms by exploiting the following properties from the Cliffor's algebra in Eq. (11.8):

$$\gamma^0\gamma^0 = \mathbb{1} \qquad \gamma^0\gamma^k + \gamma^k\gamma^0 = 0 \qquad \gamma^j\gamma^k + \gamma^k\gamma^j = 2\delta^{jk}\mathbb{1}\,. \tag{11.21}$$

Clearly, when $p_k a_0 \ll 1$, then $\sin(p_k a_0) \sim p_k a_0$ and the energy reduces to the familiar dispersion relation of relativistic particles: $E(\mathbf{p}) = \pm\sqrt{c^2\hbar^2\mathbf{p}^2 + m_0^2c^4}$. However, the sine function $\sin(p_k a_0)$ has an additional zero at $p_k = \pi/a_0$, which yields another low-energy solution at the border of the BZ. Therefore, in D spatial dimensions, we obtain 2^D fermion species. The extra degrees of freedom affect the extrapolation to the continuum limit such that the correct continuum results can not be recovered.

11.1.3.2 Nielsen and Ninomiya Theorem

Handling and removing fermion doublers is neither easy nor without cost. Indeed, according to the Nielsen and Ninomiya no-go theorem [410, 411], any *local, hermitian*, lattice fermion theory, with *translational invariance* and *chiral symmetry*, necessary displays fermion doublers. Hence, any attempt to remove doublers requires violating at least one of the four hypotheses of the theorem. Choosing which of these characteristics to compromise allows for different strategies, with the Wilson fermion [366] and staggered fermion [409] methods being some of the most well-known approaches. While the former one being already studied for quantum simulations [412–415], throughout the whole thesis, we will rely on the second solution, staggered fermions, introduced by Kogut and Susskind in [405], which has been widely exploited in numerical and quantum simulations both the electric and magnetic gauge contributions play a non-trivial role. [282, 397–402, 416].

11.1.3.3 Staggered Fermions

Staggered fermions do not completely resolve the Fermion doubling problem, yet they represent a compromise that reduces the number of doublers by distributing the components of the Dirac spinor across multiple lattice sites. This doubles the periodicity of the lattice, from a_0 to $2a_0$, and halves the Brillouin zone to $[-\frac{\pi}{2a_0}, \frac{\pi}{2a_0}]^D$. Such a solution looses locality, but maintains a modified translational invariance with a checkerboard pattern and a remnant of chiral symmetry, which is important for maintaining some of the physical characteristics of massless lattice fermions [415].

The practical realization of staggered fermions strongly depends on the spatial dimension of the lattice, as it affects the representation of spinors and gamma matrices [417]. In detail, we need to adjust the matrices γ^0 and $\gamma^0\gamma^k$ for $k = 1 \ldots D$ while maintaining all the Clifford's algebra properties.

(1+1)D Case —In one spatial dimension, $\mathbf{n} = n_1$, and the four-component spinor reduces to a two-spinor with the 0th (time) component and the 3rd (space) component: $\begin{pmatrix} \hat{\psi}_0 \\ \hat{\psi}_3 \end{pmatrix}$. Correspondingly, γ^0 and $\gamma^0\gamma^1$ can be any two Pauli matrices, for example σ^z and σ^x. We have then:

$$\hat{H}^{1\text{D}}_{\mathbf{n}} = \begin{pmatrix} \hat{\psi}^\dagger_{0,\mathbf{n}} & \hat{\psi}^\dagger_{3,\mathbf{n}} \end{pmatrix} \begin{pmatrix} m_0c^2 & ic\hbar\partial_1 \\ ic\hbar\partial_1 & -m_0c^2 \end{pmatrix} \begin{pmatrix} \hat{\psi}_{0,\mathbf{n}} \\ \hat{\psi}_{3,\mathbf{n}} \end{pmatrix} . \tag{11.22}$$

Then, in the staggered fermion solution, we decompose the spinor components by placing $\hat{\psi}_0$ on *even* sites (where $(-1)^{n_1} = +1$) and $\hat{\psi}_3$ on *odd* sites (where $(-1)^{n_1} = -1$). Then, for *even* sites, we have:

$$\hat{H}^{1\text{D}}_{\text{even}} = m_0c^2 \sum_{\mathbf{n}} \hat{\psi}^\dagger_{\mathbf{n}} \hat{\psi}_{\mathbf{n}} + \frac{ic\hbar}{2a_0} \sum_{\mathbf{n}} \hat{\psi}^\dagger_{\mathbf{n}} (\hat{\psi}_{\mathbf{n}+\mu_1} - \hat{\psi}_{\mathbf{n}-\mu_1}) , \tag{11.23}$$

whereas, for *odd* sites, we have

$$\hat{H}^{1\text{D}}_{\text{odd}} = +\frac{ic\hbar}{2a_0} \sum_{\mathbf{n}} \hat{\psi}^\dagger_{\mathbf{n}} (\hat{\psi}_{\mathbf{n}+\mu_1} - \hat{\psi}_{\mathbf{n}-\mu_1}) - m_0c^2 \sum_{\mathbf{n}} \hat{\psi}^\dagger_{\mathbf{n}} \hat{\psi}_{\mathbf{n}} . \tag{11.24}$$

Notice that each of these two contributions contains the hermitian conjugate of the hopping term of the other one. Therefore, combining even and odd sites, the (1+1)D Dirac Hamiltonian via staggered fermions reads:

$$\hat{H}^{1\text{D}}_{\text{Dirac}} = \sum_{\mathbf{n}} \left[\frac{ic\hbar}{2a_0} \hat{\psi}^\dagger_{\mathbf{n}} \hat{\psi}_{\mathbf{n}+\mu_1} + \text{h.c} \right] + m_0c^2 \sum_{\mathbf{n}} (-1)^{\mathbf{n}} \hat{\psi}^\dagger_{\mathbf{n}} \hat{\psi}_{\mathbf{n}} , \tag{11.25}$$

(2+1)D Case —The case of two-spatial dimensions is less trivial, as there are multiple ways of defining the matrices $(\gamma^0, \gamma^0\gamma^1, \gamma^0\gamma^2)$. For instance, using two-spinors, there are two solutions: $(\gamma^0, \gamma^0\gamma^1, \gamma^0\gamma^2) \to (\sigma^z, \sigma^x, \pm\sigma^y)$. Choosing the case with $+\sigma^y$, we obtain:

$$\hat{H}^{2\text{D}}_{\mathbf{n}} = \begin{pmatrix} \hat{\psi}^\dagger_{0,\mathbf{n}} & \hat{\psi}^\dagger_{3,\mathbf{n}} \end{pmatrix} \begin{pmatrix} m_0c^2 & ic\hbar(\partial_1 - i\partial_2) \\ ic\hbar(\partial_1 + i\partial_2) & -m_0c^2 \end{pmatrix} \begin{pmatrix} \hat{\psi}_{0,\mathbf{n}} \\ \hat{\psi}_{3,\mathbf{n}} \end{pmatrix} . \tag{11.26}$$

The decomposition of the spinor components along the lattice is similar to the (1+1)D case: we place $\hat{\psi}_0$ in *even* sites (where $(-1)^{n_1+n_2} = +1$) and $\hat{\psi}_3$ in *odd* ones (where $(-1)^{n_1+n_2} = -1$). Then, for *even* sites, we have:

$$\begin{aligned} \hat{H}^{2\text{D}}_{\text{even}} &= m_0c^2 \sum_{\mathbf{n}} \hat{\psi}^\dagger_{\mathbf{n}} \hat{\psi}_{\mathbf{n}} + \frac{ic\hbar}{2a_0} \sum_{\mathbf{n}} \hat{\psi}^\dagger_{\mathbf{n}} (\hat{\psi}_{\mathbf{n}+\mu_1} - \hat{\psi}_{\mathbf{n}-\mu_1}) \\ &+ \frac{c\hbar}{2a_0} \sum_{\mathbf{n}} \hat{\psi}^\dagger_{\mathbf{n}} (\hat{\psi}_{\mathbf{n}+\mu_2} - \hat{\psi}_{\mathbf{n}-\mu_2}) \end{aligned} \tag{11.27}$$

As for the *odd* sites, we have

$$\hat{H}^{2D}_{\text{odd}} = -m_0c^2 \sum_{\mathbf{n}} \hat{\psi}^\dagger_{\mathbf{n}} \hat{\psi}_{\mathbf{n}} + \frac{ic\hbar}{2a_0} \sum_{\mathbf{n}} \hat{\psi}^\dagger_{\mathbf{n}} (\hat{\psi}_{\mathbf{n}+\mu_1} - \hat{\psi}_{\mathbf{n}-\mu_1})$$
$$- \frac{c\hbar}{2a_0} \sum_{\mathbf{n}} \hat{\psi}^\dagger_{\mathbf{n}} (\hat{\psi}_{\mathbf{n}+\mu_2} - \hat{\psi}_{\mathbf{n}-\mu_2}) \tag{11.28}$$

Combining the two cases and noting the hermitian conjugate of the hopping terms, we find

$$\begin{aligned} \hat{H}^{2D} &= \frac{c\hbar}{2a_0} \sum_{\mathbf{n}} \left[\left[i\hat{\psi}^\dagger_{\mathbf{n}} \hat{\psi}_{\mathbf{n}+\mu_1} + (-1)^{n_1+n_2} \hat{\psi}^\dagger_{\mathbf{n}} \hat{\psi}_{\mathbf{n}+\mu_2} \right] + \text{h.c} \right] \\ &+ m_0c^2 \sum_{\mathbf{n}} (-1)^{\mathbf{n}} \hat{\psi}^\dagger_{\mathbf{n}} \hat{\psi}_{\mathbf{n}} \, . \end{aligned} \tag{11.29}$$

(3+1)D Case —In the three-dimensional case, we have to use the original 4-component spinor $\hat{\psi}_{\mathbf{n}} = (\hat{\psi}_{0,\mathbf{n}}, \hat{\psi}_{1,\mathbf{n}}, \hat{\psi}_{2,\mathbf{n}}, \hat{\psi}_{3,\mathbf{n}})$ and the original definition of the gammma matrices:

$$\gamma^0 = \begin{pmatrix} \mathbb{1}_2 & \\ & -\mathbb{1}_2 \end{pmatrix} \qquad \text{and} \qquad \gamma^k = \begin{pmatrix} & \sigma^k \\ \sigma^k & \end{pmatrix} . \tag{11.30}$$

Correspondingly, we have:

$$\hat{H}^{3D}_{\mathbf{n}} = \hat{\psi}^\dagger_{\mathbf{n}} \begin{pmatrix} m_0c^2 & & ic\hbar\partial_3 & ic\hbar(\partial_1 - i\partial_2) \\ & m_0c^2 & ic\hbar(\partial_1 + i\partial_2) & -ic\hbar\partial_3 \\ ic\hbar\partial_3 & ic\hbar(\partial_1 - i\partial_2) & -m_0c^2 & \\ ic\hbar(\partial_1 + i\partial_2) & -ic\hbar\partial_3 & & -m_0c^2 \end{pmatrix} \hat{\psi}_{\mathbf{n}}. \tag{11.31}$$

In this case, decomposing the spinor components along the lattice is less intuitive, but effordable. Namely, we place the component $\hat{\psi}_0$ at site $\mathbf{n} = (0, 0, 0)$: it couples with $\hat{\psi}_3$ at sites where $n_1 + n_2$ is odd and n_3 is even, with $\hat{\psi}_2$ at sites where $n_1 + n_2$ is even and n_3 is odd, with $\hat{\psi}_1$ at sites where $n_1 + n_2$ is odd and n_3 is odd, and finally with $\hat{\psi}_0$ at sites where $n_1 + n_2$ is even and n_3 is even. Correspondingly, $\hat{H}^{3D}$ decomposes into four separate sub-theories, with each theory touching a single fermion specie per site. In particular, $\hat{\psi}_0$ and $\hat{\psi}_1$ occupy even sites, while $\hat{\psi}_2$ and $\hat{\psi}_3$ occupy odd sites. As for the mass term, it acts positively on $\hat{\psi}_0$ and $\hat{\psi}_1$ (even sites), and negatively on $\hat{\psi}_2$ and $\hat{\psi}_3$ (odd sites). As for the x-derivative, it acts positively everywhere, while the y- and z-derivatives act positively on $\hat{\psi}_0$ and $\hat{\psi}_2$ (where $n_1 + n_2$ is even) and negatively on $\hat{\psi}_1$ and $\hat{\psi}_3$ (where $n_1 + n_2$ is odd). Summarizing, we obtain:

$$\hat{H}^{3D}_{\text{Dirac}}$$
$$= +\frac{c\hbar}{2a_0}\sum_{\mathbf{n}}\Big[\Big[i\hat{\psi}^{\dagger}_{\mathbf{n}}\hat{\psi}_{\mathbf{n}+\mu_1} + (-1)^{n_1+n_2}\hat{\psi}^{\dagger}_{\mathbf{n}}\hat{\psi}_{\mathbf{n}+\mu_2} + i(-1)^{n_1+n_2}\hat{\psi}^{\dagger}_{\mathbf{n}}\hat{\psi}_{\mathbf{n}+\mu_3}\Big] + \text{h.c}\Big]$$
$$+ m_0c^2\sum_{\mathbf{n}}(-1)^{\mathbf{n}}\hat{\psi}^{\dagger}_{\mathbf{n}}\hat{\psi}_{\mathbf{n}}, \tag{11.32}$$

which perfectly matches [409]. We stress that the staggered phases in the hopping and the mass term are fundamental for recovering correct resultss.

11.1.4 Discretization of Gauge Fields

In QFT, the principle of gauge invariance underlies the interactions between matter fields (such as fermions) and gauge fields (such as bosons, e.g. photons or gluons). Such an interaction is ruled by local (gauge) transformations from a corresponding gauge group.

11.1.4.1 Basics of Group Theory

Let us consider a generic compact Lie group $\mathcal{G}$ (e.g., SU(N) or U(1)), whose algebra has generators T^a, where $a \in \{1 \dots \dim \mathcal{G}\}$. In the fundamental representation of non-Abelian groups $\mathcal{G}$ like SU(N), these generators could be the Gell-Mann matrices for SU(3), or Pauli matrices for SU(2). In general, the generators T^a satisfy the following commutations rules

$$[T^a, T^b] = if^{abc}T^c\,, \tag{11.33}$$

where f^{abc} are the *structure constants* of the gauge group $\mathcal{G}$. For a non-Abelian group, these constants are fully antisymmetric in the indices a, b, c, encoding the non-commuting relation between the generators (and correspondingly between the resulting transformations). For instance, in the SU(2) case, f^{abc} is the Levi-Civita symbol ϵ^{abc}. Conversely, in the Abelian scenario, like U(1), the generator is unique and scalar $T = 1$ and $f^{abc} = 0$.

Then, any group element $\Omega(r)$ can vary from point to point in spacetime and is defined as

$$\Omega(r) = \exp(i\theta^a(r)T^a) \quad \text{with} \quad \Omega(r)\Omega^{\dagger}(r) = \Omega^{\dagger}(r)\Omega(r) = \mathbb{1}\,, \tag{11.34}$$

where $\theta^a(r)$ are the local parameters of the gauge transformation. For example, in the case of SU(2), $\Omega(r)$ could be a 2×2 unitary matrix with determinant 1. For U(1) (electromagnetism), $\Omega(r)$ would simply be a phase factor $\exp(i\theta(r))$.

11.1.4.2 Matter Fields

Assuming a single matter flavor in the fundamental representation of the gauge group $\mathcal{G}$, gauge transformations of spinors are generated by the charge operator $\hat{Q}^a(r)$, defined as:

$$\hat{Q}^a(r) = \sum_{\alpha,\beta} \hat{\psi}^\dagger_\alpha(r) T^a_{\alpha\beta} \hat{\psi}_\beta(r). \tag{11.35}$$

This operator represents how the matter fields interact with the gauge fields and underlies the dynamics of the gauge theory when multiple flavors are present. Correspondingly, under a local gauge transformation $\Omega(r)$, matter fields transform as follows:

$$\hat{\psi}(r) \to \hat{\psi}'(r) = \Omega(r)\hat{\psi}(r). \tag{11.36}$$

11.1.4.3 The Covariant Derivative

The original Dirac equation of Eq. (11.7) is clearly not invariant under local gauge transformations such as the ones in Eq. (11.36); indeed when transforming the derivative, we find:

$$\partial_\mu \hat{\psi}(r) \to \partial_\mu \hat{\psi}'(r) = \partial_\mu \left(\Omega(r)\hat{\psi}(r) \right) = (\partial_\mu \Omega(r))\hat{\psi}(r) + \Omega(r)\partial_\mu \hat{\psi}(r), \tag{11.37}$$

so that the transformed Dirac equation reads

$$\begin{aligned} 0 &= i\hbar\gamma^\mu \Big[(\partial_\mu \Omega(r))\hat{\psi}(r) + \Omega(r)\partial_\mu \hat{\psi}(r) \Big] - m_0 c \Omega(r)\hat{\psi}(r) \\ &= \Omega(r)\Big(i\hbar\gamma^\mu \partial_\mu \hat{\psi}(r) - m_0 c \hat{\psi}(r) \Big) + i\hbar\gamma^\mu (\partial_\mu \Omega(r))\hat{\psi}(r). \end{aligned} \tag{11.38}$$

To impose invariance under this transformation, the extra term $i\hbar\gamma^\mu(\partial_\mu\Omega(r))\psi(r)$ must vanish. This would spontaneously happen just in case $\Omega(r)$ is constant (which corresponds to a *global*, rather than local, gauge transformation). In general, to achieve gauge invariance, the ordinary derivative ∂_μ must be replaced by the covariant derivative $\hat{D}_\mu$, which transforms covariantly under gauge transformations thereby canceling the extra terms introduced by the local gauge transformation. Namely, we have:

$$\hat{D}_\mu \hat{\psi}(r) = \left(\partial_\mu - i\frac{q}{\hbar}\hat{A}_\mu(r) \right)\hat{\psi}(r) = \left(\partial_\mu - i g_0 \hat{A}_\mu(r) \right)\hat{\psi}(r), \tag{11.39}$$

where q is the charge, $g_0 = q/\hbar$ is the gauge coupling, while $\hat{A}_\mu(r) = \hat{A}^a_\mu(r)T^a$ is the full gauge field operator obtained by contracting the gauge field components $\hat{A}^a_\mu(r)$ to the generators T^a via the gauge group index a. The gauge field operator transforms as follows:

$$\hat{A}_\mu(r) \to \hat{A}'_\mu(r) = \Omega(r)\hat{A}_\mu(r)\Omega^\dagger(r) + \frac{i}{g_0}\Omega(r)\partial_\mu \Omega^\dagger(r). \tag{11.40}$$

Correspondingly, the continuum Dirac Hamiltonian in Eq. (11.11) is made covariant as follows:

$$\hat{H}_{\text{Dirac}} = \int d\mathbf{r}\Big[\hat{\psi}^{\dagger}(\mathbf{r})(ic\hbar\gamma^{0}\gamma^{k}(\partial_k - ig_0\hat{A}_k) + m_0c^2\gamma^0)\hat{\psi}(\mathbf{r})\Big]. \tag{11.41}$$

11.1.4.4 The Pure Gauge Hamiltonian

To make the Dirac equation gauge invariant, we needed to introduce an extra d.o.f., a gauge field, whose dynamics needs to be properly accounted and added to the Dirac Hamiltonian. In the continuum, the dynamics of gauge fields is captured by the field strength tensor $\hat{F}_{\mu\nu}(r)$, which for a generic non-Abelian gauge field $\hat{A}_\mu(r)$ is given by:

$$\begin{aligned}\hat{F}_{\mu\nu}(r) &= \partial_\mu\hat{A}_\nu(r) - \partial_\nu\hat{A}_\mu(r) - ig_0\Big[\hat{A}_\mu(r), \hat{A}_\nu(r)\Big]\\ \hat{F}^a_{\mu\nu}(r)T^a &= \Big[\partial_\mu\hat{A}^a_\nu(r) - \partial_\nu\hat{A}^a_\mu(r) + g_0 f^{abc}\hat{A}^b_\mu(r)\hat{A}^c_\nu(r)\Big]T^a.\end{aligned} \tag{11.42}$$

To move towards the Hamiltonian formalism, we separate its electric and magnetic components:

$$\hat{E}^a_k(r) = \hat{F}^a_{0k}(r) = \frac{1}{c}\partial_t\hat{A}^a_k(r) - \partial_k\hat{A}^a_0(r) - g_0 f^{abc}\hat{A}^b_0(r)\hat{A}^c_k(r) \tag{11.43a}$$

$$\hat{B}^a_i(r) = -\frac{1}{2}\epsilon_{ijk}\hat{F}^a_{jk}(r) = \epsilon_{ijk}\Big(\partial_j\hat{A}^a_k(r) - \frac{g_0}{2}f^{abc}\hat{A}^b_j(r)\hat{A}^c_k(r)\Big), \tag{11.43b}$$

where i, j, k are spatial indices. In the Coulomb gauge, where $\hat{A}^0(r) = 0$, the electric field reduces to $\hat{E}^a_k(r) = \hat{F}^a_{0k}(r) = \partial_t\hat{A}^a_k(r)/c$. This choice does not change the physics, as it simply exploits an extra freedom in defining the gauge fields.

The corresponding Hamiltonian of the pure gauge fields is

$$\hat{H}_{\text{gauge}} = \int d\mathbf{r}\Big[\frac{\epsilon_0}{2}\hat{E}^a_k(r)\hat{E}^a_k(r) + \frac{1}{2\mu_0}\hat{B}^a_k(r)\hat{B}^a_k(r)\Big]. \tag{11.44}$$

where ϵ_0 and $\mu_0 = (c^2\epsilon_0)^{-1}$ correspond to the vacuum permittivity and permeability respectively. In dimensioned units (such as SI), the physical dimensions of the constants read

$$\begin{aligned}[\epsilon_0] &= (\text{charge})^2(\text{length})^{2-D}(\text{energy})^{-1}\\ [\mu_0] &= (\text{charge})^{-2}(\text{length})^{D-2}(\text{energy})\cdot(\text{length})^{-2}(\text{time})^2\\ &= (\text{charge})^{-2}(\text{length})^{D-4}(\text{energy})(\text{time})^2.\end{aligned} \tag{11.45}$$

Correspondingly, the electric and magnetic fields have the following dimensions:

$$\begin{aligned}\Big[\hat{E}\Big] &= (\text{charge})^{-1}(\text{length})^{-1}(\text{energy})\\ \Big[\hat{B}\Big] &= (\text{charge})^{-1}(\text{length})^{-2}(\text{energy})(\text{time}).\end{aligned} \tag{11.46}$$

As for the coupling constant g_0, its dimensional analysis reads

$$[g_0] = (\text{charge})(\text{energy})^{-1}(\text{time})^{-1}\,, \tag{11.47}$$

while for the vector potential we have:

$$\left[\hat{A}\right] = (\text{charge})^{-1}(\text{length})^{-1}(\text{energy})(\text{time})\,. \tag{11.48}$$

11.1.4.5 Discretized Gauge Fields

When moving from the continuum to the lattice, each spatial component k of the covariant derivative in Eq. (11.39) is replaced by the *parallel transport*[1] $\hat{U}_{\mathbf{n},\mu_k}(t)$, a dimensionless link variable reflecting the gauge field's influence along the link connecting neighboring lattice sites $\mathbf{n}$ and $\mathbf{n} + \boldsymbol{\mu}_k$. Omitting again the time dependence, we define the parallel transport as:

$$\hat{D}_k = (\partial_k - i g_0 \hat{A}^k) \quad \longrightarrow \quad \hat{U}_{\mathbf{n},\mu_k} = \exp(i g_0 \int_{\mathbf{n}}^{\mathbf{n}+\mu_k} \hat{A}^k(\mathbf{r}) d\ell) \sim \exp(i g_0 a_0 \hat{A}^k)\,, \tag{11.49}$$

where in the second step we assumed $\hat{A}^k(r) \sim \hat{A}^k_{\mathbf{n}}$ to be almost constant along the lattice link $(\mathbf{n}, \boldsymbol{\mu}_k)$ of length a_0. Assuming a_0 to be sufficiently small, we can further expand the parallel transporter up to the first order and notice that:

$$\begin{aligned}
\left[\frac{i\hat{\psi}^\dagger_{\mathbf{n}}\hat{U}_{\mathbf{n},\mu_k}\hat{\psi}_{\mathbf{n}+\mu_k}}{2a_0} + \text{h.c}\right] &= i\left[\frac{\hat{\psi}^\dagger_{\mathbf{n}}\hat{U}_{\mathbf{n},\mu_k}\hat{\psi}_{\mathbf{n}+\mu_k} - \hat{\psi}^\dagger_{\mathbf{n}+\mu_k}\hat{U}^\dagger_{\mathbf{n},\mu_k}\hat{\psi}_{\mathbf{n}}}{2a_0}\right] \\
&= i\left[\frac{\hat{\psi}^\dagger_{\mathbf{n}}\hat{U}_{\mathbf{n},\mu_k}\hat{\psi}_{\mathbf{n}+\mu_k} - \hat{\psi}^\dagger_{\mathbf{n}}\hat{U}^\dagger_{\mathbf{n}-\mu_k,\mu_k}\hat{\psi}_{\mathbf{n},-\mu_k}}{2a_0}\right] \\
&= i\hat{\psi}^\dagger_{\mathbf{n}}\frac{\left[\hat{U}_{\mathbf{n},\mu_k}\hat{\psi}_{\mathbf{n}+\mu_k} - \hat{U}^\dagger_{\mathbf{n}-\mu_k,\mu_k}\hat{\psi}_{\mathbf{n},-\mu_k}\right]}{2a_0} \\
&\sim \frac{i\hat{\psi}^\dagger_{\mathbf{n}}}{2a_0}\left[(1 - i g_0 \hat{A}^k_{\mathbf{n}} a_0)\hat{\psi}_{\mathbf{n}+\mu_k} - (1 + i g_0 \hat{A}^k_{\mathbf{n}-\mu_k} a_0)\hat{\psi}_{\mathbf{n},-\mu_k}\right] \\
&= i\hat{\psi}^\dagger_{\mathbf{n}}\left[\frac{\hat{\psi}_{\mathbf{n}+\mu_k} - \hat{\psi}_{\mathbf{n}-\mu_k}}{2a_0} - i g_0 \hat{A}^k_{\mathbf{n}}\hat{\psi}^\dagger_{\mathbf{n}+\mu_k}\right] \\
&\xrightarrow{a_0\to 0} i\hat{\psi}^\dagger(\mathbf{r})(\partial_k - i g_0 \hat{A}^k(\mathbf{r}))\hat{\psi}(\mathbf{r}).
\end{aligned} \tag{11.50}$$

[1] Parallel transports have several names in literature. Sometimes they are also called *connections* referring to a differential geometry framework. Sometimes, they are called *comparators*.

Then, the lattice version of the covariant Dirac Hamiltonian with staggered fermions is

$$\hat{H}_{\text{Dirac}} = \frac{c\hbar}{2a_0} \sum_{\mathbf{n},\mu} \left[s_{\mathbf{n},\mu} \hat{\psi}_{\mathbf{n}} \hat{U}_{\mathbf{n},\mu} \hat{\psi}_{\mathbf{n}+\mu} + \text{h.c} \right] + m_0 c^2 \sum_{\mathbf{n}} s_{\mathbf{n}} \hat{\psi}^{\dagger}_{\mathbf{n}} \hat{\psi}_{\mathbf{n},\mu}, \tag{11.51}$$

where $s_{\mathbf{n},\mu}$ and $s_{\mathbf{n}}$ are phases that arise when using staggered fermions.

Similarly to the vector potential, gauge transformations of the parallel transporter read:

$$\hat{U}_{\mathbf{n},\mu_k} \to \Omega_{\mathbf{n}} \hat{U}_{\mathbf{n},\mu_k} \Omega^{\dagger}_{\mathbf{n}+\mu_k}, \tag{11.52}$$

where $\Omega_{\mathbf{n}}$ is the gauge transformation matrix at site $\mathbf{n}$. In general, due to the non-commuting algebra in Eq. (11.136), the left $\Omega_{\mathbf{n}}$ and the right $\Omega^{\dagger}_{\mathbf{n}+\mu_k}$ gauge transformations are different and generated by the dimensionless conjugate fields $\hat{L}^a_{\mathbf{n},\mu}$ and $\hat{R}^a_{\mathbf{n},\mu}$ respectively, which satisfy the following relations:

$$`\hat{L}^a_{\mathbf{n},\mu} \hat{U}^{\alpha\beta}_{\mathbf{n}',\mu'} = -\delta_{\mathbf{n}\mathbf{n}'} \delta_{\mu\mu'} \sum_{\gamma} T^a_{\alpha\gamma} \hat{U}^{\gamma\beta}_{\mathbf{X},\mu}, \tag{11.53a}$$

$$[\hat{R}^a_{\mathbf{n},\mu}, \hat{U}^{\alpha\beta}_{\mathbf{n}',\mu'}] = +\delta_{\mathbf{n}\mathbf{n}'} \delta_{\mu\mu'} \sum_{\gamma} \hat{U}^{\alpha\gamma}_{\mathbf{X},\mu} T^a_{\gamma\beta}, \tag{11.53b}$$

$$\hat{R}^a_{\mathbf{n},\mu} = \hat{U}^{\dagger}_{\mathbf{n},\mu} \hat{L}^a_{\mathbf{n},\mu} \hat{U}_{\mathbf{n},\mu}, \tag{11.53c}$$

where $\alpha, \beta, \gamma, \delta$ span the matrix indices of the group generators T^a. In the Abelian U(1) case, where there is only one generator, there is only one electric field and the parallel transporter acts as a raising operator. Namely, we have:

$$\hat{E}_{\mathbf{n},\mu} = \hat{L}_{\mathbf{n},\mu} = \hat{R}_{\mathbf{n},\mu} \qquad [\hat{E}_{\mathbf{n},\mu}, \hat{U}_{\mathbf{n},\mu}] = \hat{U}_{\mathbf{n},\mu}. \tag{11.54}$$

11.1.4.6 The Pure Gauge Lattice Hamiltonian

Once we have discretized all the gauge fields, we can build the corresponding lattice version of the pure gauge Hamiltonian in Eq. (11.44). The electric energy density is given by the Casimir operator:

$$\hat{E}^2_{\mathbf{n},\mu} = \hat{L}^a_{\mathbf{n},\mu} \hat{L}^a_{\mathbf{n},\mu} = \hat{R}^a_{\mathbf{n},\mu} \hat{R}^a_{\mathbf{n},\mu}. \tag{11.55}$$

To make the Casimir operator adimensional, the electric Hamiltonian is redefined as follows:

$$\begin{aligned} \hat{H}_{\text{elec}} = \frac{\epsilon_0}{2} \int d^D \mathbf{r} \hat{E}_k(r) \hat{E}_k(r) \quad \to \quad \hat{H}^{\text{latt}}_{\text{elec}} &= a_0^D \frac{\epsilon_0}{2} \sum_{\mathbf{n},k} \frac{q^2 a_0^{2-2D}}{\epsilon_0^2} \hat{E}^2_{\mathbf{n},\mu_k} \\ &= \sum_{\mathbf{n},k} \frac{q^2 a_0^{2-D}}{2\epsilon_0} \hat{E}^2_{\mathbf{n},\mu_k}. \end{aligned} \tag{11.56}$$

We can recast the latter in dimensionless units, by redefining the gauge coupling constant as

$$g = g_0 \sqrt{\frac{\hbar}{c\epsilon_0}} a_0^{\frac{3-D}{2}} = q \frac{a_0^{\frac{3-D}{2}}}{\sqrt{c\hbar\epsilon_0}} = q a_0^{\frac{3-D}{2}} \sqrt{\frac{c\mu_0}{\hbar}} . \tag{11.57}$$

By doing so, we obtain

$$\hat{H}_{\text{elec}}^{\text{latt}} = \sum_{\mathbf{n},k} \frac{g^2 c\hbar}{2a_0} \hat{E}^2_{\mathbf{n},\mu_k} . \tag{11.58}$$

Notice that the new gauge coupling constant g is dimensionless yet depends on the lattice spacing. As discussed for the matter fields, such an hidden relation becomes important when discussing the continuum limit.

Correspondingly, the simplest way to describe the magnetic energy density on the lattice is by using Wilson loops [366, 405], i.e. gauge-invariant loops made out of parallel transporters $\hat{U}$. On a hypercubic lattice, the smallest non-trivial loop is represented by a plaquette operator

$$\hat{U}_{\square} = \sum_{\alpha,\beta,\gamma,\delta} \hat{U}^{\alpha\beta}_{\mathbf{n},\mu} \hat{U}^{\beta\gamma}_{\mathbf{n}+\mu,\mu'} \hat{U}^{\gamma\delta\dagger}_{\mathbf{n}+\mu',\mu} \hat{U}^{\delta\gamma\dagger}_{\mathbf{n},\mu'} , \tag{11.59}$$

where $\boldsymbol{\mu}$ and $\boldsymbol{\mu}'$ span the plaquette's plane. Since the parallel transporters in Eq. (11.49) commutes when acting on different links (and so the vector potentials do), one can notice that:

$$\begin{aligned} \hat{U}_{\square} =& \exp\left(i g_0 \oint_{\partial\square} \hat{A}_k d\ell_k \right) \underset{*}{=} \exp\left(i g_0 \oint_{\square} \nabla \times \hat{\mathbf{A}} \cdot d\mathbf{s} \right) \\ =& \exp\left(i g_0 \oint_{\square} \hat{\mathbf{B}} \cdot d\mathbf{s} \right) \sim \exp\left(i g_0 a_0^2 \hat{B} \right) . \end{aligned} \tag{11.60}$$

where in the $*$ we used the Stokes theorem, while in the last step we assumed the magnetic field to be almost constant in the plaquette of area a_0^2. Given these assumptions, we have:

$$\text{Tr}\left(\hat{U}_{\square} + \hat{U}^{\dagger}_{\square} \right) = 2\cos\left(g_0 a_0^2 \hat{B} \right) \sim 2\left[1 - \frac{1}{2} g_0^2 a_0^4 \hat{B}^2 \right], \tag{11.61}$$

from which we obtain that

$$\hat{B}^2 \sim -\frac{1}{g_0^2 a_0^4} \text{Tr}\left(\hat{U}_{\square} + \hat{U}^{\dagger}_{\square} \right) . \tag{11.62}$$

Then, the magnetic Hamiltonian can be discretized as follows:

$$\begin{aligned}\hat{H}_{\text{magn}} &= \frac{1}{2\mu_0}\int d^D\mathbf{r}B_k(r)B_k(r) \rightarrow \hat{H}^{\text{latt}}_{\text{magn}} \\ &= -\frac{1}{2\mu_0 g_0^2 a_0^{4-D}}\sum_{\square}\text{Tr}\left(\hat{U}_{\square}+\hat{U}^{\dagger}_{\square}\right).\end{aligned} \tag{11.63}$$

Finally, by using the dimensionless gauge coupling in Eq. (11.57), we obtain:

$$\hat{H}^{\text{latt}}_{\text{magn}} = -\frac{c\hbar}{2g^2a_0}\sum_{\square}\text{Tr}\left(\hat{U}_{\square}+\hat{U}^{\dagger}_{\square}\right). \tag{11.64}$$

Notice that plaquette terms only exist in $D > 1$, contributing to the increased complexity of quantum and TN simulations of LGTs in higher dimensions [418].

11.1.4.7 The Lattice Gauge Hamiltonian with Staggered Fermions

Summarizing and combining all the ingredients (matter and gauge fields), the general *Kogut-Susskind Hamiltonian* via staggered fermions [354] reads:

$$\begin{aligned}H_{\text{LGT}} =& \frac{c\hbar}{2a_0}\sum_{\mathbf{n},\mu}\sum_{\alpha,\beta}\left[s_{\mathbf{n},\mu}\hat{\psi}_{\mathbf{n},\alpha}\hat{U}^{\alpha\beta}_{\mathbf{n},\mu}\hat{\psi}_{\mathbf{n}+\mu,\beta}+\text{h.c}\right]+m_0c^2\sum_{\mathbf{n},\alpha}s_{\mathbf{n}}\hat{\psi}^{\dagger}_{\mathbf{n},\alpha}\hat{\psi}_{\mathbf{n},\mu} \\ &+\frac{c\hbar g^2}{2a_0}\sum_{\mathbf{n},\mu}\hat{E}^2_{\mathbf{n},\mu}-\frac{c\hbar}{2g^2a_0}\sum_{\square}\text{Tr}\left(\hat{U}_{\square}+\hat{U}^{\dagger}_{\square}\right),\end{aligned} \tag{11.65}$$

where we expressed the dependence on the matrix indices α, β of the gauge algebra generators.

11.2 Dressed-Site Model for Hamiltonian Lattice Gauge Theories

We have seen all the ingredients for discretizing an LGT with dynamical matter. However, there remain fundamental challenges to be faced for attacking Hamiltonians like Eq. (11.65) with numerical methods.

First of all, we need to achieve a finite-dimensional encoding of the continuous gauge fields such as U(1) or SU(N), especially beyond one spatial dimension, where decoupling the gauge field's longitudinal component is required [419]. Among known truncation recipes are Quantum Link Models (QLM) [420–424], finite subgroups [73, 425, 426], digitization of gauge fields [427], and fusion-algebra deformation [428]. Whatever the adopted solution, further effort is required to satisfy gauge symmetry at each local site, which involves the concurrent evaluation of all the gauge links and the attached matter site. In practice, an extensive number of local constraints (energy penalties), which corresponds to QMB operators concurrently acting on 2D+1 sites (2D gauge links and 1 matter site), where D is the number of spatial dimensions. Moreover, since the Gauss law is the highest energy scale of the model, any (even small) violation of these constraints could lead to a substantial deviation from the expected physical behavior.

Another important issue to be faced in simulating LGTs is to account for the Fermi statistics of matter fields. In several TN methods as well on well-established conventional digital quantum simulation platforms (e.g. superconducting qubits, trapped-ions, Rydberg arrays, quantum dots) [429–433], fermionic algebra (mutually-anticommuting operations) must be encoded into a genuinely local algebra (mutually-commuting operations) of qubits. Standard fermion-to-qubit encodings, such as the Jordan-Wigner (JW) transformation [434], and other modern approaches [435–440] make any fermionic Hamiltonian interaction inherently long-range and expensive from a computational perspective.

In numerical methods such as TNs, as well as in quantum computations, all these requirements significantly impact the computational resources needed for the simulations, as well as their efficiency. In this section, we present a theoretical scheme which is able to order achieve a finite and controllable gauge field truncation, avoid the Fermi statistics of matter fields and directly access the gauge invariant subspace, revealing suitable for TN methods as well as for quantum simulations of LGTs. Building on the work of [441, 442], we *dress* every physical matter site with the information related to its adjacent gauge links. A pictorial scheme of this approach is outlined in Fig. 11.1[Right]: (a) starting from the original description matter fields on sites and gauge fields on links in Eq. (11.65), we truncate the gauge group imposing an energy cut-off on the Casimir operator. Then, (b)

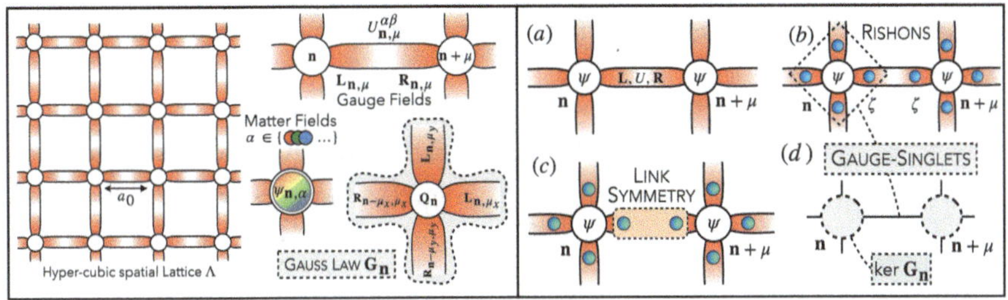

Fig. 11.1 [Left] Graphical representation of the degrees of freedom of a 2D LGT: fermionic matter fields, defined on lattice sites, and gauge fields (the parallel transporter, and the chromo-electric fields, $\hat{L}_{\mathbf{n},\mu}$ and $\hat{R}_{\mathbf{n},\mu}$), living on lattice links. A local gauge transformation at n acts on a matter site and all its attached links. [Right] Pictorial representation of the dressed site formalism adopted for TN simulations of LGTs: (**a**) we truncate the gauge link fields with an energy cutoff in the irreducible representation basis; (**b**) the truncated gauge link is split into two representations, one per half-link; each is equipped with a proper fermionic rishon mode $\hat{\zeta}$. (**c**) All the half-links are absorbed into the attached matter site, forming a gauge-invariant dressed site (**d**) whose Hilbert space spans all the possible gauge singlets. (Figure taken from [396])

we express each truncated gauge link as a pair of fermionic rishon modes $\hat{\zeta}$ and (c) constrain their link dynamics according to the original gauge group algebra. Ultimately, (d) we merge each of these modes to its adjacent matter site, ending up in a compact *dressed-site* formalism [441–443]. The resulting effective Hamiltonian is made out of only bosonic operators and directly acts on the gauge invariant Hilbert sub-space. Correspondingly, as done in Loop String Hadrons methods [444], the original (Abelian/non-Abelian) gauge invariance is exactly rewritten into an Abelian, nearest-neighbor, diagonal selection rule, and the explicit dependence on the fermionic matter is eliminated [282, 443, 445]. Such an approach is especially suitable when dealing with LGTs in high-dimensional lattices.

We will practically apply the dressed-site formalism on two paradigmatic examples that have been exploited for numerical simulations [396, 399, 401]: U(1) in Sect. 11.3 and SU(2) in Sect. 11.4.

11.2.1 Gauge Field Truncation

The link Hilbert space is the space of square-integrable functions on the gauge group $\mathcal{G}$, $L^2(\mathcal{G})$, which is infinite-dimensional for continuous groups such as U(1) or SU(N) [446].

In one space dimension, gauge degrees of freedom are unphysical (absence of transverse polarizations) and can thus be integrated out, albeit at the price of introducing non-local interactions [447]. Beyond one dimension, the removal is much more delicate, because it requires first decoupling the gauge field's longitudinal component [419]. When impossible or inconvenient to remove, gauge degrees of freedom might have to be truncated to perform TN or quantum simulation. Among known truncation recipes are Quantum Link Models (QLM) [420–424], an approach already considered for practical quantum simulation of LGTs [418, 448–457], finite subgroups [73, 425, 426], digitization of gauge fields [427], and fusion-algebra deformation [428].

Another approach is to truncate the spectrum of the electric energy density operator $\left\| \hat{E}^2 \right\| \leq \Theta$ on each link [399, 400]. The cutoff is conveniently imposed in the irreducible representation (irrep) basis $\{|jm_{\mathrm{L}}m_{\mathrm{R}}\rangle\}$ [446] of $L^2(\mathcal{G})$, where $\hat{E}^2$ is diagonal:

$$\hat{E}^2 \, |jm_{\mathrm{L}}m_{\mathrm{R}}\rangle = C_2(j) \, |jm_{\mathrm{L}}m_{\mathrm{R}}\rangle \ . \tag{11.66}$$

Here, m_{L} and m_{R} are indices in the j-irrep of $\mathcal{G}$ and $C_2(j)$ is the quadratic Casimir of j [446]. In the strong coupling limit, where the electric energy term dominates H_{LGT}, this truncation is equivalent to an energy cutoff [400].

11.2.2 Gauss Law and the Dressed Site

The most distinctive feature of gauge theories is arguably the presence of local constraints, analogous to the Gauss law of classical electrodynamics, relating the configuration of the gauge field to the spatial distribution of charges [458]. At the quantum level, Gauss law is the statement that only gauge invariant states are physical, namely, $\hat{G}^a_{\mathbf{n}} |\Psi_{\text{phys}}\rangle = 0 \; \forall \mathbf{n}, a$, where $\hat{G}^a_{\mathbf{n}}$ are the generators of local gauge transformations at $\mathbf{n}$:

$$\hat{G}^a_{\mathbf{n}} = \hat{Q}^a_{\mathbf{n}} + q^a_{\mathbf{n}} + \sum_{\mu} [\hat{L}^a_{\mathbf{n},\mu} + \hat{R}^a_{\mathbf{n}-\mu,\mu}] , \tag{11.67}$$

with $q^a_{\mathbf{n}}$ representing eventual static background charges (typically vanishing). Lattice Gauss law provides a set of *vertex constraints*, each involving a lattice site and its 2D neighboring links.

Due to Gauss law, the physical Hilbert space of LGTs is much smaller than the tensor product of all local sites and link Hilbert spaces. Properly exploiting gauge symmetries can thus significantly speed up numerical simulations [442]. Strategies that solve Gauss law by eliminating (partially or entirely) either the gauge fields or the matter fields have been developed. Nonetheless, such approaches come with specific limitations: the range of interaction has to be extended, moreover, integrating-out gauge fields become problematic in $D > 1$ [419], while the recipe for removing matter is a model (matter content) dependent [445].

Another possibility is to enforce Gauss law using a dressed site construction [441–443, 445], sketched in Fig. 11.1 and outlined below. Dressed sites have local dimensions typically larger than those resulting from the aforementioned approaches, but they are obtained from a model-independent prescription which has the advantage of preserving the locality of the interactions [400, 442].

As a first step, we factorize each gauge link in a pair of modes, living at its edges. Namely, every link is decomposed into a pair of left (L) and right (R) *rishon* d.o.f., each associated with a Hilbert space spanned by the basis states $|jm\rangle$, identifying $|jm_{\text{L}}m_{\text{R}}\rangle \hookrightarrow |jm_{\text{L}}\rangle \otimes |jm_{\text{R}}\rangle$, and writing parallel transporters as rishon bilinears [399]:

$$\hat{U}^{\alpha\beta}_{\mathbf{n},\mu} \to \sum_i \hat{\zeta}^{L(i)\alpha}_{\mathbf{n},+\mu} \hat{\zeta}^{R(i)\beta\,\dagger}_{\mathbf{n}+\mu,-\mu} . \tag{11.68}$$

Physical gauge link configurations, i.e. those with the left and right rishons in the same irrep, are selected introducing a local link symmetry at the TN simulation level [400, 442, 459]. Notice that such a constraint is always Abelian, regardless of the Abelian or non-Abelian nature of the gauge group.

11.2.3 Gauge-Invariant Dressed Site Operators

Crucially, the gauge generators $\hat{G}^a_{\mathbf{n}}$ now involve only the matter site at **n** and its $2D$ neighboring rishons. Fusing these degrees of freedom in a composite site, Gauss law becomes an internal constraint that singles out the dressed site Hilbert space as its gauge invariant subspace:

$$\mathcal{H}_{\text{dress}} = \ker G \subset \mathcal{H}_{\text{matt}} \otimes (\mathcal{H}_{\text{rish}})^{\otimes 2D} \,. \tag{11.69}$$

Therefore, the effective operators of the resulting dressed-site Hamiltonian should be obtained by projecting the obtained ones on the subspace generated by a gauge-invariant basis M. Namely, for any dressed-site operator O among the previously defined, the corresponding effective operator O^{eff} acting on gauge-invariant states reads:

$$O^{\text{eff}} = M^{\dagger} \cdot O \cdot M \,, \quad \text{where} \quad M^{\dagger} \cdot M = \mathbb{1} \quad \& \quad \left(M \cdot M^{\dagger}\right)^2 = \left(M \cdot M^{\dagger}\right) . \tag{11.70}$$

The practical computation of the gauge-invariant basis M can be determined as the kernel of the Gauss Law operator of the corresponding LGT (see Sect. 11.4 for SU(2) and Sect. 11.3 for U(1)). The expansion of the gauge singlet basis states of $\mathcal{H}_{\text{dress}}$ in terms of the matter and rishon bases is computed via Clebsch-Gordan decomposition [400].

The resulting operators O^{eff}, for any spatial dimension D and any value of the gauge truncation j_{max}, are available for U(1) and SU(2) in the GitHub repository ed-lgt [459], which also allows for simulations of these LGTs via Exact Diagonalization.

11.2.4 Defermionization

As aforementioned in the introduction, an important challenge to be faced in simulating LGTs is to account for the Fermi statistics of matter fields. In several TN methods and conventional digital quantum simulation platforms [429–433], fermionic (anticommuting) algebra must be encoded into a genuinely local (commuting) algebra of qub(d)its. Standard fermion-to-qubit encodings, such as the Jordan-Wigner (JW) transformation [434], and other modern approaches [435–437] make any fermionic Hamiltonian interaction inherently long-range and expensive from a computational perspective [460], especially for higher-dimensional LGTs.

Within the dressed-site formalism, it is additionally possible resolve this problem and effectively eliminate the Fermi statistics of matter fields. This is possible for any gauge theory where the gauge field has a well-defined parity. Specifically, a local parity operator $\hat{P}_{\mathbf{n},\mu} = \hat{P}^{\dagger}_{\mathbf{n},\mu}$ such that $\hat{P}^2_{\mathbf{n},\mu} = 1$ must satisfy $\{\hat{U}_{\mathbf{n},\mu}, \hat{P}_{\mathbf{n},\mu}\} = 0$, as it happens for $\mathbb{Z}_{2N}$, U(N), and SU(2N) [399, 443, 445]. In these cases, it is possible to consider rishons with a Fermi statistics, and, as a result, all physical (gauge invariant) dressed site operators are genuinely local, i.e. they mutually commute at a nonzero distance (as spins or bosons) [403].

As the resulting Hamiltonian in the dressed site formalims is made out of different fermions (matter fields and gauge rishons), it is convenient to rule the tensor product of general fermion operators. For a fermionic QMB system with particles arbitrarily sorted along a certain path, any tensor product of fermionic operators should take into consideration the proper anti-commutation rules. Namely, a generic fermionic operator $\hat{F}_{\mathbf{n}}$ acting on the $\mathbf{n}^{th}$ position along the path reads:

$$\hat{F}_{\mathbf{n}} = \begin{pmatrix} \hat{f}_{11} & \dots & \hat{f}_{1N} \\ \vdots & & \vdots \\ \hat{f}_{N1} & \dots & \hat{f}_{NN} \end{pmatrix}_F = \dots \otimes \hat{P}_{\mathbf{n}-\mu} \otimes \hat{F}_{\mathbf{n}} \otimes \mathbb{1}_{\mathbf{n}+\mu} \otimes \dots , \tag{11.71}$$

where $\hat{P}_{\mathbf{n}} = \hat{P}_{\mathbf{n}}^{\dagger} = \hat{P}_{\mathbf{n}}^{-1}$ is a fermion parity operator that gets inverted after the action of a fermionic operator:

$$\left\{\hat{P}_{\mathbf{n}}, \hat{F}_{\mathbf{n}}\right\} = 0 \qquad \left[\hat{P}_{\mathbf{n}}, \hat{F}_{\mathbf{n}'\neq\mathbf{n}}\right] = 0 \qquad \forall \mathbf{n}, \mathbf{n}' \in \Lambda \,. \tag{11.72}$$

Therefore, matter fields admit a notion of parity satisfying Eq. (11.72). For Dirac fermions

$$\hat{\psi}_{\text{Dirac}} = \begin{pmatrix} 0 & 1 \\ 0 & 0 \end{pmatrix}_F \qquad \hat{P}_{\text{Dirac}} = \begin{pmatrix} +1 & 0 \\ 0 & -1 \end{pmatrix}, \tag{11.73}$$

where the subscript F is a reminder that the $\hat{\psi}$ matrix is meant 'as a fermion', with the global action in Eq. (11.71). Similarly, as for Majorana fermions, we have:

$$\hat{\gamma}_{\text{Majorana}} = \begin{pmatrix} 0 & 1 \\ 1 & 0 \end{pmatrix}_F \qquad \hat{P}_{\text{Majorana}} = \begin{pmatrix} +1 & 0 \\ 0 & -1 \end{pmatrix}. \tag{11.74}$$

Similarly, as for the incoming (fermion) rishon-operators, we will provide an adequate notion of parity (see Eq. (11.134) for SU(2) and Eq. (11.87) for U(1)).

11.3 U(1) Lattice Gauge Theory

In this section, we apply the dressed-site formalism developed in Sect. 11.2 to an Abelian scenario, focusing on the U(1) LGT including dynamical matter, i.e. the lattice realization of quantum-electrodynamics (QED). Again, for simplicity we will consider a (2+1)D system, where all the gauge (electric and magnetic) contributions play a non-trivial role.

11.3.1 The Model

Within the Wilson and Kogut-Susskind formulation of LGTs [366, 405], the U(1) version of the general Hamiltonian in Eq. (11.65) reads:

$$\begin{aligned}\hat{H}_{\mathrm{U}(1)} =& \frac{c\hbar}{2a_0}\sum_{\mathbf{n}}\left[-i\hat{\psi}^{\dagger}_{\mathbf{n}}\hat{U}_{\mathbf{n},\mu x}\hat{\psi}_{\mathbf{n}+\mu x} - (-1)^{n_x+n_y}\hat{\psi}_{\mathbf{n}}\hat{U}_{\mathbf{n},\mu y}\hat{\psi}_{\mathbf{n}+\mu y} + \mathrm{h.c}\right] \\ &+ m_0c^2\sum_{\mathbf{n}}(-1)^{\mathbf{n}}\hat{\psi}^{\dagger}_{\mathbf{n}}\hat{\psi}_{\mathbf{n}} + \frac{g^2c\hbar}{2a_0}\sum_{\mathbf{n},\mu}\hat{E}^2_{\mathbf{n},\mu} - \frac{c\hbar}{2g^2a_0}\sum_{\square}\mathrm{Tr}\left(\hat{U}_{\square}+\hat{U}^{\dagger}_{\square}\right),\end{aligned} \tag{11.75}$$

where, since there is only one generator, the electric field is unique $\hat{L} = \hat{R} = \hat{E}$ and the parallel transporter acts as a raising operator:

$$\left[\hat{U}, \hat{E}\right] = \hat{U} \qquad \hat{U}^{\dagger}\hat{U} = \mathbb{1} \qquad \left[\hat{U}, \hat{U}^{\dagger}\right] = 0\,. \tag{11.76}$$

In the electric basis representation, where $\hat{E}$ is diagonal, Eq. (11.76) can be obtained assuming the following behaviors: given the $|\ell\rangle$ electric state of the gauge link Hilbert space,

$$\begin{cases}\hat{E}\,|\ell\rangle = \ell\,|\ell\rangle \\ \hat{U}\,|\ell\rangle = |\ell-1\rangle \\ \hat{U}^{\dagger}\,|\ell\rangle = |\ell+1\rangle\end{cases} \quad \text{so that} \quad \begin{cases}\hat{E}\hat{U}\,|\ell\rangle = (\ell-1)\,|\ell-1\rangle \\ \hat{U}\hat{E}\,|\ell\rangle = \ell\,|\ell-1\rangle\end{cases}\,. \tag{11.77}$$

Correspondingly, U(1) Gauss law requires that $G_{\mathbf{n}}\left|\Psi_{\mathrm{phys}}\right\rangle = 0\ \forall\mathbf{n}$, where $G_{\mathbf{n}}$ is the local generator of the U(1) gauge symmetry, satisfies $[G_{\mathbf{n}}, \hat{H}] = 0$, and is defined as follows:

$$\hat{G}_{\mathbf{n}} = \hat{E}_{\mathbf{n},\mu} - \hat{E}_{\mathbf{n}-\mu,\mu} - \hat{\psi}^{\dagger}_{\mathbf{n}}\hat{\psi}_{\mathbf{n}} + \frac{1-(-1)^{\mathbf{n}}}{2}\,. \tag{11.78}$$

11.3.2 Truncating the U(1) Gauge Group

To perform numerical simulations of Eq. (11.75), we need to perform a truncation of the infinite algebra of the U(1) gauge fields. As discussed in Sect. 11.2.1, one possible solution is to express the gauge operators $\hat{E}$ and $\hat{U}$ directly in terms of spin operators in the j representation of SU(2):

$$\hat{E} = \hat{S}^z(j) \qquad \hat{U} = \frac{1}{j}\hat{S}^-(j)\,. \tag{11.79}$$

Correspondingly, the gauge link Hilbert space $\mathcal{H}_{\text{link}}$ is made out of $(2j+1)$ states $|\ell\rangle$, which can be labeled as $|j, m\rangle$, where $-j \leq m \geq j$ is the third component of the spin momentum j. According to the rules of angular momentum summation, we would have:

$$\begin{cases} \hat{E}\,|j,m\rangle = S^z\,|j,m\rangle = m\,|j,m\rangle \\ \hat{U}\,|j,m\rangle = \frac{S^-}{j}\,|j,m\rangle = \sqrt{(1-\frac{1}{j}) - \frac{m}{j^2}(m-1)}\,|j,m-1\rangle \\ \hat{U}^\dagger\,|j,m\rangle = \frac{S^+}{j}\,|j,m\rangle = \sqrt{(1+\frac{1}{j}) - \frac{m}{j^2}(m+1)}\,|j,m+1\rangle\ , \end{cases} \tag{11.80}$$

which yields to the following commutation relations:

$$\begin{aligned} \left[\hat{U}, \hat{E}\right] &= \frac{1}{j}\left[\hat{S}^-(j), \hat{S}^z(j)\right] = \frac{\hat{S}^-(j)}{j} = \hat{U} \\ \left[\hat{U}, \hat{U}^\dagger\right] &= \frac{1}{j^2}\left[\hat{S}^-(j), \hat{S}^+(j)\right] = -\frac{2\hat{S}^z(j)}{j^2} \underset{j\to\infty}{\longrightarrow} 0. \end{aligned} \tag{11.81}$$

Such a definition of parallel transport provides a uniform convergence to Eq. (11.76) in the large-j limit. Similarly, we can define $\hat{U} = \text{diag}_{-1}(+1, \ldots, +1)$ as a ladder operator, which is perfectly unitary in the core, yet badly converges in the (upper and lower) truncated states. In the limit of $j \to \infty$, both solutions recover Eq. (11.77). As we will discuss in Sect. 11.4.6, the error induced by the truncation is particularly relevant in the small coupling limit, while remaining negligible in most scenarios [426, 460]. Choosing the ladder operator can reveal optimal for qubit base quantum hardware [461], while Eq. (11.79) is preferable in case of quantum systems that allow for the implementation of interacting spin chains with large spins [462]. For completeness, we mention that there are mappings of the parallel transporter preserving unitarity [463–466].

11.3.3 Rishon Decomposition of U(1) Gauge Fields

The next step towards the dressed-site formalism is decomposing the $(2j+1)$ states of the link basis as a combination of two states, one per half-link: $|j, m\rangle_{\mathbf{n},\mu} = \left|j, m_{\mathbf{n},+\mu}, m_{\mathbf{n}+\mu,-\mu}\right\rangle$. Omitting the j-label, the truncated link Hilbert space reads:

$$\mathcal{H}_j = \{|+j,-j\rangle, |+j-1,-j+1\rangle, \ldots, |m,m\rangle, \ldots |-j+1,+j-1\rangle, |-j,+j\rangle\}. \tag{11.82}$$

In the Euclidean basis, these gauge link states would read:

$$|+j,-j\rangle = \begin{pmatrix}1\\0\\\vdots\\0\\0\end{pmatrix} \quad |+j-1,-j+1\rangle = \begin{pmatrix}0\\1\\\vdots\\0\\0\end{pmatrix} \quad \dots \quad |-j,+j\rangle = \begin{pmatrix}0\\0\\\vdots\\0\\1\end{pmatrix} \tag{11.83}$$

Then, we express the parallel transport $\hat{U}$ as the product of two rishon modes, $\hat{\zeta}_{L(R)}$:

$$\hat{U}_{\mathbf{n},\mu} = \hat{\zeta}_{A,\mathbf{n}}\hat{\zeta}^{\dagger}_{B,\mathbf{n}+\mu,-\mu} \tag{11.84}$$

where, in full generality, we assumed the two rishon modes of the $(\mathbf{n},\mu)$-link to belong to two different species, A and B.[2] A general definition for the rishon operators is

$$\hat{\zeta}_A = \begin{pmatrix} 0 & a_1 & & & \\ & 0 & a_2 & & \\ & & 0 & a_3 & \\ & & & 0 & \ddots \end{pmatrix}_F \qquad \hat{\zeta}_B = \begin{pmatrix} 0 & b_1 & & & \\ & 0 & b_2 & & \\ & & 0 & b_3 & \\ & & & 0 & \ddots \end{pmatrix}_F , \tag{11.85}$$

where $\{a_i\}_{i=1\dots 2s}$ and $\{b_i\}_{i=1\dots 2s}$ are complex numbers to be determined in the specific spin-s representation of U(1). To obtain an operative Hamiltonian that is fully bosonic in every term, we require the rishon modes to satisfy a fermionic algebra. Namely, $\hat{\zeta}$-rishons must anti-commute among themselves and with matter fields:

$$\left\{\hat{\zeta}_{A,\mathbf{n},\mu}, \hat{\zeta}_{B,\mathbf{n}+\mu,-\mu}\right\} = 0 \qquad \left\{\hat{\zeta}_{\mathbf{n},\mu}, \hat{\psi}_{\mathbf{n}}\right\} = 0. \tag{11.86}$$

Being fermions, $\hat{\zeta}$-rishon must also satisfy (11.72). We define the rishon parity operator as

$$P^{\mathrm{U}(1)}_{\hat{\zeta}} = \mathrm{diag}(+1,-1,+1,-1,\dots), \tag{11.87}$$

where by convention we establish the first rishon state to be *even*, while the remaining ones are determined by alternating the sign of parity. Then, the last state is even (odd) if the spin-j representation is *integer* (*semi-integer*).

[2] It is possible to prove that, in *semi-integer* spin-representations, the two species coincide.

We can then properly express the parallel transport in (11.84) as follows:

$$\begin{aligned}\hat{U}_{\mathbf{n},\mu} =& \hat{\zeta}_{A,\mathbf{n},+\mu}\hat{\zeta}^{\dagger}_{B,\mathbf{n}+\mu,-\mu} = \left[\hat{\zeta}_{A,\mathbf{n},+\mu}\otimes \mathbb{1}_{\mathbf{n}+\mu,-\mu}\right]\times\left[\hat{P}_{\zeta,\mathbf{n},+\mu}\otimes\hat{\zeta}^{\dagger}_{B,\mathbf{n}+\mu,-\mu}\right]\\ =& \hat{\zeta}_{A,\mathbf{n},+\mu}\cdot\hat{P}_{\zeta,\mathbf{n},+\mu}\otimes\hat{\zeta}^{\dagger}_{B,\mathbf{n}+\mu,-\mu}\\ =& \underbrace{\begin{pmatrix} 0 & -a_1 & & & \\ & 0 & +a_2 & & \\ & & 0 & -a_3 & \\ & & & & \ddots \end{pmatrix}}_{\mathbf{n}} \otimes \underbrace{\begin{pmatrix} 0 & & & & \\ \overline{b}_1 & 0 & & & \\ & \overline{b}_2 & 0 & & \\ & & \overline{b}_3 & 0 & \\ & & & & \ddots \end{pmatrix}}_{\mathbf{n}+\mu}.\end{aligned} \tag{11.88}$$

The correct matrix elements $\{a_i\}_{i=1\ldots 2j_{\mathrm{max}}}$ and $\{b_i\}_{i=1\ldots 2j_{\mathrm{max}}}$ of the two $\hat{\zeta}$-rishon modes in the specific spin-j_{max} representation for $U(1)$ are the ones matching the original choice of the parallel transporter (spin operator as in Eq. (11.79) or ladder operator).

11.3.4 Constructing the Dressed-Site Operators

We stress that the rishon decomposition is a general approach and can be applied to lattices of whatever dimension D. However, as discussed for SU(2), for a fixed D, the construction of the Hamiltonian operators requires attention in choosing an internal ordering of the dressed-site basis. Working in a $D = 2$-dimensional lattice, the single dressed-site basis can be sketched as:

$$\left|\begin{matrix} & \hat{\zeta}_{A,\mathbf{n},+\mu y} & \\ \hat{\zeta}_{B,\mathbf{n},-\mu x} & \psi_{\mathbf{n}} & \hat{\zeta}_{A,\mathbf{n},+\mu x} \\ & \hat{\zeta}_{B,\mathbf{n},-\mu y} & \end{matrix}\right\rangle \quad \text{whose d.o.f. are ordered as:} \quad \left|\begin{matrix} & 4 & \\ 1 & 0 & 3 \\ & 2 & \end{matrix}\right\rangle. \tag{11.89}$$

Every dressed-site operator is then constructed as in (11.71) according to the previous order. We are then ready to express each Hamiltonian term via dressed-site operators.

11.3.4.1 Hopping Operators

In the hopping Hamiltonian, the matter-gauge interaction (apart from the staggered factors) along the link $(\mathbf{n}, \boldsymbol{\mu})$ can be rewritten in terms of two bosonic *arrival* operators. Namely

$$\begin{aligned}\hat{H}^{\text{hopping}}_{\mathbf{n},\mu} &= \left[\hat{\psi}^{\dagger}_{\mathbf{n}}\hat{U}_{\mathbf{n},\mu}\hat{\psi}_{\mathbf{n}+\mu} + \text{h.c}\right] \\ &= \left[\hat{\psi}^{\dagger}_{\mathbf{n}}\hat{\zeta}_{A,\mathbf{n},+\mu}\hat{\zeta}^{\dagger}_{B,\mathbf{n}+\mu,-\mu}\hat{\psi}_{\mathbf{n}+\mu} + \text{h.c}\right] \text{ where } \quad \begin{aligned}\hat{Q}^{\dagger}_{\mathbf{n},+\mu} &= \hat{\psi}^{\dagger}_{\mathbf{n}}\hat{\zeta}_{A,\mathbf{n},+\mu} \\ \hat{Q}^{\dagger}_{\mathbf{n},-\mu} &= \hat{\psi}^{\dagger}_{\mathbf{n}}\hat{\zeta}_{B,\mathbf{n},-\mu}.\end{aligned} \\ &= \left[\hat{Q}^{\dagger}_{\mathbf{n},+\mu}\hat{Q}_{\mathbf{n}+\mu,-\mu} + \text{h.c}\right]\end{aligned} \tag{11.90}$$

Then, according to the internal ordering of dressed site defined in (11.89), we have for instance:

$$\begin{aligned}\hat{Q}^{\dagger}_{\mathbf{n},+\mu x} = \hat{\psi}^{\dagger}_{\mathbf{n}}\hat{\zeta}_{A,\mathbf{n},+\mu x} = \quad & \hat{\psi}^{\dagger}_{\mathbf{n}} \otimes \mathbb{1}_{\mathbf{n},-\mu x} \otimes \mathbb{1}_{\mathbf{n},-\mu y} \otimes \mathbb{1}_{\mathbf{n},+\mu x} \otimes \mathbb{1}_{\mathbf{n},+\mu y} \\ & \times \hat{P}_{\psi,\mathbf{n}} \otimes \hat{P}_{\zeta,\mathbf{n},-\mu x} \otimes \hat{P}_{\zeta,\mathbf{n},-\mu y} \otimes \hat{\zeta}_{A,\mathbf{n},+\mu x} \otimes \mathbb{1}_{\mathbf{n},+\mu y} \\ & = \hat{\psi}^{\dagger}_{\mathbf{n}} \cdot \hat{P}_{\psi,\mathbf{n}} \otimes \hat{P}_{\zeta,\mathbf{n},-\mu x} \otimes \hat{P}_{\zeta,\mathbf{n},-\mu y} \otimes \hat{\zeta}_{A,\mathbf{n},+\mu x} \otimes \mathbb{1}_{\mathbf{n},+\mu y}.\end{aligned} \tag{11.91}$$

11.3.4.2 Number Density Operators

Similarly, number density operators can be expressed as:

$$\hat{N}_{\mathbf{n}} = \hat{\psi}^{\dagger}_{\mathbf{n}}\hat{\psi}_{\mathbf{n}} = \hat{\psi}^{\dagger}_{\mathbf{n}} \cdot \hat{\psi}_{\mathbf{n}} \bigotimes_{\pm\mu} \mathbb{1}_{\mathbf{n},\mu}. \tag{11.92}$$

11.3.4.3 Magnetic Operators

Let us rewrite the single plaquette interaction in terms of rishon modes. The idea is to perform a series of fermionic swaps to make the rishons of the same dressed site close in the internal ordering of the plaquette. We have then:

$$\begin{aligned}\hat{U}_{\square} =& \left[\hat{U}_{\mathbf{n},\mu_x}\hat{U}_{\mathbf{n}+\mu_x,+\mu_y}\hat{U}^{\dagger}_{\mathbf{n}+\mu_y,+\mu_x}\hat{U}^{\dagger}_{\mathbf{n},\mu_y}\right] \\ =& \left[\left(\hat{\zeta}_{A,\mathbf{n},+\mu_x}\hat{\zeta}^{\dagger}_{B,\mathbf{n}+\mu_x,-\mu_x}\right)\left(\hat{\zeta}_{A,\mathbf{n}+\mu_x,+\mu_y}\hat{\zeta}^{\dagger}_{B,\mathbf{n}+\mu_x+\mu_y,-\mu_y}\right)\times\right. \\ & \left.\left(\hat{\zeta}_{A,\mathbf{n}+\mu_y,+\mu_x}\hat{\zeta}^{\dagger}_{B,\mathbf{n}+\mu_x+\mu_y,-\mu_x}\right)^{\dagger}\left(\hat{\zeta}_{A,\mathbf{n},+\mu_y}\hat{\zeta}^{\dagger}_{B,\mathbf{n}+\mu_y,-\mu_y}\right)^{\dagger}\right] \\ =& \left[\left(\hat{\zeta}_{A,\mathbf{n},+\mu_x}\hat{\zeta}^{\dagger}_{B,\mathbf{n}+\mu_x,-\mu_x}\right)\left(\hat{\zeta}_{A,\mathbf{n}+\mu_x,+\mu_y}\hat{\zeta}^{\dagger}_{B,\mathbf{n}+\mu_x+\mu_y,-\mu_y}\right)\times\right. \\ & \left.\left(\hat{\zeta}_{B,\mathbf{n}+\mu_x+\mu_y,-\mu_x}\hat{\zeta}^{\dagger}_{A,\mathbf{n}+\mu_y,+\mu_x}\right)\left(\hat{\zeta}_{B,\mathbf{n}+\mu_y,-\mu_y}\hat{\zeta}^{\dagger}_{A,\mathbf{n},+\mu_y}\right)\right] \\ =& \left[\left(\hat{\zeta}_{A,\mathbf{n},+\mu_x}\hat{\zeta}^{\dagger}_{A,\mathbf{n},+\mu_y}\right)\left(\hat{\zeta}^{\dagger}_{B,\mathbf{n}+\mu_x,-\mu_x}\hat{\zeta}_{A,\mathbf{n}+\mu_x,+\mu_y}\right)\times\right. \\ & \left.\left(\hat{\zeta}^{\dagger}_{B,\mathbf{n}+\mu_x+\mu_y,-\mu_y}\hat{\zeta}_{B,\mathbf{n}+\mu_x+\mu_y,-\mu_x}\right)\left(\hat{\zeta}^{\dagger}_{A,\mathbf{n}+\mu_y,+\mu_x}\hat{\zeta}_{B,\mathbf{n}+\mu_y,-\mu_y}\right)\right] \\ =& \left[-\hat{C}_{\mathbf{n},+\mu_x,+\mu_y}\hat{C}_{\mathbf{n}+\mu_x,+\mu_y,-\mu_x}\hat{C}_{\mathbf{n}+\mu_x+\mu_y,-\mu_x,-\mu_y}\hat{C}_{\mathbf{n}+\mu_y,-\mu_y,+\mu_x}\right],\end{aligned} \tag{11.93}$$

where we merged rishon modes belonging to the same dressed site into *corner* operators defined as $\hat{C}_{\mathbf{n},\mu_1,\mu_2} = \hat{\zeta}_{\mathbf{n},\mu_1}\hat{\zeta}^{\dagger}_{\mathbf{n},\mu_2}$. For instance, from the internal order of the dressed-site basis in (11.89), we have for instance:

$$\begin{aligned}\hat{C}_{\mathbf{n},-\mu y,+\mu x} &= \hat{\zeta}_{B,\mathbf{n},-\mu y}\hat{\zeta}^{\dagger}_{A,\mathbf{n},+\mu x} \\ &= \hat{P}_{\psi,\mathbf{n}} \otimes \hat{P}_{\hat{\zeta},\mathbf{n},-\mu x} \otimes \hat{\zeta}_{B,\mathbf{n},-\mu y} \otimes \mathbb{1}_{\mathbf{n},+\mu x} \otimes \mathbb{1}_{\mathbf{n},+\mu y} \\ &= \times\, \hat{P}_{\psi,\mathbf{n}} \otimes \hat{P}_{\hat{\zeta},\mathbf{n},-\mu x} \otimes \hat{P}_{\hat{\zeta},\mathbf{n},-\mu y} \otimes \hat{\zeta}^{\dagger}_{A,\mathbf{n},+\mu x} \otimes \mathbb{1}_{\mathbf{n},+\mu y} \\ &= \mathbb{1}_{\mathbf{n}} \otimes \mathbb{1}_{\mathbf{n},-\mu x} \otimes \hat{\zeta}_{B,\mathbf{n},-\mu y} \cdot \hat{P}_{\hat{\zeta},\mathbf{n},-\mu y} \otimes \hat{\zeta}^{\dagger}_{A,\mathbf{n},+\mu x} \otimes \mathbb{1}_{\mathbf{n},+\mu y}.\end{aligned} \tag{11.94}$$

11.3.4.4 Electric Field Operators

In the dressed-site formalism, as with the parallel transporter, the electric field contribution is equally divided across the two half-links, with each half contributing half the total electric energy. Namely, from (11.79) we have:

$$\hat{E}^2_{\mathbf{n},\mu} = \frac{1}{2}\left[\hat{S}^z_{\mathbf{n},\mu}(j)^2 + \hat{S}^z_{\mathbf{n}+\mu,-\mu}(j)^2\right]. \tag{11.95}$$

In the operative Hamiltonian, we combine the $2D$ half-links contribution to the electric energy onto the same single dressed-site operator:

$$\hat{\Gamma}_{\mathbf{n}} = \frac{1}{2}\sum_{\pm\mu} \hat{S}^z_{\mathbf{n},\mu}(j)^2. \tag{11.96}$$

11.3.4.5 Operative Dressed-Site U(1) Hamiltonian

We can then rewrite the (2+1)D U(1) truncated Hamiltonian in Eq. (11.75) as follows:

$$\begin{aligned}H_{\mathrm{U}(1)} =& \frac{c\hbar}{2a_0}\sum_{\mathbf{n}}\left[-\mathrm{i}\hat{Q}^{\dagger}_{\mathbf{n},+\mu x}\hat{Q}_{\mathbf{n}+\mu x,-\mu x} - (-1)^{\mathbf{n}}\hat{Q}^{\dagger}_{\mathbf{n},+\mu y}\hat{Q}_{\mathbf{n}+\mu y,-\mu y} + \mathrm{h.c}\right] \\ &+ m_0c^2\sum_{\mathbf{n}}(-1)^{\mathbf{n}}\hat{N}_{\mathbf{n}} + \frac{g^2c\hbar}{4a_0}\sum_{\mathbf{n}}\hat{\Gamma}_{\mathbf{n}} - \frac{c\hbar}{2g^2a_0}\sum_{\Box}\mathrm{Tr}\left(\hat{U}_{\Box} + \hat{U}^{\dagger}_{\Box}\right).\end{aligned} \tag{11.97}$$

11.3.4.6 U(1) Link-Symmetry Operators

The equivalence between Eqs. (11.75) and (11.97) is correct as long as the link symmetries are satisfied, i.e., as long as the rishon states of each link have opposite electric fields. This constraint reduces to a $\mathbb{Z}_2$ symmetry on each link that can be easily resolved with dedicated libraries [459, 467], or imposed as two-body penalty terms (one per link) in the Hamiltonian, in such a way to behave as the largest energy scale of the problem.

11.3.4.7 Open Boundary Conditions and Half-Integer Representations

In the case of open boundary conditions (OBC), we may want to freeze the half links on the bordering dressed sites of the lattice to the ground of the electric field. For *integer* j-representations, this can be easily achieved by selecting the rishon configurations with $|j, 0\rangle$ on sites along the corresponding border $\boldsymbol{\mu}$. Equivalently, for each of these sites, we can add a penalty term like $\hat{H}^{\text{border}}_{\mathbf{n},\mu} = \alpha\left(\hat{E}_{\mathbf{n},\mu} - j\right)^2$. As for *semi-integer* representations, where the minimal electric field is degenerate, we can choose between one of the following penalties:

$$\hat{H}^{\text{border}}_{\mathbf{n},-\mu} = \alpha\left(\hat{E}_{\mathbf{n},-\mu} - \lceil j \rceil\right)^2 \quad \text{and} \quad \hat{H}^{\text{border}}_{\mathbf{n},+\mu} = \alpha\left(\hat{E}_{\mathbf{n},+\mu} - \lfloor j \rfloor\right)^2, \tag{11.98}$$

where $\lceil j \rceil$ ($\lfloor j \rfloor$) corresponds to the minimum upper (maximum lower) integer representation.

11.3.4.8 Projecting Operators on the U(1) Gauge-Invariant Dressed-Site Basis

As discussed in Sect. 11.2.3, all the previous operators need to be projected in the U(1) gauge-invariant subspace. Since we have chosen the staggered fermions formulation, we expect two gauge invariant bases, one for *even* and one for *odd* sites, respectively M_+ and M_-. Each one coincides with the kernel of the corresponding U(1) Gauss Law operator $\hat{G}_{\mathbf{n},p}$:

$$\hat{G}_{\mathbf{n},p} M_p = \left[\hat{N}_{\mathbf{n}} + \sum_{\mu} \hat{E}_{\mu} - \frac{1-p}{2}\mathbb{1}_{\mathbf{n}}\right] M_p = 0, \tag{11.99}$$

where $p = \pm 1$ is the parity of the lattice site.

11.3.5 Lattice QED with Minimally Truncated Gauge Fields

As an example, here we consider the smallest integer $j=1$ spin-representation of the U(1) gauge fields. Such a truncation is enough to guarantee the contribution of all the terms in the Hamiltonian equation (11.75) while maintaining a non-degenerate minimal value of the Casimir. Moreover, it provides a good approximation of the untruncated low energy physics in the strong coupling limit $g \gg 1$, where the gauge field energy term dominates the Hamiltonian. Remarkably, it has been succesfully adopted in gives access to non trivial statical and dynamical features of the U(1) LGT in different spatial dimensions from (1+1)D [398], to (2+1)D [468] and (3+1)D [397] (see Fig. 11.2).

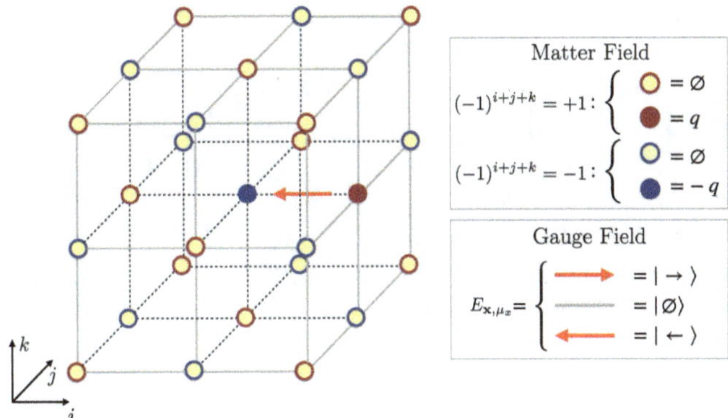

Fig. 11.2 Scheme of the (3+1)D QED LGT in the minimal integer spin ($j = 1$) truncation of the electric fields. Fermionic degrees of freedom are represented by staggered fermions on sites with different parity: on the even (odd) sites, a full red (blue) circle corresponds to a particle (antiparticle) with a positive (negative) charge. As an illustrative example, it is shown a gauge-invariant configuration of matter and gauge fields with one particle and one antiparticle in the sector of zero total charge. (Figure from [397])

Within this truncation, the gauge fields can be described by the following three-dimensional operators:

$$\hat{E} = \hat{S}^z = \begin{pmatrix} +1 & & \\ & 0 & \\ & & -1 \end{pmatrix} \qquad \hat{U} = \hat{S}^- = \sqrt{2}\begin{pmatrix} 0 & & \\ 1 & 0 & \\ & 1 & 0 \end{pmatrix}. \tag{11.100}$$

The corresponding link Hilbert space $\mathcal{H}^{j=1}$ defined in (11.83) is made out of the following states:

$$|1,1\rangle = \begin{pmatrix} 1 \\ 0 \\ 0 \end{pmatrix} \qquad |0,0\rangle = \begin{pmatrix} 0 \\ 1 \\ 0 \end{pmatrix} \qquad |-1,-1\rangle = \begin{pmatrix} 0 \\ 0 \\ 1 \end{pmatrix}. \tag{11.101}$$

Then, from the definition of $\hat{U}$ in (11.100), the only relevant matrix elements of $\hat{\zeta}_A\hat{\zeta}_B^\dagger$ are the ones that correspond to the action on sites belonging to (11.83):

$$\begin{aligned} \hat{\zeta}_A\hat{\zeta}_B^\dagger\,|1,1\rangle &= +a_2\overline{b}_1\,|0,0\rangle \\ \hat{\zeta}_A\hat{\zeta}_B^\dagger\,|0,0\rangle &= -a_1\overline{b}_2\,|-1,-1\rangle \qquad \text{whose solution is} \qquad \begin{aligned} a_1 &= a_2 = 1 \\ b_1 &= +\sqrt{2} = -b_2. \end{aligned} \\ \hat{\zeta}_A\hat{\zeta}_B^\dagger\,|-1,-1\rangle &= 0 \end{aligned} \tag{11.102}$$

Summarizing, we have found

$$\hat{\zeta}_A = \begin{pmatrix} 0 & +1 & \\ & 0 & +1 \\ & & 0 \end{pmatrix} \qquad \hat{\zeta}_B = \begin{pmatrix} 0 & +\sqrt{2} & \\ & 0 & -\sqrt{2} \\ & & 0 \end{pmatrix}. \tag{11.103}$$

By following the dressed-scheme developed so far, Ref [397] was able to simulate the challenging (3+1)D version of the U(1) Hamiltonian in Eq. (11.75), built upon a gauge-invariant local Hilbert space of dimension 267 (see Table 11.1 and [396]).

$$\begin{aligned} \hat{H}^{(3+1)D}_{\mathrm{U}(1)} = &-\frac{c\hbar}{2a_0}\sum_{\mathbf{n}}\Big[i\hat{\psi}^{\dagger}_{\mathbf{n}}\hat{U}_{\mathbf{n},\mu x}\hat{\psi}_{\mathbf{n}+\mu x} + (-1)^{n_x+n_y}\hat{\psi}_{\mathbf{n}}\hat{U}_{\mathbf{n},\mu y}\hat{\psi}_{\mathbf{n}+\mu y} \\ &\qquad + i(-1)^{n_x+n_y}\hat{\psi}_{\mathbf{n}}\hat{U}_{\mathbf{n},\mu z}\hat{\psi}_{\mathbf{n}+\mu z} + \mathrm{h.c}\Big] \\ &+ m_0c^2\sum_{\mathbf{n}}(-1)^{\mathbf{n}}\hat{\psi}^{\dagger}_{\mathbf{n}}\hat{\psi}_{\mathbf{n}} + \frac{g^2c\hbar}{2a_0}\sum_{\mathbf{n},\mu}\hat{E}^2_{\mathbf{n},\mu} \\ &- \frac{c\hbar}{2g^2a_0}\sum_{\square}\mathrm{Tr}\left(\hat{U}_{\square} + \hat{U}^{\dagger}_{\square}\right). \end{aligned} \tag{11.104}$$

By developing a 3D realization of binary TTN methods described in Chap. 8, the authors of Ref. [397] succesfully simulate the model at zero and finite charge densities, while address fundamental questions such as characterizing collective phases (see for example Fig. 11.3), identifying a confining phase at large gauge coupling, and studying charge-screening effects.

11.4 SU(2) Yang-Mills Lattice Gauge Theory

In this section, we apply the dressed-site formalism to a non-Abelian scenario, focusing on the SU(2) Yang-Mills (YM) LGT with dynamical matter. For simplicity we will focus on a (2+1)D system, where both the electric and magnetic gauge contributions play a non-trivial role. Generalizations to the (3+1)D are trivial, provided to consider the correct phase factors in Eq. (11.32) due to staggered fermions. Restrictions to the (1+1)D case are considered in Sect. 11.4.5 and applied in [401, 402, 416].

11.4.1 The Model

Before introducing the SU(2) YM Hamiltonian, let us recall the main properties of the SU(2) gauge theory. First, the SU(2) algebra generators are the Pauli matrices

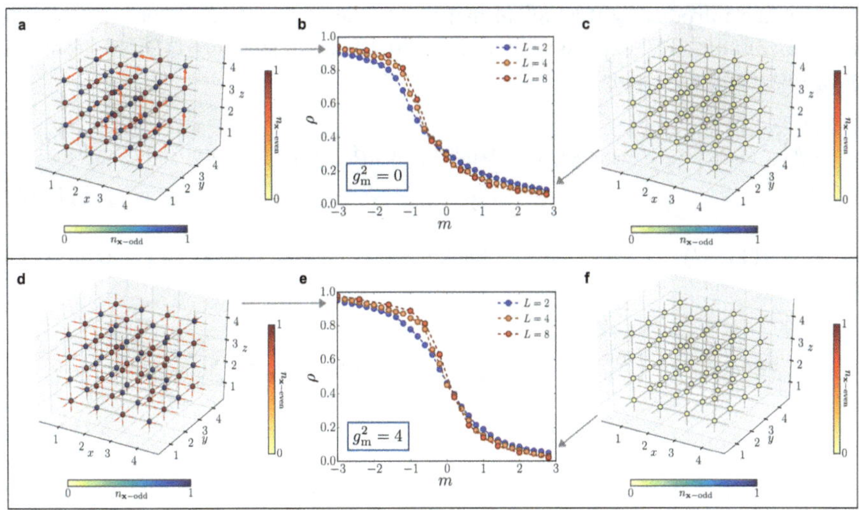

Fig. 11.3 Transition at zero charge density in the (3+1)D QED LGT with minimally truncated gauge fields. (Top) Ground state charge occupation and electric field on links for (**a**) $m = -3.0$, (**c**) $m = 3.0$ without magnetic interaction: $g_m^2 = 0$. (**b**) Particle density as a function of m, for different system sizes L and $g_m^2 = 0$. (Bottom) Ground state charge occupation and electric field on links for (**d**) $m = -3.0$ and (**f**) $m = 3.0$ in the presence of magnetic interactions $g_m^2 = 4$. (**e**) Particle density as a function of m, for different system size L and $g_m^2 = 4$. (Figure from [397])

$\hat{T}^a = \hat{\sigma}^a/2$, while the structure constants are equal to the Levi-Civita tensor $f^{abc} = \varepsilon^{abc}$. Correspondingly, left/right generators of SU(2) gauge transformations $\hat{L}_{\mathbf{n},\mu}/\hat{R}_{\mathbf{n},\mu}$ are hermitian and related each other as follows: for $a, b, c \in \{x, y, z\}$

$$\left[\hat{L}^a_{\mathbf{n},\mu}, \hat{R}^b_{\mathbf{n},\mu}\right] = 0 \quad \left[\hat{L}^a_{\mathbf{n},\mu}, \hat{L}^b_{\mathbf{n}',\mathbf{n}'+\mu'}\right] = i\delta_{\mathbf{n}\mathbf{n}'}\delta_{\mu\mu'}\epsilon_{abc}\hat{L}^c_{\mathbf{n},\mu} \quad \text{(same with } \hat{R}\text{)}. \tag{11.105}$$

Their trace yields the Casimir operator

$$\hat{E}^2_{\mathbf{n},\mu} = \sum_a \hat{L}^a_{\mathbf{n},\mu}\hat{L}^a_{\mathbf{n},\mu} = \sum_a \hat{R}^a_{\mathbf{n},\mu}\hat{R}^a_{\mathbf{n},\mu}\,, \tag{11.106}$$

and the commutations rules with the parallel transporter $\hat{U}^{\alpha\beta}$ read

$$\begin{aligned} \left[\hat{L}^a_{\mathbf{n},\mu}, \hat{U}^{\alpha\beta}_{\mathbf{n}',\mu'}\right] &= -\delta_{\mathbf{n}\mathbf{n}'}\delta_{\mu'}\sum_\gamma \frac{\sigma^a_{\alpha\gamma}}{2}\hat{U}^{\gamma\beta}_{\mathbf{n},\mu}, \\ [\hat{R}^a_{\mathbf{n},\mu}, \hat{U}^{\alpha\beta}_{\mathbf{n}',\mu'}] &= +\delta_{\mathbf{n}\mathbf{n}'}\delta_{\mu\mu'}\sum_\gamma \hat{U}^{\alpha\gamma}_{\mathbf{n},\mu}\frac{\sigma^a_{\gamma\beta}}{2}\,. \end{aligned} \tag{11.107}$$

Then, when introducing dynamical matter, we focus on single-flavor matter fields with SU(2)-color 1/2, expressed in terms of SU(2)-color staggered (Dirac) fermions $\hat{\psi}_{\mathbf{n},\alpha}$ [409] and satisfying

$$\left\{\hat{\psi}^{\dagger}_{\mathbf{n},\alpha}\hat{\psi}_{\mathbf{n}',\beta}\right\} = i\hbar\delta_{\mathbf{n},\mathbf{n}'}\delta_{\alpha,\beta}, \qquad \text{where} \qquad \alpha,\beta\in\{r,g\} \qquad \text{are SU(2)-colors}. \tag{11.108}$$

From Eq. (11.67) and assuming no background charges, Gauss Law operators are defined as

$$\hat{G}^{a}_{\mathbf{n}} = \hat{Q}^{a}_{\mathbf{n}} + \sum_{\mu}\left(\hat{R}^{a}_{\mathbf{n}-\mu,\mathbf{n}} + \hat{L}^{a}_{\mathbf{n},\mu}\right), \qquad \text{where} \qquad \hat{Q}^{a}_{\mathbf{n}} = \sum_{\alpha\beta}\hat{\psi}^{\dagger}_{\mathbf{n},\alpha}\frac{\sigma^{a}_{\alpha\beta}}{2}\hat{\psi}_{\mathbf{n},\beta} \tag{11.109}$$

rotates the quark field at **n**, and, if squared, it yields the *matter color density*, i.e. the quadratic Casimir operator of the matter field gauge group transformations:

$$\hat{S}^{2}_{\text{matt},\mathbf{n}} = \sum_{a}\hat{Q}^{a}_{\mathbf{n}}\hat{Q}^{a}_{\mathbf{n}}. \tag{11.110}$$

Correspondingly, $\hat{L}^{a}_{\mathbf{n}-\mu,\mathbf{n}}$ and $\hat{R}^{a}_{\mathbf{n},\mu}$ account for the transformation of the gauge links at its left and its right respectively [469].

The corresponding (2+1)D Hamiltonian with staggered fermions (see Sect. 11.1.3) reads:

$$\begin{aligned}\hat{H}^{\text{YM}}_{\text{SU(2)}} = &+\frac{c\hbar}{2a_0}\sum_{\alpha,\beta}\sum_{\mathbf{n}}\Big[\text{-i}\hat{\psi}^{\dagger}_{\mathbf{n},\alpha}\hat{U}^{\alpha\beta}_{\mathbf{n},\mu_x}\hat{\psi}_{\mathbf{n}+\mu_x,\beta}\\ &-(-1)^{n_x+n_y}\hat{\psi}^{\dagger}_{\mathbf{n},\alpha}\hat{U}^{\alpha\beta}_{\mathbf{n},\mu_y}\hat{\psi}_{\mathbf{n}+\mu_y,\beta}+\text{h.c}\Big]\\ &+m_0c^2\sum_{\mathbf{n}}(-1)^{n_x+n_y}\sum_{\alpha}\hat{\psi}^{\dagger}_{\mathbf{n},\alpha}\hat{\psi}_{\mathbf{n},\alpha}+\hat{H}_{\text{pure}}.\end{aligned} \tag{11.111}$$

As for the pure gauge Hamiltonian, we have

$$\hat{H}_{\text{pure}} = g^2\frac{c\hbar}{2a_0}\sum_{\mathbf{n},\mu}\hat{E}^{2}_{\mathbf{n},\mu} - \frac{c\hbar}{2a_0 g^2}\sum_{\square}\text{Tr}\left(\hat{U}_{\square}+\hat{U}^{\dagger}_{\square}\right), \tag{11.112}$$

where $\hat{U}_{\square}$ is the plaquette operator defined in Eq. (11.59), while the gauge coupling $g(q,a_0)\propto q\sqrt{a_0}$ defined Eq. (11.57) is dimensionless, but scales nonetheless with the lattice spacing a_0 to ensure that the color charge q of a quark stays finite in the continuum limit.

11.4.2 Truncating the SU(2) Gauge Group

As discussed in Sect. 11.2.1, to make LGT Hamiltonians in Eqs. (11.111) and (11.112) suitable for TN methods and quantum hardware, a finite-dimensional gauge-link Hilbert space is typically required. Here, we propose an effective truncation for the SU(2) gauge fields that can be applied to lattices of any spatial dimension and scalable to arbitrarily large truncations. In the limit of an infinitely large spin irreducible representation of the SU(2) gauge group, it recovers all the properties of the original SU(2) Yang-Mills LGT. Remarkably, the use of this approach in the smallest non-trivial truncation of SU(2), labeled as *hardcore-gluon* approximation, has been used to achieve non-trivial results as the ones discussed in [399] and [401, 402]. Of course, to accurately represent the full theory, for weak-g, larger gauge representations are required: this increases the computational challenges but it is still potentially accessible via TNs. Anyway, such a description of the gauge group is the first building block for building logical dressed-sites discussed in Sect. 11.2, where Gauss Law is already satisfied (see Fig. 11.4).

To truncate the continuous SU(2) gauge group, we express it in terms of the irreducible representation (irrep) basis [446, 470]. As SU(2) admits a *quasi-real*-representation, where the fundamental and the anti-fundamental representations coincide, $\forall \mathbf{n}, \boldsymbol{\mu} \in \Lambda$, the gauge Hilbert space of the $(\mathbf{n}, \boldsymbol{\mu})$ link can be written as:

$$\mathcal{H}_{\text{gauge}} = \{|j, m_{\text{L}}, m_{\text{R}}\rangle\}, \qquad \text{where} \qquad -j \leq m_{L(R)} \leq j \tag{11.113}$$

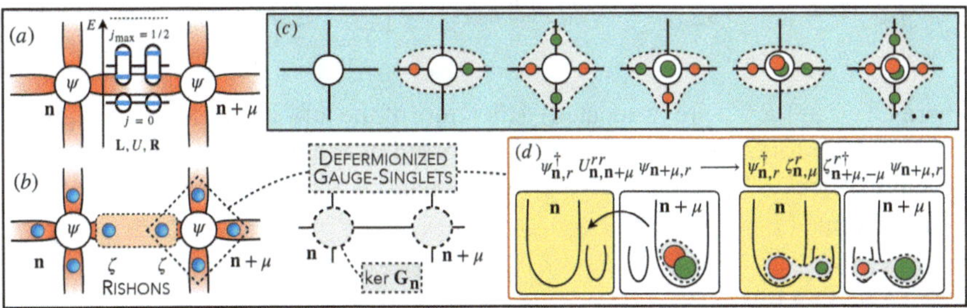

Fig. 11.4 Sketched representation of the dressed-site model developed in Sect. 11.4 for SU(2): (**a**) starting from the formulation with matter sites and truncated SU(2) gauge links, (**b**) we split the latter in pairs of rishon modes $\hat{\zeta}$ defined in Eq. (11.131) and constrain their dynamics with the SU(2) link symmetry in Eq. (11.140). We then (**c**) merge them with matter fields into dressed SU(2) gauge-singlets with a bosonic statistics. As a consequence, within the dressed-site formalism, the hopping always involves an even number of fermions, i.e. the matter field plus a rishon mode (**d**). (Figure inspired by [396, 399])

is the corresponding third spin component associated with the left (right) side of the SU(2) link-irrep j. Then, the parallel transporter $\hat{U}^{\alpha\beta}$ is given in terms of Clebsh-Gordan coefficients:

$$\langle j'm'_{\mathrm{L}}m'_{\mathrm{R}}|U\rangle^{\alpha\beta}|jm_{\mathrm{L}}m_{\mathrm{R}}\rangle = \sqrt{\frac{2j+1}{2j'+1}}C^{jm_{\mathrm{L}}}_{\frac{1}{2},\alpha;j'm'_{\mathrm{L}}}C^{j'm'_{\mathrm{R}}}_{jm_{\mathrm{R}};\frac{1}{2},\beta}. \tag{11.114}$$

Correspondingly, left and right generators are defined as:

$$\langle j'm'_{\mathrm{L}}m'_{\mathrm{R}}|\hat{L}^{a}|jm_{\mathrm{L}}m_{\mathrm{R}}\rangle = \delta_{j,j'}S^{(j)a}_{m'_{\mathrm{L}},m_{\mathrm{L}}}\delta_{m'_{\mathrm{R}},m_{\mathrm{R}}}, \tag{11.115a}$$

$$\langle j'm'_{\mathrm{L}}m'_{\mathrm{R}}|\hat{R}^{a}|jm_{\mathrm{L}}m_{\mathrm{R}}\rangle = \delta_{j,j'}\delta_{m'_{\mathrm{L}},m_{\mathrm{L}}}S^{(j)a}_{m'_{\mathrm{R}},m_{\mathrm{R}}}. \tag{11.115b}$$

Then, the chromo-electric energy operator, i.e. the quadratic Casimir in Eq. (11.106) reads [469]:

$$\hat{E}^2\,|jm_{\mathrm{L}}m_{\mathrm{R}}\rangle = j(j+1)\,|jm_{\mathrm{L}}m_{\mathrm{R}}\rangle\,. \tag{11.116}$$

Then, the truncation of the gauge fields is applied by imposing a local energy cutoff Θ on the $\hat{E}^2$ spectrum, keeping the irreps $j \leq j_{\max}$ such that $C_2(j_{\max}) \leq \Theta$. The truncation keeps the electric field operator $\hat{E}$ hermitian and protects the SU(2) gauge algebra rules of Eqs. (11.105) and (11.107). Namely, truncated $\hat{U}^{\alpha\beta}$ fields satisfy the following left and right gauge transformations:

$$\left[\hat{L}^{a}_{j_{\max}}, \hat{U}^{\alpha\beta}\right] = -\sum_{\gamma}\frac{\sigma^{a}_{\alpha\gamma}}{2}\hat{U}_{\gamma\beta} \quad \left[\hat{R}^{a}_{j_{\max}}, \hat{U}_{\alpha\beta}\right] = +\sum_{\gamma}\hat{U}_{\alpha\gamma}\frac{\sigma^{a}_{\gamma\beta}}{2}, \tag{11.117}$$

where $\hat{L}^{a}_{j_{\max}}$ and $\hat{R}^{a}_{j_{\max}}$ are (truncated) generators of the left- and right-handed groups of SU(2) transformations. They can be expressed as the block-diagonal direct sum of spin matrices S^{a}_{j} in consecutive j-representations from the smallest ($j = 0$) to the largest one ($j = j_{\max}$):

$$\begin{aligned}\hat{L}^{a}_{j_{\max}} &= \bigoplus_{j=0}^{j_{\max}}\left(S^{a}_{j}\otimes\mathbb{1}_{j}\right) = \mathrm{diag}(S^{a}_{0}\otimes\mathbb{1}_{0},\ldots,S^{a}_{j_{\max}}\otimes\mathbb{1}_{j_{\max}})\\ \hat{R}^{a}_{j_{\max}} &= \bigoplus_{j=0}^{j_{\max}}\left(\mathbb{1}_{j}\otimes S^{a}_{j}\right) = \mathrm{diag}(\mathbb{1}_{0}\otimes S^{a}_{0},\ldots,\mathbb{1}_{j_{\max}}\otimes S^{a}_{j_{\max}}).\end{aligned} \tag{11.118}$$

Unfortunately, the truncation makes $\hat{U}^{\alpha\beta}$ unitary only as long as it acts on spin shells with $j < j_{\max}$ (it loses norm on the largest spin shell). Correspondingly, Wilson loops stay unitary as long as the outer spin shell $j = j_{\max}$ is nowhere populated. Namely, it implies that, $\forall j < j_{\max}$

$$\sum_{\beta\gamma} \hat{U}^{\alpha\beta} |0;0,0\rangle \otimes \hat{U}_{\beta\gamma} |j; m_{\mathrm{L}}, m_{\mathrm{R}}\rangle \otimes \hat{U}_{\gamma\delta} |0;0,0\rangle \tag{11.119}$$

displays the same norm of $\hat{U}_{\alpha\delta}\,|000\rangle\ \forall\alpha,\delta$, that is $1/\sqrt{2}$. Ultimately, we require the parallel transporter to display *spatial reflection symmetry*. Namely:

$$\hat{U}^{\dagger}_{\alpha\beta} = -\hat{\mathcal{F}}\hat{U}_{\beta\alpha}\hat{\mathcal{F}}, \tag{11.120}$$

where $\hat{\mathcal{F}}$ is the *swap operator* on a gauge link:

$$\hat{\mathcal{F}} = \sum_{j, m_{\mathrm{L}}, m_{\mathrm{R}}} |j; m_{\mathrm{L}}, m_{\mathrm{R}}\rangle\langle j; m_{\mathrm{R}}, m_{\mathrm{L}}| = \hat{\mathcal{F}}^{-1} = \hat{\mathcal{F}}^{\dagger}. \tag{11.121}$$

Whatever the gauge field truncation, in hopping terms, the action of $\hat{U}$ has to match the fundamental irrep of the matter field, whose Hilbert space in the Fock space, can be written as:

$$\mathcal{H}_{\text{site}} = \Big\{ |0\rangle = |\Omega\rangle\,,\ |r\rangle = \psi_r^{\dagger}\,|\Omega\rangle\,,\ |g\rangle = \psi_g^{\dagger}\,|\Omega\rangle\,,\ |2\rangle = \psi_r^{\dagger}\psi_g^{\dagger}\,|\Omega\rangle \Big\}, \tag{11.122}$$

whose action in the irrep basis $|j, m\rangle$ corresponds to the following SU(2) charges:

$$\mathcal{H}_{\text{site}} = \Big\{ |0, 0\rangle\,, \left|\tfrac{1}{2}, \tfrac{1}{2}\right\rangle, \left|\tfrac{1}{2}; -\tfrac{1}{2}\right\rangle, |0, 0\rangle \Big\}. \tag{11.123}$$

11.4.3 Rishon Decomposition of SU(2) Gauge Fields

With all these premises, the resulting truncated SU(2) Yang-Mills Hamiltonian of Eqs. (11.111)–(11.112) will act on the Hilbert space resulting from the tensor product of all the single matter Hilbert spaces and the corresponding ones of each truncated gauge field:

$$\mathcal{H} = \bigotimes_{\mathbf{n}} \mathcal{H}_{\text{matt}} \bigotimes_{\mathbf{n},\mu} \mathcal{H}_{\text{tr. link}}\,. \tag{11.124}$$

Notice that matter and gauge fields display different statistics. In these terms, we expect $\hat{U}^{\alpha\beta}_{\mathbf{n},\mu}$ to be *mutually bosonic*, as it commutes with all the matter-fields operators:

$$\left[\hat{U}^{\alpha\beta}_{\mathbf{n},\boldsymbol{\mu}}, \hat{\psi}^{(\dagger)}_{\mathbf{n},\alpha}\right] = 0 \qquad \forall \mathbf{n}, \forall \boldsymbol{\mu}, \forall \alpha, \beta \tag{11.125}$$

and *purely local*, as its link-algebra commutes with the one of any other link:

$$\left[\hat{U}^{\alpha\beta}_{\mathbf{n},\boldsymbol{\mu}}, \hat{U}^{\gamma\delta}_{\mathbf{n}',\mathbf{n}'+\boldsymbol{\mu}'}\right] = 0 \qquad \forall \mathbf{n} \neq \mathbf{n}', \boldsymbol{\mu} \neq \boldsymbol{\mu}', \forall \alpha, \beta, \gamma, \delta\,. \tag{11.126}$$

As anticipated in Sect. 11.2, in the electric basis, the parallel transporter $\hat{U}^{\alpha\beta}$ admits a rishon-decomposition to arbitrary truncation of the maximum allowed spin shell $j_{\max}$. Starting from a given spin shell j, we have to separately account for the action when both rishons are increased to shell $j+\frac{1}{2}$, and both are decreased to shell $j-\frac{1}{2}$. We can then decompose $\hat{U}^{\alpha\beta}$ as follows:

$$\hat{U}^{\alpha\beta}_{\mathbf{n},\mathbf{n}+\boldsymbol{\mu}} = \hat{\zeta}^{\alpha}_{A,\mathbf{n},\boldsymbol{\mu}}\hat{\zeta}^{\beta\dagger}_{B,\mathbf{n}+\boldsymbol{\mu},-\boldsymbol{\mu}} + \hat{\zeta}^{\alpha}_{B,\mathbf{n},\boldsymbol{\mu}}\hat{\zeta}^{\beta\dagger}_{A,\mathbf{n}+\boldsymbol{\mu},-\boldsymbol{\mu}}, \tag{11.127}$$

where the two $\hat{\zeta}$-rishon species, A and B, act respectively as raising and lowering the spin shell of the SU(2) gauge irrep. Interestingly, they are related to each other as:

$$\hat{\zeta}^{\alpha}_{A} = i\sigma^{y}_{\alpha,\beta}\hat{\zeta}^{\beta\dagger}_{B} \qquad\qquad \hat{\zeta}^{\alpha\dagger}_{A} = i\sigma^{y}_{\alpha,\beta}\hat{\zeta}^{\beta}_{B}\,. \tag{11.128}$$

We can then rewrite Eq. (11.127) in terms of one species, e.g. B. Renaming $\hat{\zeta}^{\alpha}_{B} = \hat{\zeta}^{\alpha}$, we obtain:

$$\hat{U}^{\alpha\beta}_{\mathbf{n},\mathbf{n}+\boldsymbol{\mu}} = i\sigma^{y}_{\alpha\gamma}\hat{\zeta}^{\gamma\dagger}_{\mathbf{n},\boldsymbol{\mu}}\hat{\zeta}^{\beta\dagger}_{\mathbf{n}+\boldsymbol{\mu},-\boldsymbol{\mu}} + i\sigma^{y}_{\beta\gamma}\hat{\zeta}^{\alpha}_{\mathbf{n},\boldsymbol{\mu}}\hat{\zeta}^{\gamma}_{\mathbf{n}+\boldsymbol{\mu},-\boldsymbol{\mu}}\,, \tag{11.129}$$

or equivalently

$$\begin{aligned}
\hat{U}^{rr}_{\mathbf{n},\mathbf{n}+\boldsymbol{\mu}} &= \hat{\zeta}^{g\dagger}_{\mathbf{n},\boldsymbol{\mu}}\hat{\zeta}^{r\dagger}_{\mathbf{n}+\boldsymbol{\mu},-\boldsymbol{\mu}} + \hat{\zeta}^{r}_{\mathbf{n},\boldsymbol{\mu}}\hat{\zeta}^{g}_{\mathbf{n}+\boldsymbol{\mu},-\boldsymbol{\mu}} \\
\hat{U}^{rg}_{\mathbf{n},\mathbf{n}+\boldsymbol{\mu}} &= \hat{\zeta}^{g\dagger}_{\mathbf{n},\boldsymbol{\mu}}\hat{\zeta}^{g\dagger}_{\mathbf{n}+\boldsymbol{\mu},-\boldsymbol{\mu}} - \hat{\zeta}^{r}_{\mathbf{n},\boldsymbol{\mu}}\hat{\zeta}^{r}_{\mathbf{n}+\boldsymbol{\mu},-\boldsymbol{\mu}} \\
\hat{U}^{gr}_{\mathbf{n},\mathbf{n}+\boldsymbol{\mu}} &= -\hat{\zeta}^{r\dagger}_{\mathbf{n},\boldsymbol{\mu}}\hat{\zeta}^{r\dagger}_{\mathbf{n}+\boldsymbol{\mu},-\boldsymbol{\mu}} + \hat{\zeta}^{g}_{\mathbf{n},\boldsymbol{\mu}}\hat{\zeta}^{g}_{\mathbf{n}+\boldsymbol{\mu},-\boldsymbol{\mu}} \\
\hat{U}^{gg}_{\mathbf{n},\mathbf{n}+\boldsymbol{\mu}} &= -\hat{\zeta}^{r\dagger}_{\mathbf{n},\boldsymbol{\mu}}\hat{\zeta}^{g\dagger}_{\mathbf{n}+\boldsymbol{\mu},-\boldsymbol{\mu}} - \hat{\zeta}^{g}_{\mathbf{n},\boldsymbol{\mu}}\hat{\zeta}^{r}_{\mathbf{n}+\boldsymbol{\mu},-\boldsymbol{\mu}}\,.
\end{aligned} \tag{11.130}$$

For a chosen truncation $j_{\max}$ of the SU(2) irreducible representation, $\hat{\zeta}$-rishons are defined as:

$$\hat{\zeta}^{g(r)} = \left[\sum_{j=0}^{j_{\max}-\frac{1}{2}}\sum_{m=-j}^{j}\chi(j,m,g(r))\,|j,m\rangle\left\langle j+\frac{1}{2}, m\mp\frac{1}{2}\right|\right]_F\,, \tag{11.131}$$

where the function $\chi(j, m, \alpha)$ reads

$$\chi(j, m, g(r)) = \sqrt{\frac{j \mp m + 1}{\sqrt{(2j+1)(2j+2)}}} . \tag{11.132}$$

It is possible to show that this construction is indeed compatible with the explicit form of the parallel transport reported in Eq. (11.114).

11.4.3.1 SU(2) Rishon Parity

Being fermions, $\hat{\zeta}$-rishons anti-commute at different orbitals and with matter fields:

$$\left\{\hat{\zeta}^{\alpha}_{\mathbf{n},\mu}, \hat{\zeta}^{\beta}_{\mathbf{n}+\mu,-\mu}\right\} = 0 \qquad \left\{\hat{\zeta}^{\alpha}_{\mathbf{n},\mu}, \hat{\psi}_{\mathbf{n},\beta}\right\} = 0 \qquad \forall \alpha, \beta . \tag{11.133}$$

To satisfy Eq. (11.133), we need to characterize them as fermion operators properly. As discussed in Sect. 11.2.4, we can then introduce the notion of a parity for the SU(2) rishon modes. Similarly to Eqs. (11.73) and (11.74), we define the SU(2) rishon parity operator $\hat{P}_{\zeta}$ with an even $(+1)$ parity sector on *integer* irreps and odd (-1) sector on *semi-integer* ones:

$$\hat{P}_{\zeta} = \mathrm{diag}(+1| -1, -1| +1, +1, +1| \dots) . \tag{11.134}$$

The action of the single rishon inverts the local parity: $\{\hat{P}, \hat{\zeta}^{\alpha}\} = 0$. Correspondingly, the parallel transporter $\hat{U}^{\alpha\beta}_{\mathbf{n},\mu}$ reads:

$$\begin{aligned}
\hat{U}^{\alpha\beta}_{\mathbf{n},\mu} =& i\sigma^{y}_{\alpha\gamma}\hat{\zeta}^{\gamma\dagger}_{\mathbf{n},\mu}\hat{\zeta}^{\beta\dagger}_{\mathbf{n}+\mu,-\mu} + i\sigma^{y}_{\beta\gamma}\hat{\zeta}^{\alpha}_{\mathbf{n},\mu}\hat{\zeta}^{\gamma}_{\mathbf{n}+\mu,-\mu} \\
=& + i\sigma^{y}_{\alpha\gamma}\left(\hat{\zeta}^{\gamma\dagger}_{\mathbf{n},\mu} \cdot \hat{P}_{\zeta,\mathbf{n},\mu}\right) \otimes \hat{\zeta}^{\beta\dagger}_{\mathbf{n}+\mu,-\mu} + i\sigma^{y}_{\beta\gamma}\left(\hat{\zeta}^{\alpha}_{\mathbf{n},\mu} \cdot \hat{P}_{\zeta,\mathbf{n},\mu}\right) \otimes \hat{\zeta}^{\gamma}_{\mathbf{n}+\mu,-\mu} .
\end{aligned} \tag{11.135}$$

Notice that the right-hand side of Eq. (11.135) preserves the SU(2) link parity, as desired.

11.4.3.2 SU(2) Rishon Algebra

Instead of relying on two separate set of SU(2) generators, $\hat{L}^{a}$ and $\hat{R}^{a}$, $\hat{\zeta}$-rishons have a unique gauge transformation algebra. The generators of SU(2) gauge rotations upon the $\hat{\zeta}$-rishon space read:

$$\hat{\mathcal{T}}^{a}_{j_{\max}} = \bigoplus_{j=0}^{j_{\max}} S^{a}_{j} = \mathrm{diag}(S^{a}_{0}, S^{a}_{1}, \dots S^{a}_{j_{\max}}) . \tag{11.136}$$

By construction, $\hat{\zeta}$ operators are SU(2) covariant, as they transform as follows:

$$\left[\hat{\zeta}^{\alpha}, \hat{\mathcal{T}}^{a}\right] = \frac{1}{2}\sum_{\beta} \hat{\sigma}^{a}_{\alpha\beta}\hat{\zeta}^{\beta} \qquad \left[\hat{\mathcal{T}}^{a}, \hat{\zeta}^{\dagger}_{\alpha}\right] = \frac{1}{2}\sum_{\beta} \hat{\zeta}^{\beta\dagger}\hat{\sigma}^{a}_{\beta\alpha} \,. \tag{11.137}$$

Moreover, $\hat{\mathcal{T}}^{a}$ is genuinely *local*: for $\forall \mathbf{n} \neq \mathbf{n}'$, $\forall \boldsymbol{\mu} \neq \boldsymbol{\mu}'$, and $\forall \alpha \in \{r, g\}$:

$$[\hat{\mathcal{T}}^{a}_{\mathbf{n},+\mu}, \hat{\zeta}^{\alpha}_{\mathbf{n}',+\mu'}] = [\hat{\mathcal{T}}^{a}_{\mathbf{n},+\mu}, \hat{\psi}_{\mathbf{n}',\alpha}] = 0 \,. \tag{11.138}$$

We easily recover the left and right generators of the gauge field at link $(\mathbf{n}, \boldsymbol{\mu})$ as:

$$\hat{L}^{a}_{\mathbf{n},+\mu} = \hat{\mathcal{T}}^{a}_{\mathbf{n},+\mu} \otimes \mathbb{1}_{\mathbf{n}+\mu,-\mu} \qquad \hat{R}^{a}_{\mathbf{n},+\mu} = \mathbb{1}_{\mathbf{n},+\mu} \otimes \hat{\mathcal{T}}^{a}_{\mathbf{n}+\mu,-\mu} \,. \tag{11.139}$$

Since in the SU(2) group the fundamental and the anti-fundamental representations coincide, the rishon formalism is meaningful as long as the quadratic Casimir operator of the two sides of the link coincide as in Eq. (11.55):

$$\left|\hat{L}^{a}_{\mathbf{n},+\mu}\right|^{2} = \left|\hat{R}^{a}_{\mathbf{n}+\mu,-\mu}\right|^{2} . \tag{11.140}$$

Thanks to Eq. (11.140), the two rishons of the link are in the same SU(2) irrep, and the parallel transport in Eq. (11.127) coincides with the one in Eq. (11.114). Correspondingly, the Casimir operator of the $(\mathbf{n}, \boldsymbol{\mu})$ link in Eq. (11.55) can be expressed as:

$$\hat{E}^{2}_{\mathbf{n},\mu} = \frac{1}{2}\left[\left|\hat{L}^{a}_{\mathbf{n},+\mu}\right|^{2} + \left|\hat{R}^{a}_{\mathbf{n}+\mu,-\mu}\right|^{2}\right] = \frac{1}{2}\left[\left|\hat{\mathcal{T}}^{a}_{\mathbf{n},+\mu}\right|^{2} + \left|\hat{\mathcal{T}}^{a}_{\mathbf{n}+\mu,-\mu}\right|^{2}\right], \tag{11.141}$$

which looks explicitly symmetric under link reversal.

11.4.4 Constructing Dressed Site Operators

We have then all the ingredients to build a *dressed-site* compact representation. In the case of a 2D lattice, one possible pictorial description of *dressed-site* states reads:

$$\left| \begin{matrix} & \hat{\zeta}_{\mathbf{n},+\mu y} & \\ \hat{\zeta}_{\mathbf{n},-\mu x} & \left(\hat{\psi}_{\mathbf{n},r}\hat{\psi}_{\mathbf{n},g}\right) & \hat{\zeta}_{\mathbf{n},+\mu x} \\ & \hat{\zeta}_{\mathbf{n},-\mu y} & \end{matrix} \right\rangle, \qquad \text{where} \qquad \left| \begin{matrix} & 5 & \\ 2 & (0,1) & 4 \\ & 3 & \end{matrix} \right\rangle \tag{11.142}$$

is a possible internal ordering to be used as in Eq. (11.71) when constructing composite operators out of matter fields and rishons inside the dressed site.

We can then rewrite the SU(2) Yang-Mills Hamiltonian in Eq. (11.111)–(11.112) via rishon modes.

11.4.4.1 Arrival Operators

Let us start with the hopping Hamiltonian term. Discarding all the pre-factors, we can focus on:

$$
\begin{aligned}
\hat{H}^{\text{hopping}}_{\mathbf{n},\boldsymbol{\mu}} &= \sum_{\alpha,\beta} \hat{\psi}^{\dagger}_{\mathbf{n},\alpha} \hat{U}^{\alpha\beta}_{\mathbf{n},\boldsymbol{\mu}} \hat{\psi}_{\mathbf{n}+\boldsymbol{\mu},\beta} \\
&= \sum_{\alpha,\beta} \hat{\psi}^{\dagger}_{\mathbf{n},\alpha} \left[\hat{\zeta}^{\alpha}_{A,\mathbf{n},\boldsymbol{\mu}} \hat{\zeta}^{\beta\dagger}_{B,\mathbf{n}+\boldsymbol{\mu},-\boldsymbol{\mu}} + \hat{\zeta}^{\alpha}_{B,\mathbf{n},\boldsymbol{\mu}} \hat{\zeta}^{\beta\dagger}_{A,\mathbf{n}+\boldsymbol{\mu},-\boldsymbol{\mu}} \right] \hat{\psi}_{\mathbf{n}+\boldsymbol{\mu},\beta} \\
&= \left[\hat{Q}^{\dagger}_{A,\mathbf{n},\boldsymbol{\mu}} \hat{Q}_{B,\mathbf{n}+\boldsymbol{\mu},-\boldsymbol{\mu}} + \hat{Q}^{\dagger}_{B,\mathbf{n}} \hat{Q}_{A,\mathbf{n}+\boldsymbol{\mu},-\boldsymbol{\mu}} \right],
\end{aligned}
$$

where we defined two species of *arrival* operators:

$$
\hat{Q}^{\dagger}_{A,\mathbf{n},\boldsymbol{\mu}} = \sum_{\alpha} \hat{\psi}^{\dagger}_{\mathbf{n},\alpha} \hat{\zeta}^{\alpha}_{A,\mathbf{n},\boldsymbol{\mu}} \qquad \hat{Q}^{\dagger}_{B,\mathbf{n},\boldsymbol{\mu}} = \sum_{\alpha} \hat{\psi}^{\dagger}_{\mathbf{n},\alpha} \hat{\zeta}^{\alpha}_{B,\mathbf{n},\boldsymbol{\mu}} . \tag{11.143}
$$

These operators' practical construction must be consistent with the internal order in Eq. (11.142).

11.4.4.2 Matter Number Density Operators

Similarly, number density operators are expressed as

$$
\hat{N}_{\mathbf{n},r} = \psi^{\dagger}_{\mathbf{n}} \psi_{\mathbf{n},r} \otimes \mathbb{1}_{\mathbf{n},g} \bigotimes_{\boldsymbol{\mu}} \mathbb{1}_{\mathbf{n},\boldsymbol{\mu}} \qquad \hat{N}_{\mathbf{n},g} = \mathbb{1}_{\mathbf{n},r} \otimes \psi^{\dagger}_{\mathbf{n}} \psi_{\mathbf{n},g} \bigotimes_{\boldsymbol{\mu}} \mathbb{1}_{\mathbf{n},\boldsymbol{\mu}} \tag{11.144}
$$

and give access to other observables like:

$$
\hat{N}_{\mathbf{n},\text{tot}} = \hat{N}_{\mathbf{n},r} + \hat{N}_{\mathbf{n},g} \qquad \hat{N}_{\mathbf{n},\text{pair}} = \hat{N}_{\mathbf{n},r} \hat{N}_{\mathbf{n},g} \qquad \hat{N}_{\mathbf{n},\text{single}} = \hat{N}_{\mathbf{n},\text{tot}} - \hat{N}_{\mathbf{n},\text{pair}} , \tag{11.145}
$$

which respectively measure the total matter density and the corresponding occupancy of pairs or single particles.

11.4.4.3 Casimir Operator

As we assumed via Eq. (11.141) that each of the two $\hat{\zeta}$-rishons equally contributes to the link-electric energy density, we can equivalently define a dressed-site operator summing the Casimir contributions from its attached rishons:

$$
\hat{\Gamma}_{\mathbf{n}} = \frac{1}{2} \sum_{\boldsymbol{\mu}} (\hat{\mathcal{T}}^{a}_{\mathbf{n},\boldsymbol{\mu}})^2 . \tag{11.146}
$$

11.4.4.4 Magnetic Operators

Expressing $\hat{U}$ as in Eq. (11.127), the plaquette term gives rise to 16 contributions:

$$\hat{U}_{\square} = \sum_{\alpha,\beta,\delta,\gamma} \left[\hat{U}^{\alpha\beta}_{\mathbf{n},\mu_x} \hat{U}^{\beta\gamma}_{\mathbf{n}+\mu_x,\mathbf{n}+\mu_x+\mu_y} \hat{U}^{\gamma\delta}_{\mathbf{n}+\mu_x+\mu_y,\mathbf{n}+\mu_y} \hat{U}^{\delta\alpha}_{\mathbf{n}+\mu_y,\mathbf{n}} \right]$$

$$= \sum_{\alpha,\beta,\delta,\gamma} \begin{pmatrix} \ulcorner & \hat{\zeta}^{\delta\dagger}_{B,+\mu x} \hat{\zeta}^{\gamma}_{A,-\mu x} & \urcorner \\ \hat{\zeta}^{\delta}_{A,-\mu y} & & \hat{\zeta}^{\gamma\dagger}_{B,-\mu y} \\ \hat{\zeta}^{\alpha\dagger}_{B,+\mu y} & & \hat{\zeta}^{\beta}_{A,+\mu y} \\ \llcorner & \hat{\zeta}^{\alpha}_{A,+\mu x} \hat{\zeta}^{\beta\dagger}_{B,-\mu x} & \lrcorner \end{pmatrix} + \dots$$

$$\underset{*}{=} -[\begin{pmatrix} C^{AA}_{-\mu y,+\mu x} & — & C^{AA}_{-\mu x,-\mu y} \\ | & & | \\ C^{AA}_{+\mu x,+\mu y} & — & C^{AA}_{+\mu y,-\mu x}] \end{pmatrix} + \dots,$$

where, in $*$, we combined rishons in pairs to form *corner operators* like the following:

$$\hat{C}^{AA}_{\mathbf{n},\mu_1,\mu_2} = \sum_{\alpha} \hat{\zeta}^{\alpha}_{A,\mathbf{n},\mu_1} \hat{\zeta}^{\alpha\dagger}_{A,\mathbf{n},\mu_2} = \sum_{\alpha,\kappa,\kappa'} i\sigma^{y}_{\alpha,\kappa} \hat{\zeta}^{\kappa\dagger}_{B,\mathbf{n},\mu_1} i\sigma^{y}_{\alpha,\kappa'} \hat{\zeta}^{\kappa'}_{B,\mathbf{n},\mu_2}$$

$$\hat{C}^{BB}_{\mathbf{n},\mu_1,\mu_2} = \sum_{\alpha} \hat{\zeta}^{\alpha}_{B,\mathbf{n},\mu_1} \hat{\zeta}^{\alpha\dagger}_{B,\mathbf{n},\mu_2} = \hat{C}^{AA}_{\mathbf{n},\mu_2,\mu_1} \tag{11.147}$$

$$\hat{C}^{AB}_{\mathbf{n},\mu_1,\mu_2} = \sum_{\alpha} \hat{\zeta}^{\alpha}_{A,\mathbf{n},\mu_1} \hat{\zeta}^{\alpha\dagger}_{B,\mathbf{n},\mu_2} = \left(\hat{C}^{BA}_{\mathbf{n},\mu_2,\mu_1} \right)^{\dagger}.$$

As for the previous dressed-site operators, the practical construction of corner operators has to be consistent with the internal ordering in Eq. (11.142).

11.4.4.5 SU(2) Link Symmetry

Within the dressed-site formalism, the condition in Eq. (11.140) requiring the two rishon of the links to display the same Casimir operator (i.e. to be in the same spin shell j) is simply an Abelian $\mathbb{Z}_2$ Link Symmetry. It can be then easily encoded in ED and TN libraries employing symmetries [32, 459].

11.4.4.6 Projecting the Operators on the SU(2) Gauge Invariant Subspace

All the dressed-site operators previously derived must be projected in the SU(2) gauge-invariant basis M. The latter is made by SU(2) singlets and can be determined as the kernel of the Gauss Law operators $\hat{\mathbf{G}}^{a}_{\mathbf{n}}$:

$$\hat{\mathbf{G}}_{\mathbf{n}} \cdot M = \left[\hat{S}^{2}_{\text{matt},\mathbf{n}} + \sum_{\mu} \sum_{a} \hat{\mathcal{T}}^{a}_{\mathbf{n},\mu} \hat{\mathcal{T}}^{a}_{\mathbf{n},\mu} \right] \cdot M = 0, \tag{11.148}$$

where $\hat{\mathcal{T}}^a_\mu$ are the ζ-rishon generators along the μth direction and defined in Eq. (11.136), while $\hat{S}^2_{\text{matt},\mathbf{n}}$ is the matter color density introduced in Eq. (11.110). In the minimal $j_{\max} = \frac{1}{2}$ truncation of SU(2), the gauge-invariant Hilbert space of every dressed site of the full Hamiltonian has 30 gauge invariant states, whereas, restricting to the pure theory, the local Hilbert space is 9-dimensional.

11.4.4.7 The Operative Defermionized Hamiltonian

We are then ready to rewrite the SU(2) lattice Yang-Mills Hamiltonian in Eq. (11.111) making use of dressed-site operators in Eqs. (11.143) and (11.147). Namely, we have:

$$
\begin{aligned}
\hat{H} =& -\frac{c\hbar}{2a_0}\sum_{\mathbf{n}}\Big[\Big[i\Big[\hat{Q}^\dagger_{A,\mathbf{n},+\mu x}\hat{Q}_{B,\mathbf{n}+\mu_x,-\mu x}\Big] \\
&+(-1)^{n_x+n_y}\Big[\hat{Q}^\dagger_{A,\mathbf{n},+\mu y}\hat{Q}_{B,\mathbf{n}+\mu_y,-\mu y}\Big]+(A\rightleftarrows B)\Big]+\text{h.c}\Big] \\
&+m_0c^2\sum_{\mathbf{n}}(-1)^{n_x+n_y}\hat{N}_{\mathbf{n},\text{tot}}+\frac{c\hbar g^2}{4}\sum_{\mathbf{n}}\hat{\Gamma}_{\mathbf{n}} \\
&-\frac{c\hbar}{2a_0g^2}\sum_{\square}\text{Tr}\left[\left[-\frac{\hat{C}^{AA}_{\ulcorner}\,\hat{C}^{AA}_{\urcorner}}{\hat{C}^{AA}_{\llcorner}\,\hat{C}^{AA}_{\lrcorner}}+\frac{\hat{C}^{BB}_{\ulcorner}\,\hat{C}^{AB}_{\urcorner}}{\hat{C}^{AA}_{\llcorner}\,\hat{C}^{AB}_{\lrcorner}}+\ldots\right]+\text{h.c}\right].
\end{aligned}
\tag{11.149}
$$

Not surprisingly, Eq. (11.149) is completely *bosonic*, as all the dressed-site operators are made out of pairs of fermions (matter field + rishon, pairs of matter fields, or rishon pairs). Then, fermionic degrees of freedom are completely hidden inside each dressed site and there is no more need to face anti-commutation rules (see Fig. 11.4d).

11.4.5 Hardcore-Gluon Model: Minimally Truncated Gauge-Fields

As an example, we consider the *hardcore-gluon* approximation, corresponding to the smallest non-trivial representations of the gauge fields, obtained truncating the Casimir up to $j_{\max} = \frac{1}{2}$. In analogy to cold quantum gases, the label *hardcore-gluon* aims to stress that the only accessible local configurations are those states reachable from the bare vacuum with at most one application of $\hat{U}^{\alpha\beta}$. Namely, we consider $(0\otimes 0)\oplus(\frac{1}{2}\otimes\frac{1}{2})$ as the gauge field space, where (j) is the irreducible representation of SU(2) [420, 421, 423, 471, 472]. Such a truncation is enough to guarantee the contribution of all the terms in the Hamiltonian Equations (11.111) and (11.112), and provides a good approximation of the untruncated low energy physics in the strong coupling limit $g \gg 1$, where the gauge field energy term dominates the Hamiltonian. Remarkably, it has been succesfully adopted in [399, 401, 402] to extract interesting physics in and out of equilibrium.

Within the *hardcore-gluon* approximation, the gauge-link Hilbert space reduces to the states $|j, m\rangle$ with $j_{\max} = \frac{1}{2}$. In analogy to the fundamental representation of SU(2), we can parametrize the $m = \pm\frac{1}{2}$ values with the colors r, g respectively:

$$\mathcal{H}_{\text{gauge}} = \{|0, 0\rangle, |r, r\rangle, |r, g\rangle, |g, r\rangle, |g, g\rangle\}. \tag{11.150}$$

We can then define the corresponding versions of the truncated gauge fields. As for the quadratic Casimir operator in Eq. (11.55), we have:

$$\hat{E}^2 = \frac{3}{4}\text{diag}(0| + 1, +1, +1, +1). \tag{11.151}$$

Correspondingly, the truncated parallel transport reduces to [446]:

$$\hat{U}^{\alpha\beta} = \frac{1}{\sqrt{2}}\left(\begin{array}{c|cccc} 0 & +\delta_{\alpha r}\delta_{\beta g} & -\delta_{\alpha r}\delta_{\beta r} & +\delta_{\alpha g}\delta_{\beta g} & -\delta_{\alpha g}\delta_{\beta r} \\ \hline -\delta_{\alpha g}\delta_{\beta r} & 0 & 0 & 0 & 0 \\ -\delta_{\alpha g}\delta_{\beta g} & 0 & 0 & 0 & 0 \\ +\delta_{\alpha r}\delta_{\beta r} & 0 & 0 & 0 & 0 \\ +\delta_{\alpha r}\delta_{\beta g} & 0 & 0 & 0 & 0 \end{array}\right), \tag{11.152}$$

where the $1/\sqrt{2}$ factor ensures that the hopping term preserves the state norm on its support. Similarly, also the rishon-decomposition in Eq. (11.127) can be simplified as follows:

$$\hat{U}^{\alpha\beta}_{\mathbf{n},\boldsymbol{\mu}} = \hat{\zeta}^{\alpha}_{\mathbf{n},\boldsymbol{\mu}}\hat{\zeta}^{\beta\dagger}_{\mathbf{n}+\boldsymbol{\mu},-\boldsymbol{\mu}}, \tag{11.153}$$

where $\hat{\zeta}$-rishons in Eq. (11.131) reduce to:

$$\begin{aligned} \hat{\zeta}_r &= \frac{|0\rangle\langle r| + |g\rangle\langle 0|}{\sqrt[4]{2}} = \frac{1}{\sqrt[4]{2}}\left(\begin{array}{c|cc} 0 & 1 & 0 \\ \hline 0 & 0 & 0 \\ 1 & 0 & 0 \end{array}\right)_F \\ \hat{\zeta}_g &= \frac{|0\rangle\langle g| - |r\rangle\langle 0|}{\sqrt[4]{2}} = \frac{1}{\sqrt[4]{2}}\left(\begin{array}{c|cc} 0 & 0 & 1 \\ \hline -1 & 0 & 0 \\ 0 & 0 & 0 \end{array}\right)_F, \end{aligned} \tag{11.154}$$

with the corresponding parity operator

$$\hat{P}_{\zeta} = +|0\rangle\langle 0| - (|r\rangle\langle r| + |g\rangle\langle g|) = \text{diag}(+1| - 1, -1) \tag{11.155}$$

and SU(2) rishon-generators

$$\hat{T}^x_{1/2} = \frac{1}{2}\left(\begin{array}{c|cc} 0 & & \\ \hline & 0 & 1 \\ & 1 & 0 \end{array}\right) \quad \hat{T}^y_{1/2} = \frac{1}{2}\left(\begin{array}{c|cc} 0 & & \\ \hline & 0 & -i \\ & i & 0 \end{array}\right) \quad \hat{T}^z_{1/2} = \frac{1}{2}\left(\begin{array}{c|cc} 0 & & \\ \hline & 1 & 0 \\ & 0 & -1 \end{array}\right). \tag{11.156}$$

By definition, the spin-Hilbert space of every side of the link $(\mathbf{n}, \boldsymbol{\mu})$ is 3-dimensional:

$$\mathcal{H}_{\mathbf{n},+\mu} = \{|0\rangle\,, |r\rangle\,, |g\rangle\} = \mathcal{H}_{\mathbf{n}+\mu,-\mu}. \tag{11.157}$$

Correspondingly, the Hilbert space of the whole link $\mathcal{H}_{\text{link}} = \mathcal{H}_{\mathbf{n},+\mu} \otimes \mathcal{H}_{\mathbf{n}+\mu,-\mu}$ has 9 states. To recover the original 5-dimensional space in Eq. (11.150), we must constrain the left and right rishons on each link to be in the same spin shell j by selecting the even parity sector of a local $\mathbb{Z}_2$ symmetry defined in Eq. (11.140).

11.4.5.1 Low-Dimensional Hardcore-Gluon SU(2) LGTs

Combining the hardcore-gluon Hilbert space with flavorless SU(2) dynamical matter, we can easily build gauge-invariant bases that are suitable for TN and quantum simulations.

For instance, the (1+1)D realization of the Hamiltonian in Eq. (11.111) in the *hardcore-gluon* approximation can be simulated with a 6-dimensional logical basis [401, 402] which is tailored for TN simulations as well as to a six-level trapped-ion qudit quantum processor [473]. In [402], the authors discuss the experimental feasibility of such a model using generalized Mølmer-Sørensen gates to simulate the dynamics with an efficient and scalable shallow circuit. Furthermore, numerical simulations of the dynamics manifest physically-relevant properties specific to non-Abelian field theories, such as baryon excitations [402], and compelling non-ergodic behaviors compatible with quantum many-body scars [401] (see also Fig. 11.5).

In the (2+1)D scenario, where the logical site is structured as in Eq. (11.142), the previous the *hardcore-gluon* approximation yields a 30-dimensional gauge-invariant local Hilbert space (see Fig. 11.6), which has been adopted for groundstate TTN simulations in [399]. In this study, the authors characterize the phase diagram

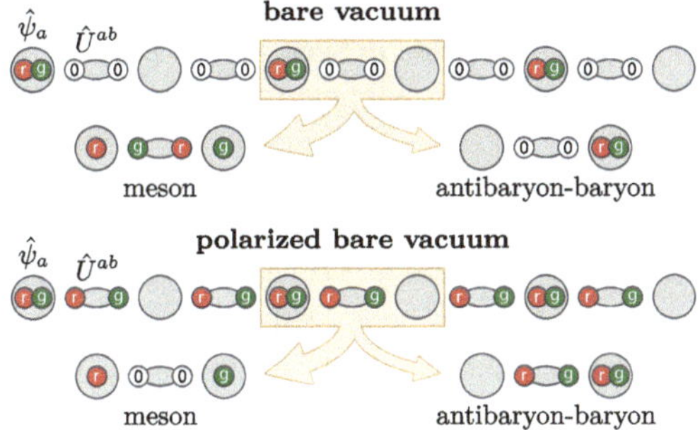

Fig. 11.5 Sketched representation of the baryon and meson formation occurring in the *hardcore-gluon* (1+1)D model developed in Sect. 11.4.5 for SU(2). Both the initial states can be easily written as product states in the corresponding 6-dimensional gauge-invariant local basis and are constructed out of a frozen staggered matter configuration with (polarized vacuum) and without (bare vacuum) gauge fields. (Figure reprinted with permission from Ref. [401] under CC-BY-4.0 license)

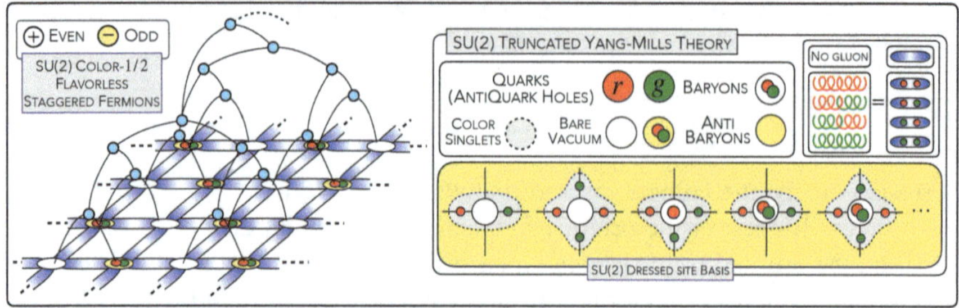

Fig. 11.6 TTN approach to (2+1)D hardcore-gluon SU(2) Yang-Mills LGT. Lattice sites host flavorless SU(2)- color-1/2 fermionic fields (red and green) in a staggered configuration (white and yellow). Lattice (blue) links describe gauge degrees of freedom from a 5-dimensional truncated Hilbert space (hardcore-gluon approximation). SU(2) Gauss Law is implemented at each lattice site. (Figure reprinted with permission from Ref. [399] under CC-BY-4.0 license)

of the model at zero and finite baryon number densities as a function of the quark bare mass and color charge. At intermediate system sizes, a liquid phase of quark-pair bound-state quasiparticles (baryons) emerges, with a finite mass towards the continuum limit. Remarkably, such a scheme allows for detecting interesting phenomena at the transition boundary where color-electric and color-magnetic terms are maximally frustrated: traces of potential deconfinement, and signatures of a possible topological order.

11.4.6 Scaling of the Local Basis Dimension

In general, the larger the gauge field truncation (i.e. the larger the spin-j representation), the larger the effective dressed-site Hilbert space. In Table 11.1, we list the dressed-site dimension $d = \dim \mathcal{H}_{\text{dress}}$ associated with the first few nontrivial gauge truncations of U(1) and SU(2), detailed in Sect. 11.4 in $D = 2, 3$ space dimensions:. Both the considered LGTs include dynamical matter, represented by one fermionic

Table 11.1 Dressed site Hilbert space dimension d for increasing number ℓ of allowed electric energy density levels in some (2+1)D and (3+1)D paradigmatic LGTs with dynamical matter and gauge groups U(1) and SU(2). Table from [396]

	d			
	$(2+1)$-dimensions		$(3+1)$-dimensions	
ℓ	U(1)	SU(2)	U(1)	SU(2)
1	35	30	267	178
2	165	168	3437	3670
3	455	600	18487	35280
4	969	1650	64953	214958
5	1771	3822	177155	967466
6	2925	7840	408421	3509062
7	4495	14688	835311	10828494
8	6545	25650	1561841	29473038

field multiplet in the fundamental representation of the gauge group. The ℓ-th row of Table 11.1 is obtained keeping only the first ℓ nonzero electric energy levels (i.e., using the ℓ-th smallest quadratic Casimir eigenvalue as cutoff Θ, see Sect. 11.2.1). As Table 11.1 shows, the local dimension increases rapidly with ℓ. For (3+1)D non-Abelian LGTs, $d \sim O(10^4)$ is reached already within the first two truncations, making the study of the untruncated limit prohibitive.

Differently from the models typically encountered in condensed matter physics, the local dimension of LGTs can thus be a limiting factor for Tensor Network simulations I especially when d becomes comparable to commonly used TN bond dimensions ($100 \lesssim \chi \lesssim 500$ for TTN). In these cases, crucially for high-dimensional LGTs, strategies aimed at further compressing the local computational basis are needed [396].

Several numerical analyses suggest that, in some cases, a small-to-moderate truncation of the gauge group is enough for accurately approximating the continuum limits, at least for low-energy states [73,398,474–477]. However, the optimal gauge truncation depends on the Hamiltonian parameters m_0 and g.

In order understand this behavior, we focus on a single QED plaquette in open boundary conditions, as it provides the minimal setting allowing for both electric and magnetic effects. Then, to characterize the convergence in the gauge truncation ℓ, we consider a candidate observable O and compute its ground state expectation value $\langle \hat{O} \rangle_\ell$ for increasing ℓ. We iterate until the relative deviation between consecutive truncations drops below some threshold ϵ_{trunc}: $|\langle \hat{O} \rangle_{\ell^*} - \langle \hat{O} \rangle_{\ell^*-1}| < \epsilon_{\text{trunc}} |\langle \hat{O} \rangle_{\ell^*}|$ for some ℓ^*. Figure 11.7a shows the minimal truncation ℓ^* at which the magnetic energy operator $\hat{O} = \text{Re}\,\hat{U}_\square$ is converged to $\epsilon_{\text{trunc}} = 10^{-5}$. We explore a grid of model parameters, whose extent has been chosen according to standard MC literature [355, 385–387, 391, 392, 407, 478–483]. $\text{Re}\,\hat{U}_\square$ is used as a benchmark due to its relevance in the weak coupling regime, where the continuum limit of $D < 3$ lattice QED is located [385]. As Fig. 11.7b shows, ℓ^* depends heavily on the coupling, growing asymptotically like $\ell^* \sim g^{-1}$ as g is decreased, while m_0 plays almost no role. An analogous inverse dependence of the minimal gauge truncation on the coupling is expected for non-Abelian LGT in arbitrary dimensions. Moreover, the continuum limit of non-Abelian LGTs in $D \leq 3$ is also expected at $g \to 0$ [385], further substantiating the need to compress the local dimension in TN simulations of LGT whenever extrapolation to the continuum is in order. Developements along this directions represent an ongoing research line we are considering towards more accurate simulations.

Apart from the growth of the local dimension, extrapolation to the continuum is further complicated by the fact that the continuum limit of a lattice quantum field theory corresponds to a critical point of the underlying lattice model [364]. Close to criticality, quantum correlations are boosted and violations of the entanglement area law are expected [46]. The higher entanglement entropy in the proximity of the continuum ($g, m_0 \ll 1$ regime) is already captured by the single-plaquette analysis of Fig. 11.7c (see [399] for the SU(2) case). Continuum limits of LGTs are thus an area of potential advantage for quantum computation over classical methods, as the former is not limited by entanglement. Nonetheless, quantum computation is also

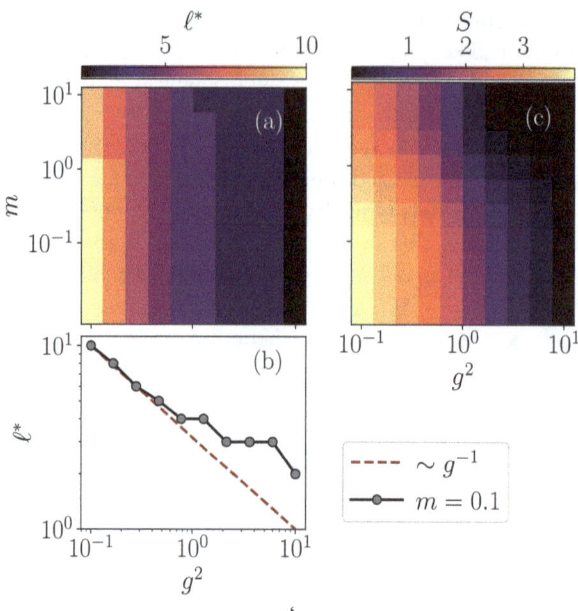

Fig. 11.7 Exact diagonalization of a QED plaquette for a grid of masses $m \in [10^{-2}, 10^1]$ and couplings $g^2 \in [10^{-1}, 10^1]$. (**a,b**) Minimal gauge truncation ℓ^* required to reach a precision $\epsilon_{\text{trunc}} = 10^{-5}$ in the magnetic energy $\langle \text{Re}\, \hat{U}_\square \rangle$. (**c**) Corresponding entanglement entropy $S(\ell^*)$ associated with a symmetric bipartition of the plaquette. (Figure from [396])

affected by the need to relax gauge truncations when $g^2 \to 0$, either by increasing the number of qubits used to encode a dressed site, which is at least $\lceil \log_2(d) \rceil$, or by using hybrid devices with both qubits and bosonic modes [484].

11.5 Summary

In this chapter, we provided a detailed exploration of Hamiltonian Lattice Gauge Theories (LGTs) with an emphasis on the discretization process for both fermionic and gauge fields. We began by reviewing the lattice discretization of fermions, highlighting the staggered fermion approach to mitigate the fermion doubling problem. The discretization of gauge fields was then discussed, where the continuous gauge fields were replaced by parallel transporters, ensuring lattice gauge invariance. We detailed how the discretization of the electric and magnetic components of the gauge fields, leading to the construction of the lattice Hamiltonian. The dimensional analysis emphasized the role of the gauge coupling and its scaling with lattice spacing.

The second part of the chapter introduced the dressed-site formalism, which effectively integrates the gauge fields into matter sites, reducing the complexity of the gauge theory and making it more amenable to TN algorithms and quantum simulations. This formalism has been thoroughly applied to two exemplary cases, such as the Abelian U(1) LGTs in Sect. 11.3 and the non-Abelian SU(2) Yang-Mills in Sect. 11.4. The techniques and concepts introduced here for these two models are foundational for understanding the numerical methods and physical results presented and investigated in several works [282, 397–403, 416] and represents a powerful scheme for any future numerical but also quantum simulations of LGTs [396].

11.6 Problems

1. Use the dressed site formalism developed in Sect. 11.3 to build the gauge invariant basis of the (1+1)D QED Hamiltonian in the first non-trivial integer truncation of the gauge fields, i.e. with $j_{\mathrm{max}} = 1$.
2. Apply the same strategy and build the corresponding basis in the (2+1)D case.
3. Compare the obtained bases with the ones available on [459].
4. Use the optimal TN ansatz available in [1] and search for the groundstate of the two Hamiltonians at $m_0 = 1$ and $g = 1$.

Out-of-equilibrium Processes 12

Simone Montangero

Out-of-equilibrium processes are ubiquitous; they surround us and life is possible because of temporary out-of-equilibrium fluctuations of the otherwise equilibrium (dead) state. Indeed, any system in the real world is in contact with the rest (the environment) and subject to changing conditions (temperature, interaction strength, etc.) which drive the system out of its equilibrium state. Moreover, any useful process, from information processing to energy harvesting, is based on the fact that a system is driven out of equilibrium and goes into another (possibly equilibrium) state. This is true also at the quantum level and, as we will see in this chapter, it is of tremendous interest to be able to study systems that change their state and to understand when this happens, how this can be efficiently driven, and to predict the result of such transformation. As we have seen, this is now possible in many scenarios for one-dimensional many-body quantum systems using TN methods introduced in the previous chapters.

In particular, the typical scenario is that of a system driven from a somehow trivial state to a more complex one. Many different scenarios that have been recently subject to intense studies can be seen as the change of a system state, from a simple, initial, given one to a more complex but significant state. The first example of such processes is the preparation of most experimental setups: typically the system is found in some equilibrium state from which another more interesting state has to be reached. A relevant instance of this scenario is any quantum information protocol and, in particular, a quantum computation: a register of qubits has to be driven from a trivial state (e.g., $|00000\rangle$) to a state which encodes the problem solution. Similarly, a quantum simulation is nothing else that a system evolving from an initial state into another less trivial one, for example from an insulating state to a superconducting one. This process is also an instance of a more general class of processes, that is, the

S. Montangero (✉)
Università di Padova, Padova, Italy
e-mail: simone.montangero@unipd.it

T. Felser, S. Montangero (eds.), *Introduction to Tensor Network Methods*,
Graduate Texts in Physics, https://doi.org/10.1007/978-3-032-17635-6_12

crossing of a quantum phase transition (an adiabatic quantum computation). As most of the hard problems in computer science can be recast in an adiabatic computation form [485] and adiabatic computation is equivalent to the circuit-base one [486], their study is of outmost importance. Finally, all these transformations can also be studied in the presence of an environment, and, in particular, it has been shown that it is possible to engineer the environments to achieve interesting states [487–489].

In this chapter, we introduce the adiabatic theorem on which adiabatic computation is based on. Then, we present the Kibble-Zurek mechanism, a powerful theory that allows predicting, under very general assumptions, the error introduced when the adiabatic condition is slightly violated. We present some approaches to go beyond the adiabatic one, and we briefly review some results on out-of-equilibrium open quantum systems. Finally, we briefly mention some other directions which have been recently explored, related to problems in high-energy physics, open systems and complexity.

12.1 Adiabatic Quantum Computation

In this section, we introduce the *adiabatic quantum computation* or *quantum annealing*. We start presenting the adiabatic theorem, the theoretical argument that enables the adiabatic quantum computation approach. We then present the simplest scenario where the adiabatic theorem can be applied, the Landau-Zener crossing and end the section with an overview of the possible applications of such method to study the quantum matter and classical hard problems.

12.1.1 Adiabatic Theorem

The adiabatic theorem states that if a time-dependent Hamiltonian $H(t)$ varies slow enough and the system is prepared in an eigenstate of the initial Hamiltonian $|\psi(0)\rangle = |E_i(0)\rangle$ (e.g. the ground state), then the system remains in the correspondent time-dependent eigenstate for the whole evolution, i.e. $|\psi(t)\rangle = |E_i(t)\rangle$, $\forall t$. The idea on which the adiabatic theorem is based is highly intuitive and is not different from a classical one: if the system is in its ground state (e.g., a particle at the minimum of a potential), slowly varying the conditions (moving the potential or deforming it) will not excite it.

A simple intuitive derivation of the above theorem can be obtained as follows [490]: given the unitary transformation $U(t)$ that diagonalizes at every time the Hamiltonian, $U^\dagger(t)H(t)U(t) = D(t)$, the time-dependent Schrödinger Eq. (3.32) can be rewritten in the diagonal basis $|\phi(t)\rangle = U^\dagger(t)|\psi(t)\rangle$ as

$$i\hbar\frac{\partial|\phi(t)\rangle}{\partial t} = D(t)|\phi(t)\rangle - i\hbar U^\dagger(t)\frac{\partial U(t)}{\partial t}|\phi(t)\rangle; \tag{12.1}$$

that is, the net effect of the time-dependent basis we have introduced is the second term on the right hand side, appearing as a correction to the Schrödinger equation. However, if the Hamiltonian is slowly varying, also the change of basis $U(t)$ will be only slowly time-dependent, and thus its derivative will be small. The adiabatic approximation consists precisely in neglecting the second term in the right hand side of Eq. (12.1), which results in a set of decoupled differential equations for the amplitude of probabilities of each eigenvector. They remain constant in time with only a varying time-dependent phase: if the system starts in a single eigenvector, its occupation probability will remain constantly one, i.e. if the system starts in $|E_0(0)\rangle$, it will remain in $|E_0(t)\rangle\ \forall t$.

Although the result presented above is already very powerful, clearly an important piece of information is missing: one should quantify what "slowly" means and possibly compute the perturbative corrections to the adiabatic approximation. This analysis can be performed [491], resulting in a bound for the probability of begin away from the ground state after starting in the initial ground state $p(t)$ of

$$p(t) \leq 4 \sum_{k \neq 0} \max_{s' \in [0, s(t)]} |a_{k0}(s')|^2 \delta^2 + O\left(\delta^3\right), \tag{12.2}$$

where the Hamiltonian $H(s(t))$ changes in time according to a parametric function $s(t)$, such that $ds/dt = \delta \cdot v(s(t))$, and δ sets the scale of parameter variation. The transition amplitudes from the ground state to the k-th ($k > 0$) normalized instantaneous eigenstate are given by

$$a_{k0}(s) = \hbar \frac{\langle \tilde{E}_k(s) | dH/ds | \tilde{E}_0(s) \rangle}{\Delta_k(s)^2}, \tag{12.3}$$

where $\Delta_k(s) = E_k(s) - E_0(s)$ and we have set all eigenvectors free phases such that the final accumulated Berry phases are zero [491]. In conclusion, parametrizing the Hamiltonian change as

$$H(s(t)) = s(t) H_0 + (1 - s(t))\, H_1 \tag{12.4}$$

with $s(0) = 0$ and $s(T_f) = 1$, one can show that the process is adiabatic (error probability $p(t) < \delta^2$) if

$$a(s) v(s) \leq 1, \tag{12.5}$$

where $a(s)$ can be bounded as

$$a(s) \leq 2\hbar \frac{||H(T_f) - H(0)||}{\min_s (\Delta_1(s))^2}; \tag{12.6}$$

which results in an adiabatic running time of

$$T_{ad} \propto \frac{1}{\min_s(\Delta_1(s))^2}. \tag{12.7}$$

The previous expression is only an upper bound for the timescale of an adiabatic process: indeed, it is possible to adapt the velocity to the instantaneous gap performing a global adiabatic optimization that allows for a speed up [492,493].

12.1.2 Applications

The simplest scenario where the adiabatic theorem can be applied is a paradigmatic scenario, the *Landau-Zener crossing* [494,495], that is, the computation of the final excitation probability in a two level quantum system with a time-dependent diagonal term and constant coupling of the form

$$H(t) = H_0 + H_1(t) = \begin{pmatrix} E_1 & V \\ V & E_0(t) \end{pmatrix} \tag{12.8}$$

where $E_0(t) = E_1 + v\,t$ for $t \in [-T; T]$. Figure 12.1 sketches the time dependence of the energy levels of such system, where easily it can be shown that the minimal gap is given by $\min_t \Delta_1(t) = 2V$. Thus, the adiabaticity condition can be expressed as

$$2V^2/\hbar v \gg 1. \tag{12.9}$$

Independently, the Landau-Zener formula predicts the final excitation probability after the crossing as

$$p = \exp\left[-\frac{2\pi V^2}{\hbar v}\right], \tag{12.10}$$

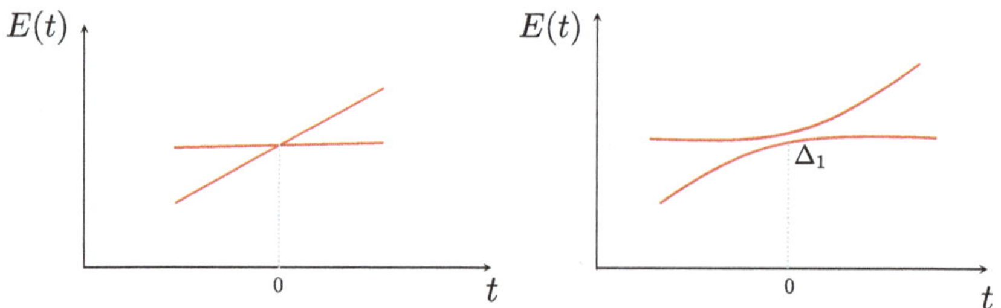

Fig. 12.1 Level crossing for $V = 0$ (left) and for $V \neq 0$ (right) in the Landau Zener crossing defined by the Hamiltonian in Eq. (12.8)

which is in perfect agreement with the condition expressed in Eq. (12.9): if the adiabaticity condition holds, the argument of the exponential is large and correspondingly the excitation probability p is small, that is, an adiabatic crossing occurred.

More generally, the adiabatic condition of Eq. (12.5) plays a crucial role in adiabatic quantum computation or simulated quantum annealing [496–498]. The scenario is similar to that presented above: a system is prepared in the ground state of an initial Hamiltonian H_0, then a competing term H_1 is switched on while the initial one is switched off, as in Eq. (12.4). The idea is that while H_0 is a Hamiltonian whose ground state is easy to prepare or compute, the ground state of the Hamiltonian H_1 encodes the solution of a difficult problem. If the process is adiabatic, the crossing defined by Eq. (12.4) returns the desired solution. In particular, it can be shown that this computation architecture is equivalent to universal quantum computing [486], and for example, that Grover's algorithm to search in unstructured databases can be recast in this language [492]. Moreover, it can be shown that most hard problems in computer science can be recast as an adiabatic quantum computation [485]. An example will serve to clarify this last point: the *partitioning problem* is an NP-complete problem defined starting from a set of numbers $n_1, \ldots n_N$. The question is if it is possible to bipartite the set of numbers in two sets of equal values. The adiabatic quantum computation version of such simple to state but extremely hard to solve problem stems from the definition of the Hamiltonian $H_1 = (\sum_i n_i \sigma_i^z)^2$, where the n_i form the set of numbers to be bipartited and σ_i^z the standard Pauli matrices. Defining, e.g., $H_0 = \sum_i \sigma_i^x$ and performing the adiabatic protocol one could find the energy E_0 of the ground state of H_1: if it is zero, which lower bounds all possible eigenenergies given that H_1 is a positive operator, the solution to that particular partitioning problem exists and can be read from the ground state. If instead $E_0 > 0$, the partition with equal values does not exist, but the mismatch has been minimized, solving an NP-hard problem.

However, what seems to be a strong attack strategy to extremely hard problems has a flaw: the adiabatic condition cannot be satisfied in the thermodynamical limit, as typically the minimal gap goes to zero either polynomially or even exponentially fast. That is, in the limit of large problem size, errors are unavoidably introduced hindering a clean solution to the problem. Nevertheless, huge efforts to exploit this strategy to the maximum have been performed, and active research both in numerical methods and in developing quantum annealers which could perform such process in the lab is under thriving activity.

Finally, another potentially huge field of application of these ideas is the crossing of a quantum phase transitions: the two competing Hamiltonians are now given by the physics of the system as in Eq. (10.9), and the parameter g is tuned in time, playing the role of the function parametrization $s(t)$. These processes naturally arise in many scenarios, in quantum science and condensed matter (e.g., during the loading of atoms in optical lattices, in superconductors or magnetic materials, etc.). Again, at the thermodynamical limit, the adiabatic condition cannot be satisfied due to the scaling of the critical gap given in Eq. (10.16). However, in this case, the structure of the problem and in particular its classification by means of the universality classes of the quantum phase transition allow us to predict the error introduced by the violation of the adiabaticity through a simple—yet extremely powerful—theory that we introduce in the next section.

12.2 Kibble-Zurek Mechanism

The Kibble-Zurek mechanism has been put forward to predict the final residual excitation energy on top of the ground state when crossing a quantum phase transition in finite time. Indeed, when crossing a gapless critical point of H(t), for any finite time, the adiabatic theorem is violated, and the final excitation probability will be non-zero and, consequently, a finite density of defects in the ordered phase will be produced [500–502]. The Kibble-Zurek mechanism provides a scaling argument for the final density of defects based on the assumption that the dynamics can be divided in either adiabatic or impulsive, according to the distance from the critical point. Its setup follows that introduced in the previous section, where a system starts in a ground state of an initial Hamiltonian which is modified in time according to Eq. (12.4), with the additional constraint that the quench is performed linearly with a typical time-scale τ_Q. In summary, we assume

$$H(t) = \varepsilon(t)H_0 + H_1, \qquad \varepsilon(t) = t/\tau_Q, \qquad t \in [-\tau_Q/2 : -\tau_Q/2]; \qquad (12.11)$$

such that $\varepsilon(t) = g(t) - g_c$ determines the distance from the critical point g_c and the crossing occurs at $t = 0$. As we have seen in the previous chapter, at the critical point the correlation length diverges as in Eq. (10.12), and near the critical point the gap scales with the distance as $\Delta \sim (g - g_c)^{\nu z}$ (see Eq. (10.16)). The system's gap introduced a natural time scale for the system relaxation time, i.e., the time needed to react to changes. For example, the variation of the distance to the critical point corresponds to a change of the correlation length. This typical timescale is nothing more than $\tau_\Delta \sim 1/\Delta = |g(t) - g_c|^{-\nu z} = |\varepsilon(t)|^{-\nu z}$. Note that τ_Δ diverges at the critical point and the thermodynamical limit, that is, it is increasingly harder for the system to adapt to the new condition (increasing its correlations). This important time scale has to be compared with the other process timescale, that is, the rate of change of the order parameter dictated by the quench time scale. The inverse of the relative rate of change of $\varepsilon(t)$ is $\tau_\varepsilon = |\varepsilon/\dot{\varepsilon}| = |\varepsilon\tau_Q|$. Thus, depending on the relation between these two timescales, different dynamics will occur:

$\tau_\Delta \gg \tau_\varepsilon$: the system can easily relax to the new ground state while the system parameter is changed. Thus, it remains in the system instantaneous ground state.
$\tau_\Delta \ll \tau_\varepsilon$: the system is not able to relax to the new ground state while the system parameter is changed, that is, it remains frozen in its initial state while $\varepsilon(t)$ evolves.

The second condition will clearly appear at some point given the divergence of τ_Δ: once again this is equivalent to stating that the adiabatic condition is violated

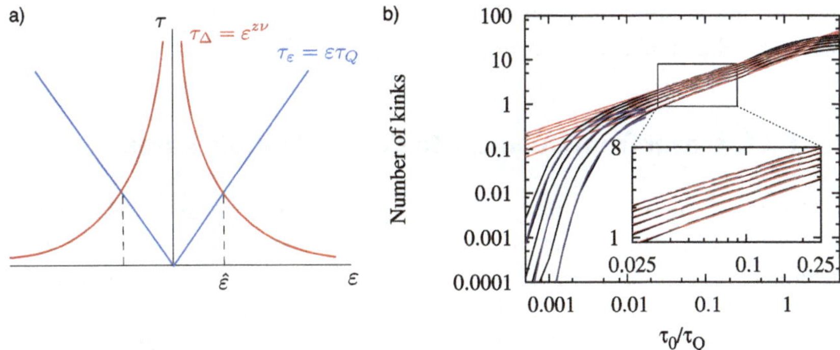

Fig. 12.2 (**a**) The Kibble-Zurek mechanism: equating the two typical system's timescales ($\tau_\varepsilon = |\varepsilon\tau_Q|$ and $\tau_\Delta = |\varepsilon(t)|^{-\nu z}$) it is possible to estimate the distance from the critical point where the system freezes $\hat{\varepsilon}$, the end of the adiabatic dynamics. (**b**) Verification of the Kibble-Zurek mechanism in the quantum Ising model in transverse field: the correlation length (proportional to the inverse of the number of kinks) as a function of the inverse of τ_Q (*black lines*, different lines reports results for different system sizes) scales as predicted by Eq. (12.13) (*red lines*). The *blue lines* follow an exponential scaling, signaling the set up of the Landau-Zener scaling of Eq. (12.10). Figure (**b**) reprinted with permission from Ref. [499].

and thus different ordered domains (of dimension given by the correlation length) will appear, e.g., forming different crystals or magnetic domains. The Kibble-Zurek mechanism is based on the assumption that the whole dynamics can be approximatively described as either adiabatic or frozen. If so, one can individuate the parameter value $\hat{\varepsilon}$ where the system changes from one to the other simply equating the two main timescales. A graphical representation of this comparison is represented in Fig. 12.2a and results in

$$\hat{\varepsilon} \sim \tau_Q^{-\frac{1}{1+\nu z}}. \tag{12.12}$$

While approaching a critical point from negative ε, the system will then be in the ground state until $-\hat{\varepsilon}$, while it will be frozen (i.e. it will not change) after that until $\hat{\varepsilon}$. Thus, around criticality, the system will achieve the maximal correlation length corresponding to that at the ground state in $-\hat{\varepsilon}$ given, using Eq. (10.12), by

$$\hat{\xi} \sim \tau_Q^{\frac{\nu}{1+\nu z}}. \tag{12.13}$$

This powerful expression allows predicting the correlation length that can be achieved driving a system at criticality in finite time only as a function of the critical exponents of the theory. The KZ mechanism has been tested in a variety of quantum toy models at zero temperature, including ordered and disordered systems, as well as for crossing isolated or extended critical regions [499, 503–518]. We report in

Fig. 12.2b the verification of the Kibble-Zurek mechanism in the Ising model in transverse field, obtained by means of DMRG simulations [499].

12.2.1 Crossover from Quantum to Classical Kibble-Zurek

The Kibble-Zurek scaling has been verified in many scenarios. However, its experimental verification in strongly correlated many-body quantum systems at zero temperature remain somehow elusive. Indeed, to achieve a clean agreement between the measured and the predicted scalings very stringent experimental constraints have to be satisfied [511,514,515,520–522]. For example, in a beautiful experiment with trapped ions [510], it has been shown that the scalings found experimentally are those of the classical theory describing the system, and not the quantum ones. Also stimulated by these findings, a full characterization and simulation of the system has been performed in [519], exploiting the fact that trapped ions undergoing the linear to zig-zag structural quantum phase transition (represented in Fig. 12.3a) can be modelled under pretty general assumptions as a ϕ^4 model [523,524],

$$H = \sum_i p_i^2 + \varepsilon(t) y_i^2 + y_i^4 + (y_{i+1} - y_i)^2; \tag{12.14}$$

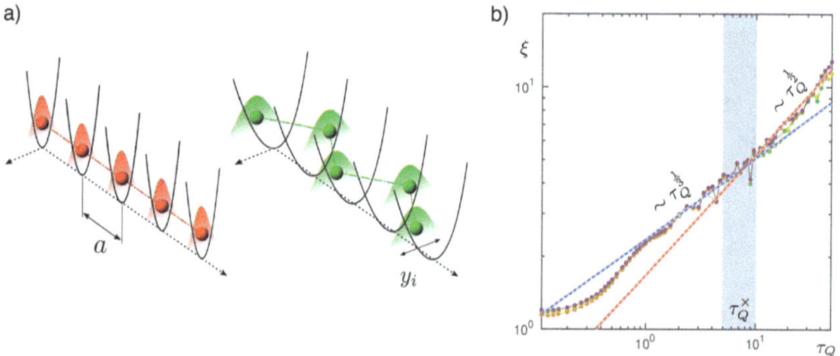

Fig. 12.3 (**a**) Linear to zig-zag quantum phase transition: N ions at fixed distance a are trapped in a linear chain (for which $\langle \sum_i (-1)^i y_i \rangle = 0$, left) for tight confinements, while releasing it beyond a critical threshold results in a zig-zag configuration of minimal energy ($\langle \sum_i (-1)^i y_i \rangle > 0$ right). The system has a $\mathbb{Z}_2$ symmetry, thus it belongs to the Ising universality class. (**b**) Verification of the Kibble-Zurek mechanism in the ϕ^4 model: final correlation length ξ as a function of the quench timescale τ_Q (full circles, different colours denoted different system sizes). For $\tau_Q < \tau_Q^\times$ the scaling is classical $\xi \propto \tau_Q^{1/3}$ (*blue dashed lines*), while for $\tau_Q > \tau_Q^\times$ the quantum scaling is found, $\xi \propto \tau_Q^{1/2}$ (*red dashed line*). Figure (**a**) reprinted with permission from Ref. [519]. Figure (**b**) reprinted with permission from Ref. [508].

where the sum runs over all ions in the chain, y_i, p_i are the transverse displacement (whose expectation value $\langle\sum_i(-1)^i y_i\rangle$ is the order parameter for the transition) and its conjugate canonical momenta such that $[y_i, p_j] = i\hbar\delta_{i,j}$. The parameter $\hbar = \hbar/\sqrt{E_0 m a^2}$ plays the role of an effective $\hbar$, with a being the lattice constant, m the particle mass, and E_0 the typical pairwise interaction energy. It is then possible to simulate its dynamics while crossing the quantum phase transition employing TN methods [508]. It has then put forward the idea that the classical scalings appear because the system freezes far away from the critical point, in the region where quantum fluctuations are still negligible. This region can be individuated by means of the Ginzburg criteria [525, 526], and used to predict where the crossing between classical and quantum scalings occurs, that is

$$\tau_Q^{\times} \sim \frac{\hbar}{\varphi}|\varepsilon|^{-1-z\nu}, \tag{12.15}$$

where φ is determined by the scaling of the gap around the critical point $\Delta_1 \simeq \varphi|\varepsilon|^{z\nu}$. A numerical verification of such relation is reported in Fig. 12.3b [508]. In conclusion, experiments willing to verify the quantum Kibble-Zurek scalings shall be devised to explore the tight region of parameters where not only quantum effects are present (i.e., low temperatures and low decoherence) but also for which the system freezes in the quantum critical region defined by Eq. (12.15).

12.3 Optimal Control of Many-body Quantum Systems

We have seen that crossing a quantum phase transition or performing adiabatic quantum computation generating the minimal amount of errors is a nontrivial task to achieve as the adiabatic strategy is hindered by the gap closure. However, are there different strategies that one can adopt? The answer is positive, and the different strategies can be grouped in two main classes as depicted in Fig. 12.4a: (1) those that aim to speed up the process remaining adiabatic and (2) those that abandon the adiabatic condition and do not require the system to remain in the instantaneous ground state but only to reach the final state with minimal energy.

The former class of fast adiabatic passage can be achieved either adapting the functional time-dependence of the changing parameter $s(t)$ to the instantaneous gap, that is, replacing the bound obtained by the minimization in Eq. (12.6) by its actual time-dependent value and thus speeding up the whole evolution [492]. Another potent strategy is to add terms to the Hamiltonian such that the nonadiabatic terms, which couple the instantaneous ground state with the excited ones, are exactly canceled [527–535]. This class of solutions, which goes under the name of *shortcuts to adiabaticity*, is very powerful and have been adopted in different situations, mostly few-body or where simple descriptions of the system exist (see [531] for a review). However, for many-body quantum systems the required additional terms quickly become pretty complicated (i.e., k-body interacting or long range) and different or hybrid strategies might be required to bring them to the lab [536]. The

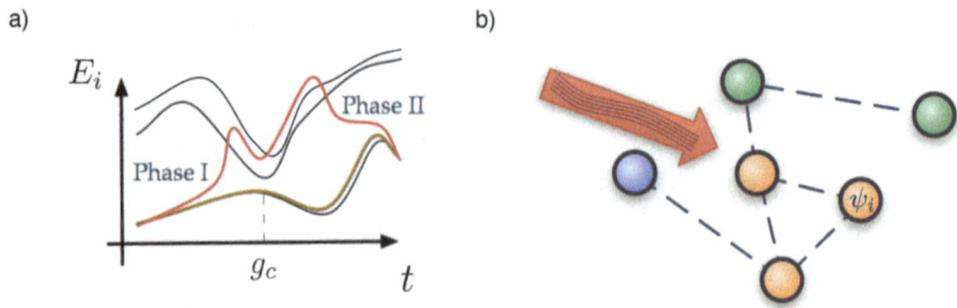

Fig. 12.4 (**a**) Eigenenergies E_i as a function of time (*black lines*) and instantaneous system's energy $E(t)$ of a system crossing a quantum phase transition via an adiabatic passage, either slow or fast by means, e.g. of counteradiabatic terms (*gold line*) or optimal passage (*red line*). (**b**) Sketch of a typical control of a many-body quantum systems problem: (**a**) an external control field (right arrow) acts on the system or a part of it, on external (e.g., particle position) or internal (e.g., atom levels) for a finite time with some finite energy inducing a transformation to a different state with different topology, particle numbers and/or internal states and that might also be a macroscopic superposition of different states of the system

second class of strategies is mainly based on optimal control theory, on which we will focus hereafter.

Quantum optimal control theory solves problems which might be recast in a functional minimization: given an initial state vector $|\psi_0\rangle$ and a dynamical law—in our case Schrödinger equation or its generalization to open and non-linear dynamics as we will see later on—depending on one (or more) external control field $\Gamma(t)$, a figure of merit F is introduced that depends on the control field. The figure of merit might include different terms usually quantifying the property of interest to be achieved and some constraints, the total field strength, the total time of the evolution, etc.. The solution of the optimal control problem is then given by the control field that minimizes the figure of merit, that is, that solves a constrained functional minimization. Finding the solution to this problem is nontrivial also when the number of degrees of freedom is limited and raises many interesting questions. Optimal control of quantum systems has been achieved in many different settings and the conditions under which this is feasible has been analyzed widely regarding controllability of the system, finite resources as for example energy-time relations or due to other physical constraints [493, 537–558].

Less explored, due to the additional complexity in their numerical and analytical descriptions, are the theory and the applications of optimal control to many-body quantum systems. Here, additional dimensions of the problem appear, as sketched in Fig. 12.4b: only recently, the role played by the new naturally arising properties of a many-particle system (e.g., correlations or entanglement in the system, the tensor structure of the system Hamiltonian and the control fields, the coordination number and topology of the connections, and of the scalability with the number of constituents of the system) has been started to be investigated [493, 535, 540, 544, 546, 549, 557–573].

In general, a quantum optimal control problem can be recast in a functional minimization,

$$\min_{\Gamma(t)} F(\Gamma(t), \psi_0, T, \ldots); \tag{12.16}$$

where F is a figure of merit that quantifies the quality of the result of the transformation, $\Gamma(t) \in \mathcal{G}$ is the control field and $\mathcal{G}$ the space of all "physically reasonable" functions defined in $t \in [0, T]$; such that the system Hamiltonian can be written as $\hat{H} = \hat{H}_D + \Gamma(t)\,\hat{H}_C$ where $\hat{H}_D$ is the drift part of the Hamiltonian that cannot be modified (e.g., an interaction between system components) and $\hat{H}_C$ the operator coupled linearly to the control field (e.g., the intensity of a laser of a magnetic field). The final value of the functional also depends on other degrees of freedom, for example, the initial state ψ_0 or the total duration of the evolution T. Moreover, additional constraints might be present such as the maximal available power of the control field or its bandwidth. Typical figures of merit are the fidelity of the final state with some desired goal state, the final expectation value of some observables or other properties such as final correlations of entanglement present in the system [538, 539, 545]. Here and in the following, we formulate the problem in terms of pure states, but it can be straightforwardly generalized to mixed states using density matrices and to (unitary) transformations [537–539, 545].

Many different algorithms have been successfully applied to minimized the functional in Eq. (12.16), mainly explicitly computing the gradient of the functional with respect to $\Gamma(t)$ and performing a gradient descent in the space of all possible control field, as the Krotov or the GRAPE algorithms: Their elegant mathematical formulations and their successful applications can be found in the literature [538, 539, 545]. Here, we concentrate on yet another approach, the dressed chopped random basis (dCRAB) for its simple mathematical formulation and because it has been used to attack optimal control problems on many-body dynamics, such as the optimal crossing of a quantum phase transition, as we will see later on [543, 561, 574].

The dCRAB optimal control algorithm is based on the idea (inspired by tensor network methods) that it might be possible to reduce the space $\mathcal{G}$ in which one search the optimal control field $\Gamma(t)$ apriori, thus extremely simplifying the search. It should be noticed that this approximation is implicitly made in all known numerical algorithms: indeed, typically, the time is discretized, and the control field becomes a piecewise constant function with the ultraviolet cutoff Δt. However, this choice is seen as an approximation that shall be taken under control, typically with the condition $\Delta t \to 0$. On the contrary, we assume that it is possible to truncate the space $\mathcal{G}$ to a small dimensional space $\mathcal{G}^* \subset \mathcal{G}$ and perform the minimization within this space to find an optimal minimum.

The first step in this program is to perform the truncation of the space: it shall be done in the most general way unless additional physical information are known: in this case, these information shall be exploited to introduce an educated guess which, however, it is typically not fundamental. An expansion into a truncated randomized basis of functions $\{h_i(t)\}_1^{N_C}$ is thus perfomed

$$\Gamma(t) = \sum_{i=1}^{N_C} c_i h_i(t), \tag{12.17}$$

The minimization defined in Eq. (12.16) is then recast in a multivariable minimization that can be attacked with standard methods, ranging from the conjugate gradient descent, genetic algorithms or, as typically done, by gradient-free minimization methods [122]. A typical choice for the basis functions $h_i(t)$ is a Fourier basis with randomized frequencies chosen within a bandwidth interval $\Delta\Omega$.

The crucial parameter here is the number of basis functions N_C needed in the expansion in Eq. (12.17) to find a satisfactory result. Early works have investigated the minimal number of independent parameters of the control function necessary to solve the optimal control problem and set a heuristic rule on the number N_C necessary to solve the optimal control problem, which scales as the Hilbert space size [575]. Recently, it has been proven that the number of coefficients necessary to reach an error $\epsilon > 0$ is a polynomial function of the dimension D_{W^+} of the space W^+ of states that can be reached in polynomial time with the Hilbert space size [552]. An upper bound for such dimension is the whole Hilbert space dimension, and for few qubits scenario, they typically coincide. However, in the many-body scenario they can drastically differ [301]. In particular, any many-body dynamics that admits an efficient numerical representation lives in a polynomial space, thus drastically reducing D_W^+ with respect to the exponentially large Hilbert space, resulting in an efficient solution of optimal control problems. Indeed, it is possible to show that, to reach a final error $\epsilon > 0$ the dimension of the space of functions to be searched in $D_{\mathcal{G}}^* \equiv N_C$ can be bounded by apolynomial function of D_W^+ [552]. It is found heuristically that typically a linear lower bound can be saturated [574,575]. In particular, exploiting a simple restarting and change of the random basis used to perform the optimization as introduced in the dCRAB formulation, allows escaping local traps with probability one for power-unconstrained problems [574]: the net results is that it is possible to solve the minimization problem with a series of one-dimensional $N_C = 1$ minimizations on different random directions in the search space. This approach has the additional advantage that minimizations in one dimension can be achieved with highly efficient simple algorithms, without the need to exploit sophisticated approaches such as evolutionary, gradient or higher orders algorithms. Finally, it has been shown that the minimal time needed to achieve an optimal transformation with finite precision is bounded by the control field bandwidth $\Delta\Omega$, i.e.

$$T \geq \frac{D_W}{\Delta\Omega}, \tag{12.18}$$

where D_W is the dimension of the set of reachable states [552].

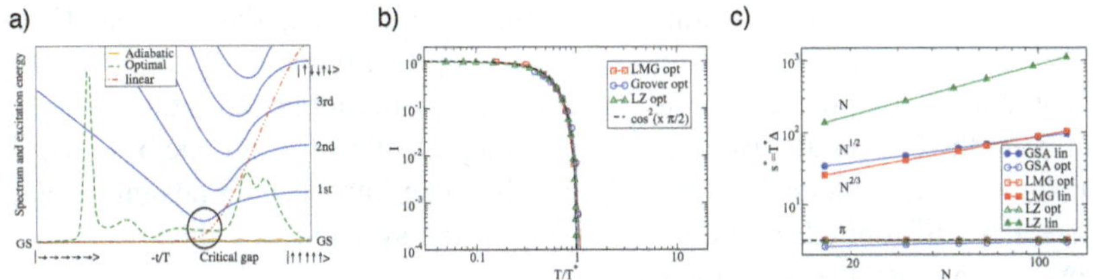

Fig. 12.5 (**a**) Adiabatic (*yellow line*), fast (*red dot-dashed line*) and optimal (*green dashed line*) passage of the quantum phase transition in LMG model. *Blue full lines* report the instantaneous eigenenergies of the system and the critical gap is highlighted. (**b**) Final infidelity as a function of the rescaled total time with $T^* = \pi/\Delta_1$ of transformation identifying the quantum speed limit of the process in different optimal passages (LMG model, Grover adiabatic algorithm, and Landau-Zener). (**c**) Scaling with system size N of the total time needed for linear and optimal annealing. For the Landau-Zener process we assumed an effective system size given by $N = \Delta_1^{-1}$. Figures reprinted with permission from [493].

We conclude this section presenting some results on the application of optimal control problem to quantum annealing. As shown in Fig. 12.4, the crossing of a quantum phase transition can also be achieved relaxing the adiabatic condition but simply imposing that the final state minimizes the energy of the final Hamiltonian. In such a way, we can aim to speed up the process not being constrained by the adiabatic condition. This analysis has been performed in [493] where it has been shown that it is indeed possible to speed up the quantum phase transition crossing: as a first test case the Lipkin-Meshkov-Glick (LMG) has been studied, a many-body model (Ising chain with infinite range interaction) that being integrable is perfectly suitable for such studies [576]. As shown in Fig. 12.5a, where the LMG model spectrum is reported as a function of time while the transverse field is varied either linearly (adiabatically and not) or optimally: it is clearly visible that while a fast naive passage ends in a highly energetic state, both the adiabatic and the optimal passage result in a good approximation of the ground state. Similar results can be obtained on different systems and protocols, such as the adiabatic Grover algorithm or the crossing of the superfluid to Mott insulator QPT in cold atoms in optical lattices [493, 540]. It is then natural to investigate how much one can compress the time necessary to cross a QPT and if a fundamental limit exists: in Fig. 12.5b we report the final error as a function of the total time of the process for the LMG model, the adiabatic Grover algorithm and, for comparison, a simple Landau-Zener crossing. As it can be clearly seen, in all these scenarios it exists a minimal time T^* below which the optimal transformation cannot be found and the error scales as $I = 1 - |\langle\psi(T)\rangle\, \psi_G|^2 = \cos^2(T/T^*)$, where $T^* = \pi/\Delta_1$ and Δ_1 is the minimal gap of the system. Here T^* is the *quantum speed limit* of the process, the minimal time necessary for the transformation to occur, a consequence of the Heisenberg indetermination principle [547, 548, 551, 554, 555, 557, 558, 572, 577–583]. This result is in perfect agreement with the analytical solutions available for the two-level systems and supports the conjecture that also in more complex scenarios we can achieve the fastest transformation allowed by nature, and that can be interpreted as an optimal two-level transition between the initial state and the final one. Finally, in Fig. 12.5c we report the scaling of the speed up as a function of the system size: the optimal crossing speeds-up the process by a square root factor in the system size (similar to the Grover algorithm gain) and by a prefactor of about one or two orders of magnitude. In conclusion, optimal annealing can introduce an advantage which however is still limited by the minimal gap of the system, even though in a more favorable way than the adiabatic passage.

Finally, we report that an optimal passage of a quantum phase transition has been experimentally verified in one-dimensional cold atoms in optical lattice undergoing the superfluid to Mott insulator transition [540].

12.4 Other Applications

In this section, we review some of the physical scenarios of out-of-equilibrium physics that can be addressed also or uniquely using TN techniques: we included them here as they have attracted a vast interest in the last years or we think that they will be subject to further investigations in the years to come.

The application of TN methods to lattice gauge theories in the context of high-energy physics is not limited to equilibrium properties as shown in Cap. 6. Indeed, they have been already applied to the study of dynamical phenomena such as the simulation of the real-time evolution of string breaking, of the Schwinger mechanism and of scattering processes. String breaking is an important phenomenon that occurs when the gauge field string connecting two charges in a gauge theory (e.g., the electric field between two charges, the gluon field between two quarks, etc.) is too costly energetically, and thus it is more convenient to break the string and produce from the vacuum particle-antiparticles pairs. The production of such pairs from the vacuum in QED is known as Schwinger mechanism [584]. Due to the sign problem in Monte Carlo methods, the study of these phenomena in the last decades has been mainly performed observing indirect features, such as the string tension or Wilson loops [334]. A semi-classical simulation of the real-time evolution of string-breaking has been presented only recently [585], followed by fully quantum simulations using TN, for a $U(1)$ and a $SU(2)$ lattice gauge theories [57, 68]. Moreover, the real-time evolution of the scattering process between to bounded particle-antiparticles states as been presented in [57], complemented by an analysis of the formation of entanglement created during the scattering process.

Another large field of the application of the techniques introduced in the previous chapters is the study of out-of-equilibrium many-body quantum systems in different forms: the field of out-of-equilibrium quantum phase transition is attracting increasing interest, that is, the characterization of the different phases a system in the presence of a bath that induces the system to equilibrate. In the presence of competing terms, different properties can arise in the fixed point of the dynamics as a function of the different weights of the unitary and/or non-unitary terms [586,587]. On the one hand, this effect can be exploited to perform the so-called *reservoir engineering*, that is, the development of a specific bath that drives the system into a state with interesting characteristics, as for example, the presence of entanglement to be exploited for quantum information protocols or the preparation of some desired state. Given that the state is a fixed point of the dissipative dynamics, it is inherently robust against perturbations, as already experimentally observed [487–489]. On the other hand, one can study the physics of such phenomena, as in the paradigmatic system of coupled cavities in presence of pumping and dissipation, that has been shown to display a quantum phase transition of light [588–590], or in the recent experimental verification in a string of trapped ions simulating the Ising model in transverse-field [591].

On the more statistical mechanics side, TN methods have also been employed to study fundamental questions related to the thermalization of many-body quantum systems, that is, to investigate if and under which conditions a closed many-body systems acts as a thermal reservoir for its subsystems, and how and if the thermalization occurs [592, 593]. More generally, one can study the effects of quenches, the onset of many-body localization and of quantum chaos [516,518,593–601].

The possible applications of tensor networks in open quantum systems go well beyond the study of the properties of the steady states of the systems: for example it has been shown that it is possible to map system-environment quantum models to effective spin chains [602, 603], or apply them to the study of atoms coupled to optical fibers, e.g. to study of chiral quantum optics and photonic circuits with time delays and quantum feedback. [604–606].

Finally, an increasing interest as grown in the application of concept of classical complexity to classify states of many-body quantum systems, also in the context of cellular automata, in particular the quantum counterpart of Conway's game of life [607–610].

12.5 Problems

1. Using t-DMRG simulate an adiabatic passage through the Ising model in transverse field quantum critical point for different system size lengths and total time of the passage. Compute the final energy and final correlation length and compare them with the values on the exact final ground state.
2. Individuate the Landau-Zener and the Kibble-Zurek regime in the residual excitation energy computed in the previous exercise. Extract the system critical exponents and discuss possible differences between the theoretical and computed ones.

Tensor Networks for Machine Learning 13

Timo Felser

Beyond the groundbreaking achievements of Tensor Networks in simulating complex quantum systems on classical computers, their applications have since transcended disciplinary boundaries. In fact, these numerical tools have found applications in diverse fields, from tackling complex optimization problems in applied mathematics to numerical analysis in quantum chemistry and material science. Among these exciting interdisciplinary applications, one area that captivates our focus in this chapter is the realm of tensor networks in machine learning.

In recent breakthroughs, tensor networks have emerged as formidable players in (un-)supervised Machine Learning (ML) tasks, showcasing performances that rival state-of-the-art Neural Networks on standard datasets like MNIST [222, 611]. Remarkably, this achievement is even more compelling given the relative youth of tensor network applications in machine learning. A distinctive advantage of TN-based machine learning lies in its capacity for real-time predictions, exemplified for instance by its accelerated analysis and classification of data produced by the Large Hadron Collider at CERN [227].

Before going into the detail on how tensor networks can perform machine learning tasks, in what follows we will briefly summarise the concepts of machine learning itself. Afterwards, we introduce the idea of using tensor networks as a classifier with its connection to quantum many-body physics. Finally, we show some state-of-the-art results and comparisons of TN-based machine learning applications.

T. Felser (✉)
Tensor AI Solutions GmbH, Pfaffenhofen, Germany
e-mail: timo.felser@tensor-solutions.com

T. Felser, S. Montangero (eds.), *Introduction to Tensor Network Methods*,
Graduate Texts in Physics, https://doi.org/10.1007/978-3-032-17635-6_13

13.1 Brief Introduction to Machine Learning

In standard software development, a programmer writes concrete rules in some programming language which the program is following upon execution. Thus, the way how the software behaves is determined by the logic of the written programming rules and the code of the programmer (as long as it is free of unintended bugs). In machine learning, however, the computer should derive the decision rules by itself from data that represents the problem at hand to solve (as shown in Fig. 13.1). The core concept of machine learning is to recognise fitting patterns in data and based thereon make predictions for new data. Therefore, we use learning methods or algorithms to derive statistical regularities from existing data, which can be represented in the form of numerical models. Using the learned patterns, these models can make predictions for new data enabling algorithm-based systems that on the one hand can support decision-making and, on the other, be used in autonomous systems.

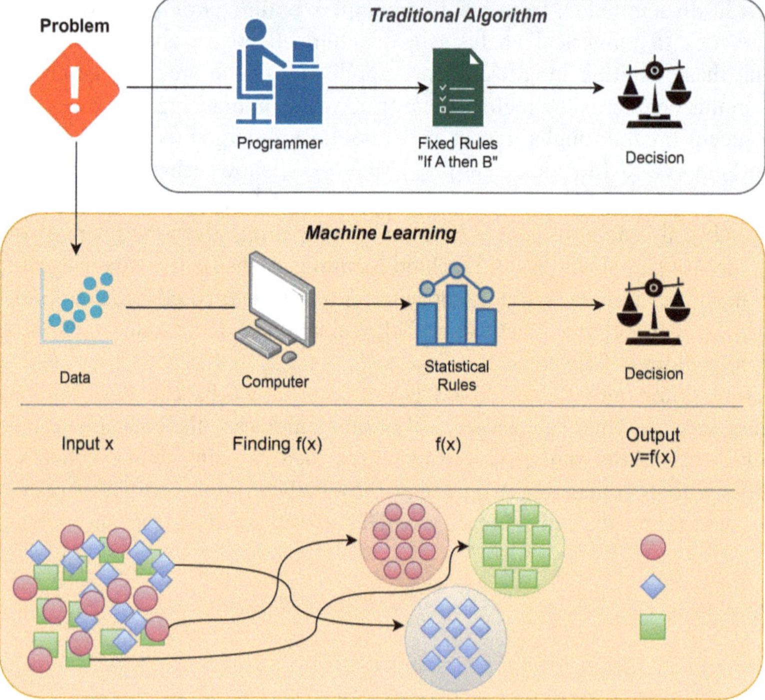

Fig. 13.1 Comparison of machine learning (ML) with standard software development (SD): (top) In SD, the programmer writes the rules for the program to execute. (bottom) In ML, however, the computer derives the decision rules (or decision function $f(x)$) in a learning process based on given data that represents the problem. Afterwards, the derived decision rules, i.e. learned patterns, are used to make predictions for new data

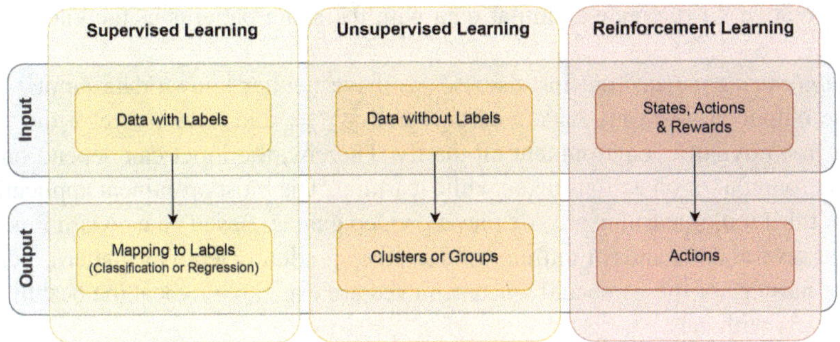

Fig. 13.2 Sketch of the different types of training procedures for machine learning: The supervised learning (left) aims to find the best mapping from known input $x_i \in X$ to known output $y_i \in Y$ (labels) and required data that is labeled accordingly. The unsupervised learning (center) aims to find the best grouping of known input $x_i \in X$, i.e. defining the labels Y. Reinforcement learning (right) aims to find the best actions when only a certain goal y_{target} is known without input data but rather having a map of possible actions to take and states to be in

Mathematically, our machine learning task is to approximate a target *decision function*

$$f : X \to Y, y = f(x) \tag{13.1}$$

which we do not know exactly or is too complex to compute exactly. Now, in order to train our model, i.e. find a good approximation for f, we have different types of training procedures depending on the data set and the specific data problem (compare Fig. 13.2):

1. *Supervised learning*: In this case, we have several data samples i in our data set from which we know the input $x_i \in X$ and the corresponding output $y_i \in Y$ of the decision function $f(x_i)$, i.e. labels. Thus, the target is to find the correct patterns—or correct function f—that produces from given samples the corresponding labels. Concretely, this task can be a *classification* in case the labels have discrete values or a *regression* in case the label space is continuous. We will describe the supervised learning in more detail in Sect. 13.1.1.
2. *Unsupervised learning*: We are talking about unsupervised learning if our data set consists of samples with known input x_i, however unknown output y_i, i.e. unlabeled data. The target in this case can be two-fold: In case we aim to have a discrete label space, we aim to group similar data samples together differentiating between data with different patterns. This procedure is also referred to as *clustering* since we aim to divide the samples in different clusters representing different distributions where each data point in a cluster follows the distribution of its cluster. Furthermore, we can as well be interested in learning the general distribution function behind the data set. This can be used for instance as

generative AI to create a similar data with the same patterns as the known data set.

3. *Reinforcement learning*: In this case, we have neither known data input x_i nor the output y_i. We only have a certain goal y_{target} and have to create the input x_i in an dynamic environment on the fly. Thereby, the input can depend on the environment or on actions taken while learning. The most prominent applications for this kind of learning are AIs playing video games. Based on a reward function and several thousands or millions of tries, they gradually learn the patterns behind the mastery of the game and when achieved are even able to beat the best human players with ease.

For sake of completeness, it is to be said that these three types of machine learning can be mixed in practice depending on the data availability and the concrete problem. For instance when having some labeled data and a lot of unlabeled data, we can perform a semi-supervised learning in which we can merge the procedures of supervised and unsupervised learning into one model.

In what follows, we will describe the supervised learning procedure in more detail. We use this type of learning to introduce further important parts of machine learning, such as *training, testing and validation*, which are good practice when performing machine learning tasks.

13.1.1 Supervised Learning

As described above, in supervised learning we have a data set consisting of input samples $x_i \in X$ and corresponding labels $y_i \in Y$. With this data, we aim to optimise the decision function f that maps the input space X into the label space Y according to Eq. (13.1). As we will see later on in Sect. 13.2, this decision function can be modeled using tensor networks. As an example, we can take the MNIST data set [613] which consists of hand-written digits (see Fig. 13.3. In this case, x_i would be the i-th picture in the data set—more concretely the vector representation of the picture—and y_i the digit it represents. The clear task is to find a classifier that is able to distinguish between the different digits, i.e. a decision function $f(x)$ that manages to map all the available pictures to the correct output. Therefore, we have two fundamental steps: The *training* and the *testing* procedure which we will explain in more detail in the following. Before, it is worth noting that in a classification, the prediction is not necessarily the decision $f(x)$ itself. In this case, the decision function of a classification

$$f(x) = \begin{pmatrix} f_0(x) \\ f_1(x) \\ \vdots \\ f_{L-1}(x) \end{pmatrix} \tag{13.2}$$

← Class "0"
← Class "1"
⋮
← Class "9"

Fig. 13.3 Example of a classification data set: The MNIST data set contains handwritten numbers as gray-scale image [612, 613]. Each image in the data set is a *sample* $x_i \in X$ for the machine learning task and has a label $y_i \in Y$ for the correct class attached

becomes a vector with the entries $f_\ell(x)$ denoting the confidence—or probability—for each of the L different classes. So, a prediction of for the MNIST data set might look like this:

$$f(x) = \begin{pmatrix} 0.0671829 \\ 0.403097 \\ 0.0223943 \\ 0.111971 \\ 0.179154 \\ 0.0447886 \\ 0.134366 \\ 0.873378 \\ 0.0223943 \\ 0.0671829 \end{pmatrix} . \tag{13.3}$$

Note, that in this case, the predicted class would be $\ell = 7$ with approximately $87,34\%$ confidence. Thus, in case the digits are labelled accordingly, this picture would be predicted as a “7”.

13.1.2 Training a Model

Now lets say we have a model $f(x, \theta)$ for our targeted decision function $f(x)$. Notice that we introduced θ for our model: This indicates the tunable parameters in our model, such as the weights in neural networks or—as we will see later on—the tensor elements in case of tensor networks. During the training, we optimise these parameters and thereby our model of the decision function $f(x, \theta)$. Therefore, we define a cost function $C(\theta)$ to minimise, such as

$$C(\theta) = \frac{1}{2}\frac{1}{N_T}\sum_{i=1}^{N_T}\sum_{\ell=1}^{L}\left(f^\ell(x_i, \theta) - y_i^\ell\right)^2 . \tag{13.4}$$

Table 13.1 Overview on the most frequently used cost functions in supervised learning: Mean absolute error, Mean square error, and Cross entropy. The Cross entropy is depicted for a binary classification but can also be generalized to more classes

Name	$C(\theta)$	$N_T \partial_\theta C(\theta)$
Mean abs. error (or *L1-error*)	$\frac{1}{N_T} \sum_{i=1}^{N_T} \sum_{\ell=1}^{L} \left\| f^\ell(x_i, \theta) - y_i^\ell \right\|$	$\sum_{i=1}^{N_T} \sum_{\ell=1}^{L} \partial_\theta f^\ell(x_i, \theta)$, if $f^\ell(x_i, \theta) > y_i^\ell$ $\sum_{i=1}^{N_T} \sum_{\ell=1}^{L} -\partial_\theta f^\ell(x_i, \theta)$, if $f^\ell(x_i, \theta) < y_i^\ell$
Mean square error (or *L2-error*)	$\frac{1}{2N_T} \sum_{i=1}^{N_T} \sum_{\ell=1}^{L} \left(f^\ell(x_i, \theta) - y_i^\ell\right)^2$	$\sum_{i=1}^{N_T} \sum_{\ell=1}^{L} \left(f^\ell(x_i, \theta) - y_i^\ell\right) \partial_\theta f^\ell(x_i, \theta)$
Cross entropy (binary class.)	$\frac{1}{N_T} \sum_{i=1}^{N_T} y_i \log\left(f(x_i, \theta)\right)$	$\sum_{i=1}^{N_T} \frac{f(x_i,\theta) - y_i}{f(x_i,\theta)(1 - f(x_i,\theta))} \partial_\theta f^\ell(x_i, \theta)$ $+ (1 - y_i) \log\left(1 - f(x_i, \theta)\right)$

Note, that i in the outer sum runs over all N_T samples in the training set and ℓ in the inner sum runs over all possible labels L, i.e. all classes, which would be all possible digits in our MNIST example. Thus, this cost-function compares the prediction $f(x_i, \theta)$ of our model for each sample i with the known output y_i^ℓ for all possible classes to compute the mean square error. Adapting the tunable parameters θ of our model, we can minimise this cost function—or the performance of the model—using optimisation techniques such as various gradient descent methods [614].

$$\min_\theta C(f(\theta)) \tag{13.5}$$

for the given data set. For the sake of completeness, we stress that the cost function in Eq. (13.4) is not the only possible cost function to be used in machine learning. In Table 13.1, we provide an excerpt of other frequently used cost functions.

Notice, that we referred to the data used in this training procedure as training set. In fact, it is critical in supervised learning to separate the available data into different sets as we will explain in what follows.

13.1.3 Testing and Validation

The fundamental idea behind the aforementioned separation of the data set is to train the model, i.e. optimise $f(\theta)$, on one data set and test it on another, for the model unknown data set. This procedure ensures that the model can in fact accurately make predictions for new data of the same type that it has not seen before. This idea

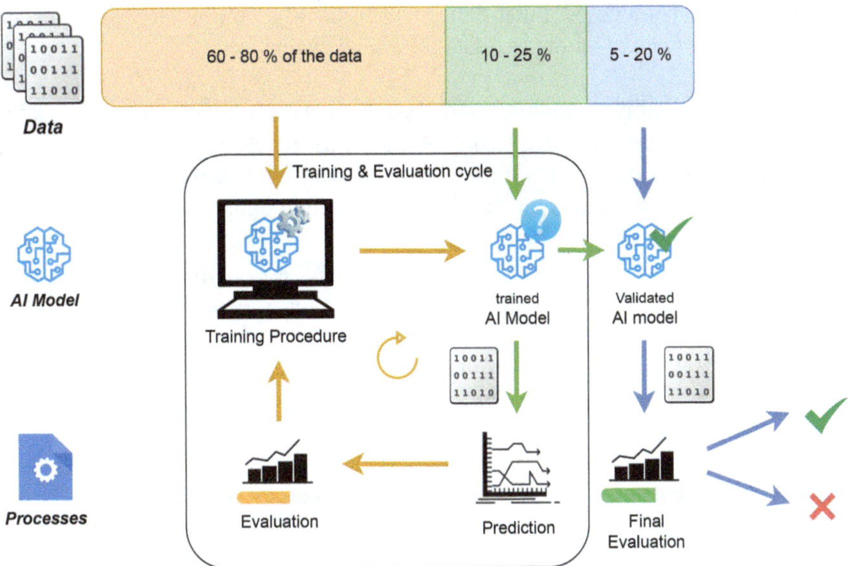

Fig. 13.4 Data separation principle for machine learning: The data is typically separated in a training set, a validation set and a testing set. The training set is used to train an AI model. The validation set is used to validate a trained AI model. Based on the predictions of the trained model on the validation set, we can evaluate its performance on data that was unseen by the model. Based on this evaluation, we can see if its learned patterns generalise well to new data or if the model needs retraining. After sufficient cycles of retraining, the final model will be benchmarked on yet another set—the testing set for a final evaluation

of keeping a good prediction performance is as well referred to as *generalisation*. Typically, in the creation of a machine learning model, one separates the entire data into a *training set*, a *validation set* and a *test set* as shown in Fig. 13.4. As a rule of thumb for large data sets, the entire procedure goes as follows:

1. Split the entire data into three groups: The training set (ca. 60–80% of the data depending on the problem at hand), the validation set (ca. 10–25%) and the test set (ca. 5–20%).[1]
2. Train, i.e. optimise according to Sect. 13.1.2, one or more models. Multiple models can be trained (as well in parallel), e.g. by using different optimisation techniques or different initial models.
3. After the training, evaluate the performance of each model on the validation data set.
4. In case the performances is not satisfying for all models, restart at 2. by keeping only the best models and discarded the poorest performing ones.

[1] These numbers usually work in case there are large enough samples available for training but are not a fixed rule.

5. In case, there is one model that performs sufficiently well on the validation set, apply this *best model* on the test data for a final evaluation.

As mentioned, this procedure is to bee seen as a rule of thumb for large enough data sets, whereas other techniques, such as *leave one out cross validation* are designed for machine learning on lower amount of data [615]. In this whole procedure it is important to know that there are multiple figures of merit to take into account when measuring the performance of a model. The most straight-forward is of course the cost-function of Eq. (13.4) itself. Another important performance indicators arise for classifications when looking at the so-called *confusion matrix* of a model. Evaluating a model on a data set, we can track how often it predicted correctly and how often it predicted incorrectly. In fact, for classification problems, such as the MNIST, we can analyse for each class the decision, and in case of an incorrect decision, we can see with which other label the model confused the data sample. For sake of simplicity, let us first look at hand-written digits with the aim to find "1"s in the data set, i.e. having two different classes distinguishing "1" from not "1". Testing a classifier, we end us with four different scenarios:

1. *True positive*: The model predicts a "1" (*positive* prediction) while the data sample is in fact a "1"
2. *False positive*: The model predicts a "1" (*positive* prediction) while the data sample is in fact not a "1"
3. *False negative*: The model predicts "not 1" (*negative* prediction) while the data sample is in fact a "1"
4. *True negative*: The model predicts "not 1" (*negative* prediction) while the data sample is in fact not a "1"

Counting each of the scenarios during the evaluation (with an evaluation set of size N), we obtain the confusion matrix as shown in Table 13.2.

With this table, we can evaluate relevant performance indicators, such as the accuracy, i.e. the number of correctly classified samples, as

$$A(f) = \frac{TP + TN}{N} . \tag{13.6}$$

Table 13.2 Representation of a binary classification in a confusion matrix

Actual label ↓	Predicted results	
	Number of predicted "1"s	Number of predicted "not 1"s
Number of actual "1"s	Number of true positives (TP)	Number of false negatives (FN)
Number of actual "not 1"s	Number of false positives (FP)	Number of true negatives (TN)

Table 13.3 Overview of the most frequently used performance metrics for evaluating a machine learning classifier

Name	Equation	Note
`Accuracy`	$\frac{TP+TN}{N}$	
`Precision`	$\frac{TP}{TP+FP}$	Fraction of correctly detected "1"s of all actual "1"s
`Recall, sensitivity` or `hit rate`	$\frac{TP}{TP+FN}$	Fraction of correctly detected "1"s of all predicted "1"s
`Balanced accuracy`	$\frac{1}{2}\left(\frac{TP}{TP+FN}+\frac{TN}{TN+FP}\right)$	Balancing the accuracy for imbalanced data sets
`F1-Score`	$\frac{2TP}{2TP+FP+FN}$	
`Informedness`	$\left(\frac{TP}{TP+FN}+\frac{TN}{TN+FP}\right)-1$	

However, depending on the data set, this figure of merit might be very contra-intuitive as can be seen in our example: For distinguishing between "1" and "not 1" in the MNIST data set, we only have 10% of the entire available data actually being a "1" while 90% of the data set would belong to the second class "not 1". Imagine now, we have a classifier that, no matter what the input, always predicts "not 1". With this strategy, the model would achieve 90% of accuracy while being completely unable to solve the task. To compensate such effects different figures of marit have been established for classification problems. The most prominent ones are listed in Table 13.3. Note, that all of these figures of merit can be generalised for a classification problem with more than two classes.

13.2 Machine Learning with Tensor Networks

Having the basic concepts of machine learning presented in the prior section in mind, we now show how tensor networks can be used as a model in machine learning. In fact, the concept of using tensor networks is increasingly investigated in applied mathematics [616–619]. The Matrix Product States, for instance, has been rediscovered as *Tensor Trains* [183] in this community. Later on, it was shown how tensor networks, in particular the MPS, can be used as a novel tool for machine learning [227, 611]. As we will see later on, the first applications of tensor networks in machine learning have already yield impressive results when performing supervised learning on standardised datasets [222, 611, 620] and even beyond [227,228]. Before we illustrate these applications, we shown in what follows how machine learning with tensor networks works.

13.2.1 Tensor Networks Predictor

A special method in machine learning is the so-called *kernel learning* [621] where the data input $x \in X$ is mapped into a higher dimension. This idea is among others the underlying success of *support vector machines* (SVM) which are designed to exploit this *kernel-trick*: By "looking" at complex non-linear data in an other, more expressive basis, it can become linearly separable and easy to classify. As an example, in Fig. 13.5 we can perfectly separate the two different data points using a linear boundary after transforming the data by $\Phi(x) = x^2$. Taking this kernel-trick into account, we can describe the decision function as introduced in Eq. (13.1) by

$$f(x) = W \times \Phi(x) \tag{13.7}$$

where $\mathbf{x}$ is the input (or feature) vector, $\Phi(x)$ the chosen feature map and W, the so-called *weight vector* which, in principle, can be exponentially large or even infinite. In fact, we can see this weight vector equivalent to a vector representing a quantum many-body system as introduced in Chap. 5. In what follows, we show how—just like in the case of quantum mechanics—we can present such a weight vector for a machine learning prediction by an order-N tensor. By then representing this tensor as a tensor network, as we did for the many-body wavefunction in quantum mechanics, we can introduce the concept of tensor networks for machine learning. For this idea to work, we need to choose $\Phi(x)$ as a separable feature map, such as

$$\Phi(\mathbf{x}) = \phi_1(x_1) \otimes \phi_2(x_2) \otimes \ldots \otimes \phi_n(x_n) \tag{13.8}$$

where we separate the complete feature map with respect to the different input features of our data set. We call Φ_k *local feature map*. Now each sample becomes a vector product—or product state in the language of quantum many-body physics. By simply contracting the entire network with the mapped input $\Phi(x)$, we can

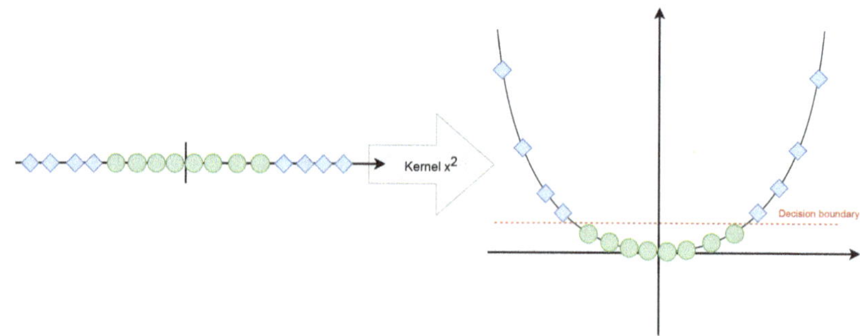

Fig. 13.5 Example for using a *kernel-trick* to separate data points of two different classes (marked with blue diamonds and green circles respectively). Transforming the linear data from the left hand side by $\Phi(x) = x^2$ it becomes easily separable by a linear function in the kernel space on the right hand side

make a tensor network prediction as we will see in more detail later on. As we will see as well, this kind of feature map leads to a deeper analogy between this machine learning approach and quantum mechanics. In fact, such separable kernels have already been investigated [622, 623] and shown to be in particular of interest for data in which we do not assume or know any particular relationships between features in the input vector before training.

For sake of completeness, we mention that for a classification problem, the *weight vector* W in Eq. (13.7) is in fact a $m \times L$ *weight matrix* where L is the number of different classes. When predicting, the classifier in this case gives the confidences for each of the classes $l \in \{1, \ldots, L\}$

$$\mathcal{P}_l(\mathbf{x}) = W_l \cdot \Phi(\mathbf{x}) \tag{13.9}$$

where we now have the additional index l, as introduced in Sect. 13.1.1.

13.2.2 Mapping Data to Quantum Spins

One of the first applications of tensor networks as a classifier in machine learning came from Ref. [611] with the idea to *map data into quantum spins*. For this, we can have a look again at the MNIST data set of hand-written digits introduced in Sect. 13.1.1. As shown in Figs. 13.3 and 13.6, the data set consists of grayscale images with n pixels where every pixel of each sample is black-white encoded, i.e. represented by a value ranging from 0.0 for white pixels to 1.0 for black ones. By choosing the local feature map introduced in Eq. (13.8) as

$$\Phi^{[i]}(x_i) = \left[\cos\left(\frac{\pi x_i}{2}\right), \sin\left(\frac{\pi x_i}{2}\right)\right] \tag{13.10}$$

we can map each feature x_i into a quantum spin. In this picture, a white pixel becomes a spin-up, a black pixel a spin-down and a gray pixel a super-position between them which depends on the grayness of the pixel as shown in Fig. 13.6 and illustrated below:

$$x_i = 0.0 \longrightarrow \Phi^{[i]}(x_i) = \begin{pmatrix} 1 \\ 0 \end{pmatrix}; \tag{13.11}$$

$$x_j = 1.0 \longrightarrow \Phi^{[j]}(x_j) = \begin{pmatrix} 0 \\ 1 \end{pmatrix}; \tag{13.12}$$

$$x_k = 0.5 \longrightarrow \Phi^{[k]}(x_k) = \frac{1}{\sqrt{2}}\begin{pmatrix} 1 \\ 1 \end{pmatrix}. \tag{13.13}$$

In fact, the idea of quantum spin is a helpful analogy to investigate further properties of tensor networks in machine learning with its connection to quantum physics and vice versa.

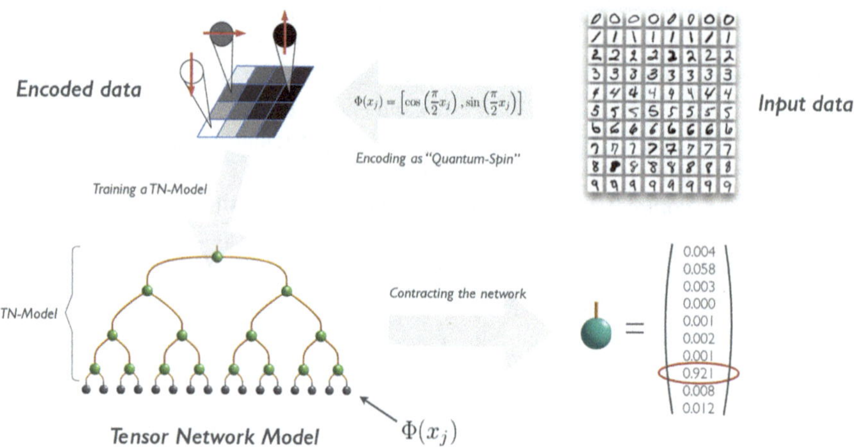

Fig. 13.6 Example of the mapping concept for Machine Learning with Tensor Networks (TN). Here, the MNIST data of handwritten digits (top right) is encoded in the basis of *quantum spins*: Each pixel (feature) of a picture will be transformed by the feature map $\Phi(x)$ into a spin-up for black pixels, a spin-down for white pixels and a superposition for gray pixels (top left). The encoded data is used for training a TN-model (bottom left). The model predicts by contracting the mapped sample with the TN-classifier resulting in a vector indicating the probabilities for each class (bottom right). Here, the TN-model predicts quite confident the number "7" as the seventh entry (the index starts at 0) is the largest in the prediction vector

13.2.3 The Matrix-Product-States as Classification Model

In what follows, we describe the training procedure, i.e. the optimisation, of a matrix-product-states (MPS) model for machine learning problems. In particular, we will focus on the classification of the MNIST-data set.

13.2.3.1 MPS Model and Predictions

We recall from Eqs. (13.7) and (13.2) that the decision function for a classification with MPS becomes

$$f_\ell(x) = W_\ell \times \Phi(x) = W_\ell^{[\mathrm{MPS}]} \times (\phi_1(x_1) \otimes \phi_2(x_2) \otimes \ldots \otimes \phi_n(x_n)) \qquad (13.14)$$

where ℓ denotes the label for the different classes and $W_\ell^{[\mathrm{MPS}]}$ the MPS representation of the weight matrix as described in Sect. 13.2.1. Note that in tensor network language the index ℓ for the label indicates an external link in the tensor network. Thus, the entire MPS has $n + 1$ outgoing links for the n input features and the label link ℓ as depicted in Fig. 13.7a. This extra link can formally added to the MPS representation introduced in Eq. (7.7) by using the form

$$W_\ell^{[\mathrm{MPS}]} = \sum_{\{i\}} A_1^{i_1} A_{i_1,2}^{i_2} \cdots A_{i_{j-1},j}^{\ell,i_j} \cdots A_{i_{n-2},n-1}^{n-1} A_{i_n,n} \left(= \left|\psi_\ell^{\mathrm{MPS}}\right\rangle \right) \qquad (13.15)$$

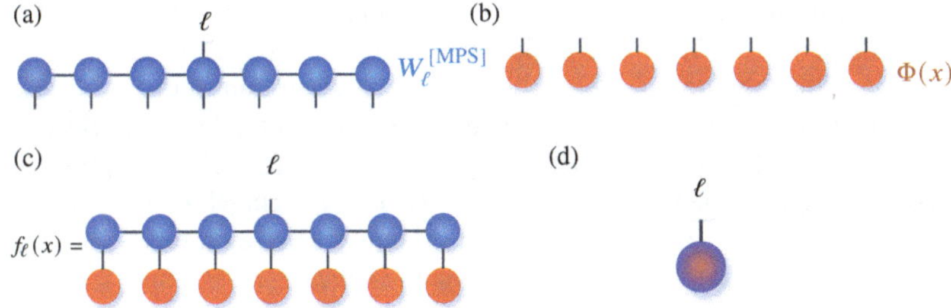

Fig. 13.7 Concept for machine learning (ML) classification with Matrix Product States (MPS). Compared to MPS for quantum systems, the MPS for classification $W_\ell^{[\mathrm{MPS}]}$ exhibits an additional link ℓ for the different classes of the ML problem (**a**). The data samples are encoded in a product state $\Phi(x)$ (**b**). The model predicts by contracting the sample with the MPS-classifier (**c**) resulting in a vector indicating the probabilities for each class (**d**)

where the link ℓ is now attached to the j-th tensor in the MPS. In brackets on the left hand side, we indicated that the MPS representation of the weight matrix can be interpreted as well as a set of wave functions for each label ℓ. Now, given the mapping $\Phi(x)$ of the input data into a quantum-like spin chain as shown above in Eq. (13.10) (see as well Fig. 13.7b), we can now make a single prediction of our MPS classifier by contracting the mapped data with the MPS over all *physical indices*. In fact, we get the equation for the contraction when replacing the MPS representation in Eq. (13.14) leading to

$$f_\ell(x) = \sum_{\{i\}} A_1^{i_1} A_{i_1,2}^{i_2} \dots A_{i_{j-1},j}^{\ell,i_j} \dots A_{i_{n-2},n-1}^{n-1} A_{i_n,n} \times (\phi_1(x_1) \otimes \phi_2(x_2) \otimes \dots \otimes \phi_n(x_n)) \ . \tag{13.16}$$

Notice that the only open link after this contraction is the label link ℓ as it should be for the classification. In fact, by using this spin map $\Phi(x)$, we can interpret each data sample, i.e. each picture in the MNIST example, as product state with the wave function $|\Phi\rangle = |\phi_1, \phi_2, \dots, \phi_n\rangle$. In this picture, the decision function $f^\ell(x)$ can be seen as calculating the overlap of $\langle \psi_\ell^{\mathrm{MPS}} | \Phi \rangle$ as illustrated in Fig. 13.7c. Consequently, the classification result corresponds to the state ψ_ℓ^{MPS} that shows the highest overlap with the "quantised" data sample. As shown in Sect. 13.1.1, we end up with vectorised decision function shown in Eq. (13.2) with the vector-index ℓ running over the different classes of the machine learning problem, i.e. a one-link tensor with index ℓ (see Fig. 13.7d).

13.2.3.2 Training an MPS Model

Having introduced the concept of predicting with an MPS, we are now moving the attention to the training process of an MPS. In this process, we adapt the tunable parameters θ of our model—in the case of an MPS, the entries of its tensors—in order to maximise the prediction accuracy or minimise a costfunction $C(\theta)$ with respect to a given data set, as described in Sect. 13.1.2. For this task, we can take inspiration of the core idea of DMRG-like variational algorithms for tensor networks used in quantum mechanics: As introduced in Sect. 7.1.3 for MPS and Sect. 8.1.4 detailed for ground state search in TTNs, the idea is to reformulate the problem on the level of each individual tensor of the network. By sweeping through the network, iteratively solving the local problems for each tensor after each other, we address the global optimisation task. Therefore, the mathematical steps required to bring the concept behind variational algorithms for tensor networks to the machine learning problem to solve are as follows:

1. Reformulate the global optimisation problem as local problem for individual tensors.
2. Optimise the individual tensors with respect to the local problem.
3. Control the global convergence when iteratively optimising the network locally.

1. Reformulating the Global Optimisation as Local Problem

Beginning with the first point, we will take the quatratic costfunction

$$C(\theta) = \frac{1}{2}\frac{1}{N_T}\sum_{\tilde{n}=1}^{N_T}\sum_{\ell=1}^{L}\left(f^{\ell}(x_{\tilde{n}},\theta) - y_{\tilde{n}}^{\ell}\right)^2 . \tag{13.17}$$

introduced in Sect. 13.1.2 as an example for demonstration. As we will see in the process, the following procedure is easily adaptable to any differentiable costfunction $C(\theta)$. In fact, when trying to optimise a local tensor in a tensor network, i.e. the parameters $\theta_k \in \theta$ of the tensor A_k, we compute the derivative of the costfunction with respect to the variable parameters θ_k while keeping all other tensors fixed. Starting with the following calculation,

$$\frac{\partial C(\theta)}{\partial \theta_k} = \frac{1}{N_T}\sum_{\tilde{n}=1}^{N_T}\sum_{\ell=1}^{L}\left(f^{\ell}(x_{\tilde{n}},\theta) - y_{\tilde{n}}^{\ell}\right)\frac{\partial f^{\ell}(x_{\tilde{n}},\theta)}{\partial \theta_k} , \tag{13.18}$$

we have to compute the derivative of the decision function $f^{\ell}(x_{\tilde{n}},\theta) = W_{\ell}(\theta) \times \Phi(x_{\tilde{n}})$ which we have already formulated in Eqs. (13.14) and (13.16) in more detail. Given that the mapped samples $\Phi(x_{\tilde{n}})$ are independent on the tunable MPS parameters θ, we only need to differentiate the MPS with respect to the tunable parameters θ_k of the tensor A_k as introduced in Sect. 5.3.1. Consequently, the derivative of the decision function becomes

$$\begin{aligned}\frac{\partial f^{\ell}(x_{\tilde{n}},\theta)}{\partial \theta_k} &= \frac{\partial W_{\ell}(\theta)}{\partial \theta_k}\Phi(x_{\tilde{n}}) \\ &= \left(\frac{\partial}{\partial \theta_k}\sum_{\{i\}} A_1^{i_1} A_{i_1,2}^{i_2} \dots A_{i_{j-1},j}^{\ell,i_j} \dots A_{i_{n-2},n-1}^{n-1} A_{i_n,n}\right)\Phi(x_{\tilde{n}})\end{aligned} \tag{13.19}$$

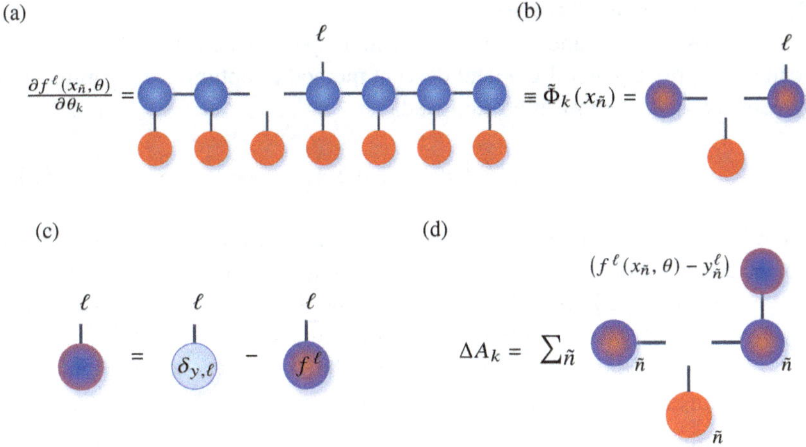

Fig. 13.8 Concept for the machine learning (ML)optimisation of Matrix Product States (MPS). To calculate the gradient for the k-th tensor A_k in the network with respect to the quadratic cost function, we require the derivative of the MPS-decision-function (**a**) which can be written in a contracted form (**b**). An auxiliary tensor representing the difference between the known, i.e. true, values and the predicted values (**c**) support the illustration of the calculation. Finally, the gradient can be calculated by weighting all the samples with the contraction shown in (**d**)

Let us now first assume, that the targeted tensor A_k does not have the label link ℓ attached, i.e. $k \neq j$, as illustrated in Fig. 13.8a). To simplify the calculation, we can define the *environment* $\tilde{\Phi}_k(x_{\tilde{n}})$ of A_k for the sample $x_{\tilde{n}}$ such that

$$f^{\ell}(x_{\tilde{n}},\theta) = \left(\sum_{\{i\}} A_1^{i_1} A_{i_1,2}^{i_2} \dots A_{i_{k-1},k}^{i_k} \dots A_{i_{j-1},j}^{\ell,i_j} \dots A_{i_{n-2},n-1}^{n-1} A_{i_n,n} \right) \Phi(x_{\tilde{n}}) \tag{13.20}$$

$$\equiv \sum_{\{i\}} \tilde{\Phi}_k(x_{\tilde{n}})_{i_{k-1}}^{\ell,i_k} A_{i_{k-1},k}^{i_k} \,. \tag{13.21}$$

In fact, this *environment* can be efficiently contracted to a large extend as shown in Fig. 13.8b). In this way, the derivative of Eq. (13.19) equals

$$\frac{\partial f^{\ell}(x_{\tilde{n}},\theta)}{\partial \theta_k} = \tilde{\Phi}(x_{\tilde{n}}) \tag{13.22}$$

and the derivative of the cost function becomes

$$\frac{\partial C(\theta)}{\partial \theta_k} = \frac{1}{N_T} \sum_{\tilde{n}=1}^{N_T} \sum_{\ell=1}^{L} \left(f^{\ell}(x_{\tilde{n}},\theta) - y_{\tilde{n}}^{\ell} \right) \tilde{\Phi}(x_{\tilde{n}}) \,. \tag{13.23}$$

2. Optimisation of One Tensors
Having the costfunction and its derivative with respect to the tensor A_k to optimise, we can now apply a general gradient descent method to optimise the parameters of the tensor. Thus the optimisation step reads

$$A_k^{[\text{new}]} = A_k + \alpha \Delta A_k \ , \tag{13.24}$$

where α denotes the step size and ΔA_k the gradient of our cost function. For the technical computation, we compute for each sample within the update batch the difference between the known, i.e. true, values and the prediction f^ℓ as an auxiliary step (see Fig. 13.8). The result will be used to compute the total gradient as shown in Fig. 13.8. Note that the contraction over the label link ℓ depends on the tensor A_k to optimise: In case the label link is attached to the tensor A_k—as $A_k = A^{\ell, i_k}_{i_{k-1}, k}$—it stays attached to A_k and not to its *environment* in Eq. (13.21). Consequently, the dimensionalities change for the optimisation of the label-link-tensor accordingly.

The gradient descent can be executed as discussed in Sect. 13.1.2. As discussed before in this section, this procedure of deriving the gradient for a single tensor and updating it with a gradient descent can be applied straightforwardly to any differentiable costfunction and any tensor network geometry.

3. Iteratively Optimising the Network Locally for Global Convergence
Once we have the ingredients for optimising every single tensor in the network individually, we can apply the idea behind a DMRG-like optimisation straightforwardly by sweeping through the network optimising all tensors after eachother. Once we went thought the network once completing a training epoch, we can iterate as many epoche as required to converge globally.

The overall optimisation scales with

$$\zeta_{opt} = O\left(N_T N m^2 N_L d\right) \ , \tag{13.25}$$

where m represents the MPS bond dimension, N is the number of tensors, N_L is the number of labels, and N_T is the number of training samples and d the dimension of the local feature maps. In practical scenarios, the predominant factor contributing to the overall cost is the substantial number of training inputs (N_T). Hence, there is a significant interest in minimizing this cost. This can be achieved by leveraging stochastic gradient descent or training in batches. Another mean of optimising the speed of the computation efficiently is paralleling the computations with respect to batches of N_T.

Lastly, it's worth mentioning that similar optimization algorithm were suggested including a space truncation in Ref. [611] and for hidden Markov models in Reference Ref. [624].

13.3 Problems

1. Exploiting the code developed in the previous exercises, extend the object MPS towards machine learning applications, i.e. include a label link to one tensor in the MPS and basic operations to encode data and compute a prediction on them.
2. With the tools developed above, write a tensor network machine learning algorithm using the MPS and apply it to the MNIST dataset [613]. Compare the resulting accuracy on the test set with those computed using other methods.

Hardware in a Nutshell for Physicists

A

Computational physics lies at the frontiers of physics in two ways: (1) from the biggest experimental efforts at CERN to the single table-top experiments in cold atoms, the support of dedicated software and hardware is fundamental. In both cases, without the automation allowed by software control most of these experiments would not be possible, and the postprocessing and the analysis of the data often requires terabytes of storage and thousands of lines of code; (2) the complexity of the software dedicated to simulate physical process has become so high that the numerical experiments now matches real ones in terms of complexity (years to be set up, thousand of lines of code), resources (supercomputers, months of computation time), not to speak of the ability of description of reality. It is then necessary to know the tools needed to attack such fascinating challenges to be able, if necessary, to modify the smallest detail of the experimental apparatus, including the software and the hardware used to run it. Hereafter, we present some fundamental concepts of computer architecture that any physicist planning to set up numerical experiments, even small, should know and might serve as a starting point for more advanced reading.

A.1 Architecture

Modern computers are built according to the von Neumann architecture, whose sketch is presented in Fig. A.1, composed of the following components each dedicated to a given task:

Memory, dedicated to data and program instructions storage.

Central Processing unit (CPU), divided in a Control Unit that fetches the instructions/data from memory and decodes the instructions and coordinates the operations to be performed, and an Arithmetic Logic Unit (ALU) dedicated to performing data processing.

T. Felser, S. Montangero (eds.), *Introduction to Tensor Network Methods*,
Graduate Texts in Physics, https://doi.org/10.1007/978-3-032-17635-6

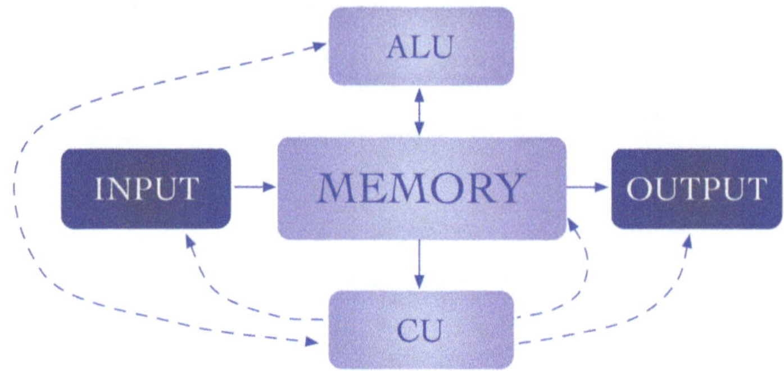

Fig. A.1 Von Neumann's computer architecture: the Control Unit (CU), Arithmetic Logical Unit (ALU), Memory, inputs and outputs devices, and their relations (*Bold lines* denote data/instruction flow, *dashed* ones control signal flow)

Input and Output (I/O) devices, the interfaces with external storage and the users.

The main innovation introduced by this architecture, which nowadays might appear somehow natural, is that it allows building a general purpose computer, while previous architectures where build to perform a specific task, i.e., the program was hardwired, not stored. The von Neumann architecture allows instead to write the computer program (the set of instructions it has to perform) in its memory, making it a machine highly versatile, as witnessed from the fact that since its introduction it has essentially remained unchanged. Indeed, in the last decades, all computer components have been extensively revised and improved, but there has been no need to rethink this basic computer architecture apart from one innovation first introduced in the PDP-11 manufactured by DEC in the late sixties. This innovation is the *system bus*, dedicated hardware that transfers data among the components along a unique path shared by them all: The uniqueness of the bus simplifies the arbitration of the possibly conflicting accesses to the memory requested concurrently by the other components.

A.2 Data and Formats

As well known, the data in modern computers is stored in bits, using binary coding. That means that there is a collection of physical entities (the memory of the system) each of them having only two possible different states (referred as 0 and 1). What seems now natural is not the only possible choice: replacing this definition results in highly interesting consequences: a quantum-bit is defined by two quantum states $|0\rangle$ and $|1\rangle$, from which all quantum technologies are nowadays stemming from [8]. Coming back to classical bits, for example, one can write the number 42 as

$$101010, \tag{A.1}$$

which should be read as $1 \times 2^5 + 0 \times 2^4 + 1 \times 2^3 + 0 \times 2^2 + 1 \times 2^1 + 0 \times 2^0$. In the jargon, 8 bits form a byte, 2 bytes form a 'word', 4 bytes a 'long word' and 8 bytes a 'long integer'. Having defined positive numbers, the elementary operations can be constructed as usual. Here we briefly report the addition of two bytes:

$$
\begin{aligned}
2 &\quad 00000001 \\
+\,3 &\quad 00000011 \\
=5 &\quad 00000101,
\end{aligned}
\tag{A.2}
$$

where the sum of each digit is done modulus two and the carry (if present) is transferred to the digit on the left.

Less straightforward is the introduction of negative numbers, however, it can naturally be introduced recalling their definition: $-n$ is the number that summed to n results in zero. It is interesting to look for a representation that does not require new circuitry to perform additions - on the other side, one needs to introduce something to change the sign to an integer. Thus, we shall look for a number that added (using the binary adder, hence $\mod 2^8$, since for simplicity we keep assuming 8 bits) to n according to the previous definition results in zero. This is achieved by the *2s-complement representation* of the integers, that keeps the usual representation of the positives in $[0, 2^7)$ and defines the opposite of a number $n \in [-2^7, 0)$ as $(2^8 - n) \mod 2^8$. Summing modulo 2^8, we have $n + 2^8 - n = 0$. A simple example with $n = 1$, hence $-n = 2^8 - 1 = 11111111$ in base 2, helps to visualize the general solution:

$$
\begin{aligned}
1 &\quad 00000001 \\
+\,-1 &\quad 11111111 \\
=\ 0 &\quad 00000000.
\end{aligned}
\tag{A.3}
$$

Here is another interesting example:

$$
\begin{aligned}
21 &\quad 00010101 \\
+\,-1 &\quad 11111111 \\
=\ 20 &\quad 00010100.
\end{aligned}
\tag{A.4}
$$

In general, the problem of finding the 2's complement representation of the opposite of n, that is computing ($2^8 - n$ modulo 8) can be implemented as $((2^8 - 1) - n) + 1$. This is convenient, since the subtraction in the first addend (in binary with 8 digits) does not require borrows, and its result can be obtained just negating each bit of the subtrahend (a zero turns into a one and viceversa). For instance, $(2^8 - 3)$ becomes $(11111111 - 00000011 = 11111100$. The operation of flipping each bit in n is often denoted by $\bar{n}$, an operation which is useful in itself, like many other elementary operations on bits (OR, AND, XOR etc...). Hence, $-n$ can be implemented as $\bar{n}+1$ almost at no cost, i.e., without requiring specific circuits.

In conclusion, with one byte, one can store the natural numbers from 0 to 255 and the integers from -128 to 127:

$$\begin{array}{rrl} \textit{Natural} & \textit{Integer} & \\ 0 & 0 & 00000000 \\ 1 & 1 & 00000001 \\ 2 & 2 & 00000010 \\ & \vdots & \\ 127 & 127 & 01111111 \\ 128 & -128 & 10000000 \\ & \vdots & \\ 253 & -3 & 11111101 \\ 254 & -2 & 11111110 \\ 255 & -1 & 11111111 \end{array} \tag{A.5}$$

To understand which is which one needs to know a-priori which representation is intended, as it happens, for instance, when a decimal input is converted in binary.

A programming language has few predefined types, one of which is INTEGER: when one defines a variable of a given kind, a slot in the memory is reserved to store its value. The type of variable defines the amount of memory reserved and the biggest and smallest number one can store safely. If one exceed these limits during a computation, unpredictable errors might occur: this is the 'overflow' error. In FORTRAN (hereafter, we concentrate on this programming language for scientific calculations. Anyway, although the examples refers specifically to it, corresponding objects can be found in any programming language) there are the following kind of integers:

TYPE	Lenght	Range	Max
INTEGER $*$ 2	2bytes	$[-2^{15} : 2^{15} - 1]$	$\approx 10^4$
INTEGER $*$ 4	4bytes	$[-2^{31} : 2^{31} - 1]$	$\approx 10^9$
INTEGER $*$ 8	8bytes	$[-2^{63} : 2^{63} - 1]$	$\approx 10^{19}$.

We now look at how real numbers are stored in a discrete binary memory according to the IEEE standard 754 [626]. We start from the fact that a real number can be written as $(-1)^S \times M \times 10^E$, where M is the mantissa, S determine the number sign, and the E the exponent specifying the order of magnitude of the number. We can then store a real number, up to some finite precision, in a binary register as follows. For sake of clarity, here we specify it for the single precision (4 bytes long or REAL*4 in FORTRAN):

Position	31	30...23	22...0
Content	S	E	M
Length	1bit	1byte	3byte − 1bit

and the corresponding number is defined as $(-1)^S \times 2^{E_8 E_7 \ldots E_0 - 127} \times 1.M_{22}M_{21}\ldots M_0$, where $E_8 E_7 \ldots E_0$ is the binary coding of the exponent E and $1.M_{22}M_{21}\ldots M_0 = 1 + M_{22}/2 + M_{21}/2^2 + M_{20}/2^4 + \ldots$ encodes the mantissa. These are the so called floating point numbers as the order of magnitude of the stored number changes according to the exponent E. Note the bias in the exponent definition, which allows encoding positive and negative exponents. In conclusion, the previous encoding allows to store safely numbers ranging from 2^{-127} to about 2^{128}, that is from about 10^{-39} to 10^{38}.

Does that mean that we have about eighty digits of precision available? A careful look at this question will result in a negative answer. Indeed, although we can store numbers that spam eighty orders of magnitude, the precision is much less as what counts is the maximum extent of numbers on which we can safely perform operations, and this is related to the amplitude of the mantissa. For example, if we want to add two floating point numbers, we shall first of all write them with the same exponent, and then perform the binary sum of the mantissas. Thus, the two most far apart numbers that we can meaningfully add are 1111...1110 and 0000...0001, that is 16772215 and 1, which correspond to eight digits of precisions. Once again, this precision is a crucial element to be taken into account while performing numerical simulations as a wrong estimate of the needed precision might lead to uncontrolled results.

To increase precision, the only possibility is to extend the memory dedicated to store each number, that is, to use double precision (REAL*8 in FORTRAN) which reserves 8 bytes for each number, divided as follows: 1 bit for the sign, 11 bits for the exponent and 52 bits for the mantissa. A straightforward calculation yields that the range of numbers which can be stored is from the order of 10^{-308} to 10^{308}, with sixteen digits of precision. Some programming languages allow defining higher precision real numbers (quadruple precision or even more); however, one should use them responsibly. Indeed, increasing the precision typically implies also reducing the computational performances of the simulation. Given that only very few physical processes can be studied or achieved with such impressive precision (atomic clocks are nowadays at about eigthteen sixteen digits of precision and struggling to improve even more [627]), the need of using higher than double precision should be carefully evaluated and motivated.

A.3 Memory and Data Processing

As stated previously, computers store data and instructions in the Memory, with the CPU in charge of the data processing. To describe the main components within a CPU, hereafter we concentrate on the old i8086 CPU—introduced at the end of the 70's and used worldwide in most desktop computers until the beginning of the 90's—which provides a perfect playground to introduce the concepts. Clearly, modern CPUs have differences with respect to what described in the following. However, the main concepts are still valid and hopefully will provide the reader with the basic knowledge to begin exploring the modern CPU components and to understand the specific description provided by each vendor.

The core of the CPU is the Control Unit (CU)—called Execution Unit (EU) in the i8086—, which implements a finite state machine (an automaton with a finite number of states and of possible transitions between them) that controls the computing processes.

The CU goes repeatedly through two steps: first it *fetches* an instruction from the *instruction queue* and then dispatches the sequence of orders (μinstructions) needed to perform it to the other involved units, e.g., the ALU for the arithmetic operations. Concurrently, the instruction queue is fed with the contents of consecutive Memory locations. Indeed, most of the time the next instruction to be executed is implicitly the next one in Memory, with the notable exception of the *jump* orders, which specify explicitly where to fetch the next instruction. When a jump is executed, the instruction queue is flushed, and feeding restart from the new location.

In von Neumann's original architecture, instructions were fetched directly from the RAM: The instruction queue was introduced as performing one operation might be much faster than retrieving the next one from memory. Thus, more consecutive instructions are loaded at the same time to avoid the control unit to remain idle. This is the first example of a design tradeoff: augmenting the complexity and cost of the circuitry to gain in performance. The same idea lays behind the introduction of the *virtual memory* and of the *memory chaches* that we discuss next.

The memory where data and instructions are stored during a computation in von Neumann's architecture is such that the time needed to fetch information does not depend on its location: hence the name, somewhat confusing, of Random Access Memory (RAM).[1]

The memory device devoted to permanently storing the data and all the software needed for the computer to run is the hard disk. However, retrieving information from the hard disk is typically much slower than accessing the RAM. Besides, the capacity of the available RAM is typically much smaller than that of the hard disk and thus data and instructions need to be swapped back and forth between them. Indeed, it has been early recognized that the set of instructions decoupled

---

[1] Indeed, previous automatic calculators used different memory structures, where in general the time needed to retrieve a piece of information depended on its location. Nowadays, the same applies to the hard disks—the mechanical ones, not the solid state ones.

from the main program and dedicated to specific tasks (as those contained in single subroutines or functions) and the use of data structures such as vectors, introduce temporal and spatial correlations in the usage of data stored in memory [628]: The *principle of locality* in computer science states that programs tend to remain most of the time in contiguous blocks of memory. Thus, having an efficient way of handling contiguous blocks of memory can be highly rewarding.

The desired information is identified in the memory by *addresses*, which give the position of a specific byte in the memory. The width in bits of the addresses defines the *address space*: e.g., with 32 bits one has an address space of 2^{32} bytes. To simplify memory management and to introduce the possibility to mimic a RAM larger than that physically present, according to the *virtual memory* concept each program is endowed with a larger address space than that of the real RAM, typically $4GB$ vs $1GB$. The virtual memory is divided into *segments* of $64KB$ each, so that any address can be seen as composed of a *base*, identifying the segment (the most significant 16 bits) and an *offset*, the least significant 16 bits, identifying a byte within the segment.

During execution, according to the locality principle, only few virtual segment are actually present in the RAM at any time: the operating system, in charge of memory management, holds a map of the virtual segments bases to the physical ones currently in the RAM. This map is exploited by the compiler to store in dedicated *registers* (fast memory slots in the CU) the actual base of the segments in use, so that the real addresses can be computed from the content of the register and the offset contained in the instruction. This is done exploiting the arithmetic capabilities of the ALU, i.e., no special hardware is needed for address calculation. The advantage of such a structure is that whenever it is needed to move blocks of contiguous memory, it is possible to specify only once the segment address, sparing half of the address length for each instruction.

In the i8086 architectures, there are five dedicated sixteen-bit registers to store the segment addresses and the instruction pointer (segment registers) plus eight sixteen-bits *main registers*: four to hold the offsets within the segments, and four to store the data locally and perform the operations, also at the single byte level. Doubling the number of bits used to specify addresses, one can access a memory of $2^{32} = 4GB$. This was considered, until recently, far than enough for any reasonable application.

To further exploit the locality principle, cache memories have also been introduced: they are intermediate blocks of memory with a higher rate of response than the RAM and physically sit near the processor, so that also the communication to and from it and the CPU is much faster than from the CPU and the RAM. The following strategy is then adopted: whenever an instruction or some data need be retrieved from the RAM, a bigger block of memory contiguous to the indicated address is copied into the cache. According to the locality principle, most of the times this will speed up the computation, since the likely next needed information will already be present in the cache. In modern computers, there may be up to three levels (referred as cache L1, L2, and L3) with smaller and faster units nearer to the CPU.

An optimizing compiler will exploit the cache as much as possible. So, the code should be written to take advantage of the cache, or at least not to row against. Typically, when working with big matrices, knowing how they are stored in the memory (either by row or by column) and writing the operations accordingly, is fundamental to achieve high performances.

Nowadays, floating point operations are performed by a *coprocessor*, i.e., a dedicated piece of hardware (the i8087, in the first series). Initially, its performances were still pretty slow as it required one hundred *clock cycles* (the unit of time of the processor, nowadays usually of the order of GHz, at the time one thousand times slower) to perform a single operation and hundreds of cycles to perform a complex trigonometric function. Nowadays, not only the clock is much faster but also the chip structures are improved, and only one clock cycle is necessary to perform standard floating point operations.

A.4 Multiprocessors

When one aims at pushing to the limit his capabilities to simulate (quantum) systems, a possible strategy is to achieve the best performance out of the available resources—the main topic of this book; the other possible strategy is to increase the available resources. Thus, one can—and eventually should—think of running the code in parallel, distributing the work to many independent processing units. For decades the possibility to run code in parallel has been possible exclusively on large supercomputers. However, nowadays there are different options we will briefly introduce in the following, together with pointers to the relevant literature to guide the interested reader.

The main difference between the possible parallel strategies relies on the interface between the RAM and the CPU. Indeed, a general parallel network is composed of different nodes of a cluster, each of them containing some local memory shared by several processing units. In modern architectures, each node might be composed of different CPUs each of them composed of independent cores, acting as an independent processing unit. Each core can process information independently from the others, executing its thread. Nowadays, in standard clusters, there are from eight to thirty-two cores in each node. The communication speed among cores and the local RAM can be considered instantaneous compared to that between nodes which introduces some overheads in the computation time as we will see later on. However, while one can add virtually an unlimited number of nodes (top500 supercomputers have typically more than ten thousand nodes [629]) and is typically limited only by practical considerations (installation and maintenance costs) scaling the numbers of cores inside the CPU is a highly nontrivial engineering challenge.

The different parallel architectures have to be addressed with the proper software tool, which we list hereafter for completeness together with some reference for further reading.

OpenMP API (Application Program Interfaces) define a set of instructions to perform multi-thread, shared memory parallel calculations on multi-core machines. There are different implementations for almost all the popular programming languages [630]. Moreover, modern compilers allow us also some automatized multi-thread optimization via specific options, typically in combination with standard mathematical libraries, e.g., LAPACK libraries in the Intel MKL (Math Kernel Libraries) implementation [130, 631]. The use of OpenMP, at least in its most straightforward applications, can be highly rewarding at a limited cost in time to exploit them, although it is always limited by the numbers of cores present in each node.

MPI (Message Passing Interface) is a library specification to perform parallel computation between different nodes also in heterogeneous networks, implemented in most of the standard programming languages [632]. Its use, however, requires a higher investment to learn the basics concerning what is required to use OpenMP. However, MPI allows the user to virtually scale the computational resources of orders of magnitudes. Unfortunately, this does not guarantee the same speedup in computational time. Indeed, the speedup strongly depends on the particular computation: Amdahl's law states that any program has a part which is inherently sequential—it cannot be parallelized—as the loading of the data; and a part which can be parallelized [633]. The total computation time on a single processor is then $T = T_s + T_p$ (where T_s and T_p are the time spent in the sequential and in the parallel part respectively), while running the code on M processors results in at least a total time of $T_M = T_s + T_p/M$. Thus, in the limit of infinite processors, we have that $T_M \to T_s$. However, if one has to decide on how many processors run code in parallel, the interesting quantity to look at is the asymptotic gain per processors, that is

$$\lim_{M\to\infty} \frac{T}{T_M} = \frac{T_s + T_p}{T_s} = 1 + \frac{T_p}{T_s}.$$

Thus, in the limit of many processors, to achieve a high gain for each additional processor, we should aim to reduce T_s as much as possible, as in the limit of vanishing T_s the gain per process diverges. So, given that the condition $T_s \ll T_p$ holds, how much gain one could expect? Mainly, the losses arise in the communication between nodes during parallel algorithms and have two different sources:

1. Latency times, the time needed to establish a connection between nodes, which sum up to a time proportional to the number of transfers.
2. Transfer times, the time needed to transfer the data between nodes limited by physical factors as the speed of transfer, the bandwidth and the physical distance between nodes, which sum up to a total time proportional to the number of bits to be transferred.

Thus, we can classify some typical linear algebra operations, which lay at the heart of most of the algorithms for simulating quantum systems, studying the ratio between communication and computation time.

Operation	# bits sent	# operations	Ratio
$a \cdot a$	$O(1)$	$O(1)$	$O(1)$
$a \cdot \vec{v}$	$O(n)$	$O(n)$	$O(1)$
$\vec{v} \cdot \vec{v}$	$O(n)$	$O(n)$	$O(1)$
$\mathrm{O} \cdot \vec{v}$	$O(n^2)$	$O(n^2)$	$O(1)$
$\mathrm{O} \cdot \mathrm{O}$	$O(n^2)$	$O(n^3)$	$O(n)$

where a, $\vec{v}$, O are a scalar, a vector and a matrix respectively and n is their size. When the ratio is of the order one, the computation speedup will be limited by the network as the number of transfers scales as the number of operation to be performed. On the contrary, when the ratio scales with the problem size, we might expect substantial speedups coming from the parallelization for large n, that is, for big matrix size and "level 3" processes (composed by three nested loops).

Finally, different algorithms can be classified in three different classes according to their behaviour under parallelization, as even in the perfect scenario the gain is typically less than T_p/T_s. In this respect, one encounters:

1. 100% algorithms, formed by independent calculations such as those due to statistical sampling or exploration of the parameter space. Here the problem can typically be trivially paralleled, and one obtains a speedup linear with the number of processors M.
2. Semi-efficient algorithms, where the gain that can be achieved is still linear with M. However, for most of the time a significative number of processing units remains idle, wasting resources.
3. Costly algorithms where some gain is obtained but overheads should be added to allow for parallel computation.

To finalize this section it is worth mentioning that other parallelization architectures can be explored that, depending on the resources at hand and on the problem to be solved, could give significant speedups. Beowulf clusters connect standard machines in parallel and are pretty straightforward to set up [634]. This can be a solution to exploit to its maximum outdated or heterogeneous hardware, a cheap solution to enter in the world of parallel computation and extremely useful for didactic purposes. Another option which is attracting increasing interested in the last years is the use of OpenCL, a programming language to program parallel code running on different architectures also integrated with Graphical Processing Units (GPUs). GPUs are very powerful dedicated hardware initially developed to match the needs of powerful graphical processing by the software game industries. They can be usefully exploited to speed up some parts of scientific computations [635, 636].

A.5 Problems

1. Integer and real numbers have a finite precision. Explore the limits of INTEGER and REAL in FORTRAN:

(a) Sum the numbers 2.000.000 and 1 with INTEGER*2 and INTEGER*4.
(b) Sum the numbers $\pi \cdot 10^{32}$ and $\sqrt{2} \cdot 10^{21}$ in single and double precision.

2. Consider a quantum system formed by N (distinguishable) subsystems (spins, atoms, particles, etc..) each described by its wave function $\psi_i \in \mathbb{C}^D$ where $\mathbb{C}^D$ is a D-dimensional Hilbert space.

(a) Write a Fortran code to describe of N non-interacting systems and for a general N-body wavefunction $\Psi \in \mathbb{C}^{D^N}$. Comment their efficiency.
(b) Write a code to compute the density matrix of a general N-body wave function $\rho = |\Psi\rangle\langle\Psi|$.
(c) Characterize the functions introduced above for different N in terms of time and memory requirements. What is the biggest N you can reach?

Software in a Nutshell for Physicists

B

The goal of computational physics is to attack problems that cannot be treated with other methods, to complement experimental and theoretical studies, building trust on their results and exploring regimes of parameters out of their reach or where the analysis is particular difficult. This is particularly true for computational quantum physics, where the field of application of analytical tools is vastly challenged, e.g., in many-body systems. Especially when aiming to perform top-class academic research, the methods and tools are not only the bare application of state-of-the-art techniques: on the contrary, they are pushed to the limit, in a fast evolving scenarios where requirements and goals change very quickly. Indeed, differently from applied and commercial fields where software is developed, the specification of what the software shall do, under which circumstances and with which resources (software requirements specification) is not—and cannot be—precisely defined once for all, but keeps evolving as the use of the software increases the understanding of the problem under attack. Thus, when planning to develop scientific software, one should be aware that no matter how hard one tries, one will eventually ends to have to fulfill three requirements, that is to

R1 Change the software adapting it to the evolving situation and the corresponding needs;
R2 Work with a lot of data;[1]
R3 Solve hard problems with either large-scale and/or inefficient algorithms.

Given that the three requirements R1, R2, and R3 above are so general and ubiquitous, professional computational physicists shall learn how to address them, as much as they learn how to solve an integral or a particular technique in the

[1] Even though a computational physicist might not work with "Big Data" in proper sense, the amount of repetitive tasks requested by careful studies is typically far from what can be done by a human being.

T. Felser, S. Montangero (eds.), *Introduction to Tensor Network Methods*,
Graduate Texts in Physics, https://doi.org/10.1007/978-3-032-17635-6

particular field of study they are devoted to. Indeed, the three aforementioned requirements call for the use of some strategies that have been developed in the last decades and in particular:

S1 Software Engineering: writing good, flexible software, easy to debug and expand;
S2 Scripting: automatize work as much as possible as pre- and post-processing will require as much work as producing the data;
S3 Optimization and HPC: be as efficient as possible, and be prepared to use "brute force" if needed.

This chapter initiates the reader into the aforementioned strategies, which shall be adopted to attack scientific computations in the most efficient way. To attack the seemingly overwhelming challenges described before, a good starting point are the set of priorities described below that anyone writing scientific software should follow as much as possible [120, 121]. Those priorities are valid in general for any scientific software (and most of them for any software), however, in the following they will be specialized for quantum physics for the sake of clearness. They are, in order or priority and accompanied by a clear motivation for such priority,

1. Correctness—to be useful;
2. Numerical Stability—to be trustful;
3. Accurate Discretization—to describe nature;
4. Flexibility—to be used;
5. Efficiency—to do a good job!

In the rest of this chapter, we introduce the reader to the most important aspects of each of them, that one should keep in mind to write good scientific software.

B.1 Correctness

The fundamental requirement for a good scientific software is obviously that they should be correct, that is, it shall perform the correct sequence of instructions, written in the correct syntax according to the programming language used and version and compiler compliant. However, even for highly skilled and experienced programmers, the probability of writing some lines of code without any error is essentially zero. Thus, actions and strategies to correct the unavoidable errors has to be developed and undertaken. The first support comes from the *compiler*, a program that translate the source code into a lower-level code. Typically, compilers manage to identify syntax errors and are constantly improved to include more and more possible error scenarios. However, even thought the compilers runs without signalling errors and produces the compiled code, the executable program, might still crash due to many different causes such as, for example, a memory allocation problem (the program tries to write in a memory position that is not allocated), an *obi-one* (Off-by-one, a recursive index start from zero instead than one or vice-versa). Finding and correcting these and other errors is the “art of debugging" and requires time and experience. However, there are some good practices that can be adopted to speed up the process, briefly introduced below:

- Use the print statement to check the variables name together with compiler options which allows one to select lines of code for debugging (as, for example, the $-d$ option of the ifort compiler together with starting the line of the debugging test with a d):

  ```
  d   PRINT *, "The value of x is", x
  ```

  This combination is a quick way to reduce the time of debugging as the lines of code written for this purpose are not cancelled but remain there, for later and repeated use when another bug appears.
- For more sophisticated checks, when multiple lines of code shall be written and/or it can be reused in other part of the program, some dedicated SUBROUTINE or FUNCTION might be defined, with an enabling global variable and extended flags and output. The simplest example of such code might assume the form of

  ```
  FUNCTION Checkpoint
  IF (DEBUG)
     PRINT *,  "The value of x is", x
  ENDIF
  END FUNCTION
  ```

- Finally, the use professional debuggers such as, for example the GNU debugger GDB, even with the help of a Graphical User Interface (DDD) [637,638] is highly recommended above a certain level of code complexity.
- Producing output that is compatible with the input is a good practice that allows to automatically generate test of increasing complexity for the program. For example, if one is working with matrices or tensors, if the input and output styles match the output can be easily used as new input without the need of time-consuming format processing.This compatibility is also extremely useful in the data acquisition stage as in many cases the program might be restarted (e.g., due to computational limit time reached, additional check of convergence etc.).
- Include in the program sections “Errors and Exceptions" for known issues: even if one is not interested in correcting them, signalling them and including explicitly a test to exclude those exceptions will save an enormous amount of time during the debugging. For example, if while implementing a given algorithm a particular case is excluded, as it would require an additional effort not needed at the moment, including an exception might be highly beneficial when the same piece of code might be reused (see later, Sect. B.4) and might need exactly the excluded scenario: after months or even years, uncovering the exception might require hours of hard debugging.
- Perform testing against random input: even though random input is typically not ill-conditioned and thus it will possibly not cover all cases (for example random matrices are never degenerate [133,639]), random testing allows one to perform a

huge number of automated testing. Again, automated testing software is available for most common programming language.

If the previous practices are followed, it should be possible to implement different strategies to have an effective and efficient debugging. In particular, one can perform *Incremental Testing*, that is to test and debug every part of the program while writing it: debugging different parts of the code separately is highly simpler than debugging them all together. However, even if all the parts are bug free, nothing assures that while combining them some errors will occur. Thus, another good strategy to adopt is *Regression Testing*, that is, to store the situations that produce bugs when they are spotted in a set of automated tests. Every time changes are made, one can rerun all tests which, provided they have been carefully planned and implemented, should highly improve the debugging speed. One can also embrace the *Extreme Programming* paradigm which relay, among others aspects, on designing and coding the tests even before writing the code itself. In such a way one should be able to easily automatize them and run them as much as possible [640].

Finally, even though the compiler compiles and the program runs without crashing, how can we be confident that the outcome is the correct one, that is, that the program is computing exactly what was written for? To increase the confidence in the output (and also to help in the debugging), it is useful to include some *proof of correctness* in the code, such for example:

- In every section of the code, even if they are trivial, include explicitly in form of comments and—if possible—of automated checks, *preconditions* and *postconditions*, with a clear message in case they are violated. This will cost the programmer some time, but will save an enormous amount of time when a bug as to be located and/or the code will return results which do not match your expectations. Typical examples of preconditions are the sign of a variable (e.g. if known to be positive), the norm of a vector, the dimensions of vectors, matrices, and tensors which typically should match. Postconditions can be individuated in some properties of the solution (e.g. eigenvector normalization, function monotonicity, expected result for some particular parameters, etc.) which should hold despite of the (unknown) value of the solution itself.
- Define loop-invariants if present as they can provide a proof for induction of the correctness of that part of the code: check the loop-invariant for a given value of the recursive index i, and check that if it is true for i then it is true also for $i + 1$.
- In physics and whenever it applies, typically one can individuate some trivial (e.g. wave function normalization, system energy, etc.) and less-trivial constants of motion which should be monitored.
- Comparisons with other programs results and with regimes where the solution of the problem is known (e.g. non-interacting case, single or few particles, low dimensional, etc.) should be performed before exploring the new regimes.

B.2 Numerical Stability

Even though all the aforementioned practices are strictly followed and the code is up and running and has passed the numerical tests—and thus one can safely assume that the program is correct and executes the algorithm has been written for—there might be cases where the output presents strange results that cannot be trusted. So, even though the program is formally correct, the results shall be discarded: this is typically due to *numerical instabilities* which are mainly the result of the difference between real-numbers and floating-point arithmetics. To detect and avoid such instabilities might not be easy but there are a number of typical scenarios—described below—that one shall avoid while writing code [121].

- Never use floating point variables to test for equality: the lines of code

```
1 REAL*8 x,y
2 x=y
3 PRINT *, (x.eq.y)
```

  might result in a FALSE output as the test for equality depends on the details of the architecture and compiler and on the check on the digits beyond the finite precision of the double precision variables. A check on the difference between the two floating point variables shall instead always used, e.g.,

```
1 REAL*8 x,y
2 err= 1d-14
3 IF (abs(x-y).lt.err) THEN
4    PRINT *, "x is equal to y"
5 ENDIF
```

- Similarly, never use floating point variables as counters. Indeed the cycle

```
1 REAL*4 x,C
2 DO (x=0.0; x<= 1.0; x = x+ 1.0/C)
```

  will increase the variable `x` by `1.0/C` every cycle. So, the cycle will loop $C - 1$ or C times depending on the round-off error on the finite precision representation of the incremental fraction, drastically and randomly affecting the result of the computation.
- Be careful with the result of the subtraction of two finite precision variables a and b if $|a - b| \ll |a|, |b|$. Indeed, if the round off errors of the two variables have opposite signs one can show that the relative error on the difference $c \equiv a - b$ is proportional to $\Delta c/c \propto (a + b)/(a - b)$, which can easily explode beyond the floating point precision.[2] This error can occur frequently as differential calculus is obviously based on differentials—very small differences between two quantities, e.g. the first order approximation of a derivative is $f'(x) \sim \Delta f/\Delta x$.

[2] To prove that simply consider the variables with their floating point errors and compute the relative error on their difference, i.e. $\hat{a} \equiv (1 + \epsilon_a)a$, $\hat{b} \equiv (1 + \epsilon_b)b$, $c \equiv (1 + \epsilon_c)c$ and compute the relative error on c given that $\epsilon_a \sim -\epsilon_b$.

When the possible values of the independent variable span different orders of magnitudes the computation of $\Delta x = x_2 - x_1$ can suffer exactly of the aforementioned problem, resulting in drastic reduction fo the result precision if not in a completely random results.

– Finally, as all scientists know, the errors propagates during arithmetical calculations and given the fast growing number of operations that modern computers can do, they shall be carefully monitored. In particular, operations like factorial and repeated multiplications shall be handle carefully. For example the cycle

```
REAL*8 f
f=F
DO (k=0; x<N; k = k+1)
   f = k*f
```

will give a completely wrong error as soon as $N \sim 20$. The reason for that is the propagation of the error in f_0 as, being a finite precision number one shall always assume $\hat{F} \equiv (1+\epsilon)F$: the error propagates rapidly (however being initially so small it will be hardly noticed) and then explodes in a few iterations corrupting the result in what seems an unpredictable behaviour.

B.3 Accurate Discretization

Typically, computational physicists develop codes to simulate physical processes, which in most cases can be described by a set of partial differential equations (PDEs). While simulating PDEs, to be sure that the developed program is providing useful information, the last check to be performed (after those described in the previous sections) is relation of the simulation results with the physical reality. Indeed, a simulation can be correct and numerically stable, however to be useful it should provide a correct description of the phenomena subject of study.

For example, the PDE can be a particular instance of the Schrödinger equation, which being a first order in time PDE, can be formally represented as

$$\begin{cases} \partial_t \psi(x,t) = \mathcal{L}[\psi(x,t)] \\ \psi(x,0) = \psi_0(x) \end{cases}, \tag{B.1}$$

where the ∂_t is the partial derivative with respect to the time variable, $\psi_0(x)$ represents the system's initial wave function and the Liouville operator $\mathcal{L} \equiv -i\hat{H}/\hbar$ is build from the system Hamiltonian $\hat{H}$. Solving Eq. (B.1) in all its possible variants is the main subject of this book, and the solution can be formally written (for time-independent Hamiltonians) as

$$\psi(x,t) = e^{\mathcal{L}t}\psi_0(x) \equiv U_t\psi_0(x), \tag{B.2}$$

where U_t is the time-ordered exponential, i.e. the time-evolution operator which maps the wave function at time zero to any following time t. Solving the problem numerically equals to computing the action of the time-evolution operator on the initial wave function. In this apparently simple operation, there is a formal step whose possible consequences shall not be underestimated: the mapping of a continuous space-time $C \equiv (x, t)$ into a discrete one necessary, due to the floating point arithmetic of computer simulations. That is, setting the precision of the program and choosing the kind of variables, e.g. REAL*4 or REAL*8, we approximate the continuos space time with a discrete one $C \approx C_\Delta$, with $\Delta \equiv (\Delta x, \Delta t)$. Typically, the first choice is to employ a linear uniform grid in space time, such that $x_\ell = \Delta x \cdot \ell$ and $t_n = \Delta t \cdot n$ with $\ell = 1, \ldots, N_x$ and $n = 1, \ldots, N_t$ even though more sophisticated choices are possible, such as the use of adaptive grids or of finite elements methods [128]. However, independently from the particular grid choice, in general the discretization introduces an error such that

$$\begin{cases} U_t = U_t^\Delta + E^\Delta \\ \psi = \psi^\Delta + e^\Delta \end{cases} ; \tag{B.3}$$

where the first addends in the right hand sides of the equations are the discretized versions of the left hand sides and E^Δ and e^Δ the corresponding errors. In conclusion, in presence of a discretization procedure, to avoid obtaining correct but not significant results, the simulation *consistency* and *accuracy* of the results shall always be carefully checked and kept under control.

In particular, the consistency of the simulation with the physical reality requests that

$$\lim_{\Delta \to 0} U_t^\Delta = U_t \text{ and } \lim_{\Delta \to 0} E^\Delta = E_0 = 0. \tag{B.4}$$

On the contrary, if $E_0 \neq 0$, then an anomaly is present which typically signals that a symmetry has been broken (parity, time-inversion, etc.) and which shall recovered. Again, the use of physical invariants might be extremely useful as physical invariants, after discretization, shall become numerical invariants, that is constant quantities within the numerical error.

Finally, the accuracy of the discretization procedure shall be individuated, that is, the scaling of the errors E_Δ with Δ. Indeed, knowing the scaling power p, such that $E_\Delta \sim \Delta^p$, allow us to maximize the resources (time and memory) needed to achieve a desired precision. A paradigmatic example of such scaling is given by the discretization procedure of the first order derivative: while approximating the first order derivative via a discrete grid, the straightforward approach (the *forward derivative* FD) is

$$FD(x) = \frac{f(x + \Delta) - f(x)}{\Delta}. \tag{B.5}$$

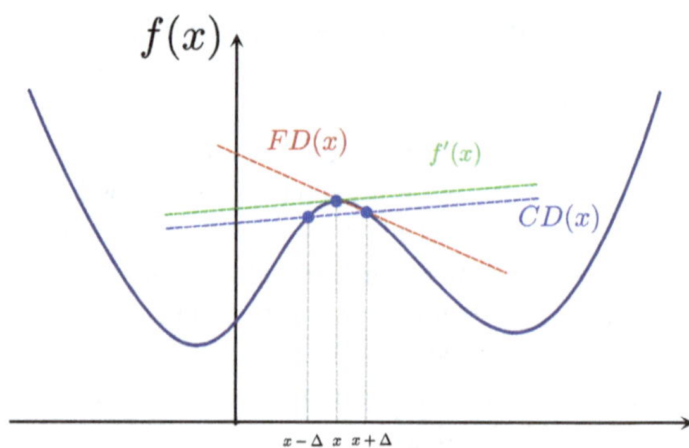

Fig. B.1 The exact derivative (*green dashed line*), the forward and the central discrete derivative (*red and blue dashed lines* respectively) according to Eqs. (B.5) and (B.6) of a function around an extremal point

The textbook definition of the derivative assures that with the above definition $\lim_{\Delta\to 0} FD(x) = f'(x)$. However, the study of the scaling of the error with Δ suggests a better strategy to compute the discrete derivative: by means of a Taylor expansion of the function $f(x) = \sum_n a_n x^n$ and $f'(x) = \sum_n a_n n x^{n-1}$, it is possible to easily show that the error introduced by the discretization is of order $O(\Delta)$, as $f(x \pm \Delta) = \sum_n a_n (x \pm \Delta)^n = f(x) \pm f'(x)\Delta + \sum_n [n(n-1)x^{n-2}/2]\Delta^2 + \ldots$ and thus $FD(x) = f'(x) + O(\Delta)$. Along the same lines, it can be shown that an alternative discretization procedure might be more favourable: the *central derivative*

$$CD(x) = \frac{f(x+\Delta) - f(x-\Delta)}{2\Delta} \tag{B.6}$$

is such that $CD(x) = f'(x) + O\big(\Delta^2\big)$. The reason for that can be easily grasped by means of the sketch presented in Fig. B.1, and lies again on the fact that the forward derivative breaks the problem symmetry with respect to the point the derivative is computed in (the grid point $x + \Delta$ assumes a prominent role with respect to the previous one $x - \Delta$). Restoring the symmetry improves drastically the precision, especially on the functions obeying to such symmetry—the even functions and in particular the lower order ones, quadratic. Notice that the two methods to compute the derivative provided above require exactly the same number of operations, thus there is no reason to use the forward derivative instead of the central one. However, typically this is not the case, and to solve the same problem there exist low- and high-order algorithms with different (decreasing) efficiencies.

Despite the paradigmatic example provide above, the estimation of the errors introduced by the discretization procedure and the search of alternative higher-order schemes is typically highly non trivial for complex algorithms and most of the results are obtained via numerical benchmarking. In general, one should find the right trade-off between high-accuracy methods (whose efficiency typically scale badly with the number of grid points) and low-accuracy but efficient methods.

B.4 Flexibility

Once the first three priorities are satisfied (guaranteeing that the code is correct, numerically stable and describing correctly the physics of interest) the written software has the potential to be used, in some cases also for years. Depending on the the specific cases, it might become a part of another more complex software and eventually might become a standard for a group of users, a community of specialists or even non-expert users. While writing a computer program, one never really knows what a potential future user might want to do this it. Here the future user might be the very same programmer one month or a year later aiming to reuse some part of the already written software; another highly skilled programmer that found your code interesting or useful; a software company that is planning to integrate it in their commercial suite; the programmer's boss; a student that is looking for software to develop a master or Ph.D. thesis; or even the IT director of a big software project or research institution that might transform the program in the core of a multi-user multi-purpose research software.

Every software undergoes what is called the *software lifecycle*, depicted in Fig. B.2: it is first described by the *requirements* or specifications that detail the properties of the software and its requested output; the specifications are then translated into a *software design* which assures that the requirements will be met; the design is then implemented in a code in a specific programming language. Finally, the software is tested, debugged and once it is reliable enough, used to produce data. Eventually, while the software usage will push the programmer to either change the specifications and/or the software design, thus closing the lifecycle. The lifecycle can go on even for decades for the more used and famous softwares, e.g.

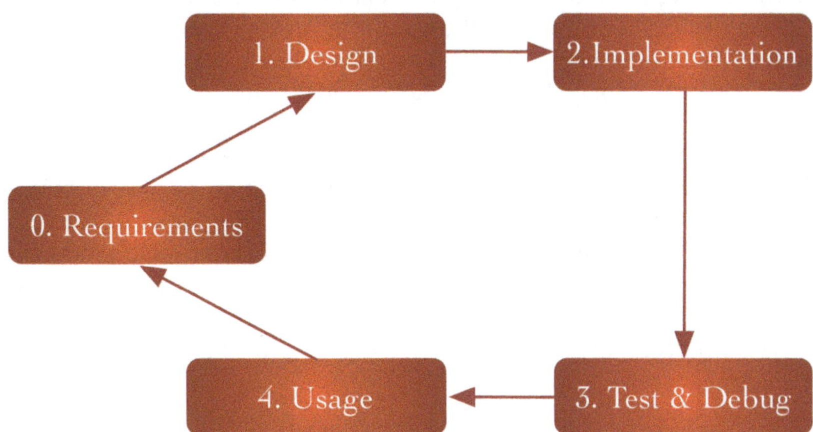

Fig. B.2 The software life cycle: After defining the software requirements, the software design phase enables the implementation phase. The source code is then tested and debugged and finally used. The code usage or changes in the environment will identify the need of additional or different requirements, which will trigger another round of the software lifecycle operative systems (Windows, OSX, Linux, etc..), large productivity (e.g. Office) or scientific (AUTOCAD, Matlab, etc.) suites. The changes in the requirements or in the software design might be triggered by different aspects that influence a software development such as, the need for new features (increasing systems size or variety, new protocols or data acquisition, etc.), bug fixes, new hardware, dependencies on other software, and changes of the targeted user (e.g. from experts to general audience).

Software engineering is the science dedicated to design and implement software, and thus to develop tools, concepts and strategies to cope with challenges that software development and its lifecycle pose. Sophisticated strategies and tools might be not relevant while programming small single-user software but they become more and more fundamental with the growing complexity of the software. As modern science is almost always based on the performances of complex software, it is highly probable that any scientists will encounter a large complex software in his academic life. Thus, it is important to know at least some basic concepts of software engineering. The first of all is that while programming one shall think generally, leave open doors and shall not specialize the software if it is not strictly necessary. If the generalization is not possible as one is writing an highly specialized part of the code, then one shall write a library that can be reused easily in the future. In general, one should aim to a good balance between being abstract and the concrete implementation that necessary are needed to write code. How to achieve that is a matter of experience, however, some general strategies shall be adopted to achieve the best possible result compatible with the developer experience and skills. Adopting them, even though they require an overhead of time while programming, will highly probably result in a huge increase of efficiency in the long run.

In particular, as said before, a software developer performs different tasks at different level of abstractions that goes from the software design to the implementation of the single line of code, and finally the interfaces between parts of program. Hereafter we briefly present some good habits that might help to focus on the important aspects of each of them and to avoid the most common errors.

Software Design When planning a novel software one shall go through the following list of points and try to answer to the related questions: depending on the complexity of the software to be developed some of them shall be carefully addressed and will influence the whole software development, while others might be skipped. In any case, spending some time on them, once again, might save a lot of time in later stages of developments and surely will help in focusing on the important aspects of the programming work while neglecting or at least postponing less important ones. A possible useful checklist is:

1. Specifications:

   Program objectives: what are the essential goals of the software, what is not going to be included?

Performances: will the performances be fundamental for the success of the project? Will the program run on a single machine, a cluster or on HPC architecture?

Input/Output: what will be the format of the inputs and outputs and their compatibility (e.g. with other programs)?

2. Data structures and functions: what are the data structures that will be needed? Is there the need to define new ones with their own functions?
3. Libraries and language: Are there already available libraries to perform some needed tasks? Which programming language provides the better tools?
4. Coordination: Is there the need to coordinate different programmers and the software development? Which tools will be used to support them?
5. Portability: Is there the need to have a portable software (i.e. on different platforms and architectures)? If so, how that will be achieved?

For a typical scientific project the answers to many of these questions are eventually quite straightforward: in particular, for most of quantum physics related projects, standard tools such as FORTRAN, C++, python—together with LAPACK (Linear Algebra PACKage [130]) mathematical libraries and some versioning and debugging tools—will do the job. However, the answering to these questions will force the scientist to better structure the software engineering part of his/her job and to avoid common errors which can undermine the success of months-long hard work.

Interfaces Once the software structure has been individuated, the data structures and the needed functions have to be coded: this means typically to write numerous parts of code (class definitions, functions, etc.) which shall be later interfaced. This operation is highly non trivial and the strategy used to attack it might influence the final result. Indeed, during this operation, one shall try to exploit as much as possible:

1. Mathematical abstractions: in scientific software typically it is straightforward, in quantum physics linear algebra and complex analysis provides the necessary tools to describe the systems of interests.
2. Information hiding: Subroutines and functions shall allow easy access to the information they contain in the interfaces and nothing more. This will avoid unclear program parts dependencies, and allow one easier debugging and software update.
3. Flexibility and modularity: Abstractions helps to increase the software flexibility as well as modularity which is the key to write good, efficient, reusable and useful software. For example, due to these characteristics, the BLAS (Basic Linear Algebra Subprograms [631]) can and are easily used as a basis for most of the scientific software available.

Hereafter we briefly describe some possible ways to organize the work, together with some comments on their effectiveness [121].

Top-down : The initial design is implemented starting from the higher levels, and refined in increasing detailed and concrete steps until the software is working. This strategy might be followed but it tends to reduce the final software modularization and increase the complexity of the debugging process.

Bottom-up : The data-structures are defined together with the functions operating on them, after that the higher-levels are coded until the whole design is complete. This is an effective strategy which however requires a very careful planning and typically high coding expertise to be exploited at best.

Middle-out : it is possible to follow an intermediate strategy that exploits the good properties of the two previous ones and reduces their risks, however it requires some courage as sometimes it requires to make some steps backwards. However, forcing the coder to refine the previously done job, tends to result in a easier to debug and higher quality code. The steps to follow are:

1. Identify the overall instructions at the higher level;
2. Identify the components needed to have a minimal working software;
3. Code the minimal components to have a working software (in their minimal version);
4. Test the software as you build it, keep the old tests and update them and re-apply them after changes;
5. Build the software components of increasing complexity starting from near the bottom level;
6. Once the minimal version of the software is working and debugged, add functionalities to reach the final goal starting from point 1.

As it can be clearly seen, this strategy will force the coder to re-design the data-structure and to rewrite parts of the code while proceeding. This should be possible (even though clearly not painless) in a well-written, test-friendly code and by construction shall result in a information-hiding, flexible, and modular code.

Coding Once the software is designed and the strategy chosen, the only thing left to do is the coding itself. Independently from the choices made, coding is a matter of knowledge, skill, patience, and experience. Hereafter some good practices are presented to guide the newcomers into the process of mastering this art.

1. A famous quote among programmers is "Premature optimization is the root of evil" which leaves little to add, if not that this is the reason why in this chapter efficiency is left as the very last point.
2. Variables and functions name shall be chosen carefully and properly. Indeed, the function

```
FUNCTION f(x)
REAL*8 x,f
f=x**2
END FUNCTION
```

is characterized by short, fast-to-write names which clearly have the advantage to increase writing speed. However, their names do not give any information about their usage or meaning, if embedded in other hundreds of such names, the coder will have hard times to remember all of them. Even more dangerous is the fact that one typically need to automatically find variables and functions in thousand lines of code: an automated search (or even worse a find and replace one, a common practice while restructuring codes) will miserably fail. On the contrary the same function written as

```
FUNCTION xsquare(xvariable)
REAL*8 xsquare,xvariable
xsquare=xvariable**2
END FUNCTION
```

slows down the coding without introducing any benefits. Thus, the best solution is to adopt a careful balance between the two extreme examples presented here, which requires the due attention in the definition of the variables name.

3. Commenting the code is always hard time for the coder, however it is a fundamental step for the aforementioned reasons. A good starting point towards good commenting practice is to look the professional libraries (e.g. BLAS libraries) and try to reproduce their standards which allow the user to compile the documenation as html from source code. However, here below some guidelines are reported:
   A. Comment the interfaces as they will be the most frequent information you will need, as for example:

```
c NORMV() computes the norm of the vector a(i) (i=1 to n)
c
c a input array
c n length of array
c norm= sqrt(sum a(i)^2) (flag=0)
c norm= -1  if n less or equal zero (flag=1)
c
  SUBROUTINE NORMV(a,n,flag)
  ...
```

   B. Declare loop invariants and conditions.
   C. State explicitly warnings for unusual behaviour.
   D. Describe what is done, not how: the implemented algorithm can be explained in the code documentation or simply refer directly to the bibliographic reference.
   E. Avoid redundant comments which increase the code length and waste time twice: to write them and to read them while looking for useful information later on.
   F. Keep the comments updated as outdated comments can drastically slow down the debugging and future reuse of the code than no comments at all.

B.5 Efficiency

The code efficiency is the last priority of the list presented in this chapter as only when all the previous ones are correctly addressed it make sense to worry about it. However, it is still a fundamental issue that might make the difference between a state-of-the-art and an amateur software and thus might open or prevent the possibility of conducting cutting edge research. Software complexity is a highly developed theory and we refer the interested reader to the literature [127]. Here we only briefly recall two results on two different definition of complexity that have implications in computational physics:

Algorithmic complexity is possibly the most widespread definition of complexity and classifies algorithms in terms of the scaling of the resources (time or memory) needed to run an algorithm as a function of the input size N. This definition allows one, beside the fundamental introduction of complexity classes [127, 641], to quantify the performances of different algorithms and distinguish between difficult ones (e.g. scaling exponentially or even super-exponentially with N) and practical ones (polynomial scaling, $O(N^p)$, with some low power p).

Smoothed complexity has been recently introduced as the scaling of the resources needed to run an algorithm as a function of the input size N *in presence of noise*. This apparent small difference is actually very important for two reasons: the first one is that noise is unavoidable in physical systems and also that finite precision arithmetic can be seen as a form of noisy arithmetic. The second one is that very recently it has been shown that many important algorithms which solve very general problems do have super-polynomial algorithmic complexity, but are also characterized by polynomial smoothed complexity [642, 643]. This implies that inefficient algorithms in terms of algorithmic complexity can be practically used as, for example, the simplex algorithm to solve linear programming problems [644].

Independently from the complexity measure chosen, in most cases there are different algorithms with different complexities to solve a given problem. The difference in complexity usually shows up in the difficulty and length of the algorithm to be coded, typically with the better scaling algorithm involving longer coding. However, the quest for efficiency calls always for a better scaling algorithm than a slower one, even if the difference is from $O(N^2)$ to $O(N \log N)$. Moreover, if the program is well organized and written, it should be easy to implement both algorithms, the first one as a quick start and as testbed for the more performing one.

Additional practices that allow boosting efficiency are the use of compiler optimization flags (e.g. the –O# and –fast options of the ifort compiler), the use of a profilers to find program bottlenecks (e.g. gprof or similar) and the carefully handling of memory allocation.

The final steps towards efficient computation are the exploitation of parallel architectures and in general of High Performance Computing tools, together with the use of supercomputers and big computational clusters at national and internationl levels [126]. However, these additional steps might require huge investments which shall be considered only when all other options have been explored and the only remaining possible step to increase efficiency is to invest in huge computational power.

A final considerations is in order: in modern professional scientific computing there is no room for the NIH (Not Invented Here) syndrome. So, there is no room for self-written libraries for standard algorithms (e.g. linear algebra). It is much more effective to spend time to learn how to use professional available libraries (e.g. BLAS, LAPACK, etc...) than starting from scratch. Indeed, usually "95% of execution time is spent in 5% of the code" and for scientific (quantum simulation) software this time is spent in basic (linear algebra) operations. Thus, the most efficient strategy is to rely on the work of years (in some cases decades!) of the maximal experts in the field: the probabilities to improve their codes without spending a similar effort in coding with similar skills is practically zero (unless of course you are part or aim to become part of the small community of high-level experts that are developing those libraries).

B.6 Problems

1. Matrix multiplication is often the bottleneck of linear algebra computations.

(a) Write explicitly the matrix-matrix multiplication loop in two different orders (column-row and viceversa):
(b) Use the FORTRAN intrinsic function.
(c) Use the LAPACK subroutine

Increase the matrix size and use the FORTRAN Function CPUTIME to monitor the code performance.

2. Include documentation, comments, pre- and post- conditions, error handling, and checkpoints in the previous exercise.

References

1. D. Bacilieri, M. Ballarin, G. Cataldi, A. Costantini, D. Jaschke, G. Magnifico, S. Montangero, S. Notarnicola, A. Pagano, L. Pavesic, D. Rattacaso, M. Rigobello, N. Reinić, S. Scarlatella, P. Silvi, D. Wanisch, *Quantum TEA: Qtealeaves* (Zenodo, Geneva, 2024). 10.5281/zenodo.13383350
2. D.J. Thouless, *The Quantum Mechanics of Many-body Systems* (Courier Dover Publications, Garden City, 2013)
3. M.D. Schwartz, *Quantum Field Theory and the Standard Model* (Cambridge University Press, Cambridge, 2014). ISBN: 978-1-107-03473-0
4. P.A. Lee, N. Nagaosa, X.-G.Wen, Doping a Mott insulator: physics of high-temperature superconductivity. Rev. Mod. Phys. (2006) 10.1103/RevModPhys.78.17
5. A. Mann, High-temperature superconductivity at 25: Still in suspense. Nature (2011) 10.1038/475280a
6. A. Szabo, N.S. Ostlund, *Modern Quantum Chemistry: Introduction to Advanced Electronic Structure Theory* (1996). ISBN: 0-48669-186-1. 10.1119/1.1973756
7. A.G.J. MacFarlane, J.P. Dowling, G.J. Milburn, Quantum technology: the second quantum revolution. Philos. Trans. A Math. Phys. Eng. Sci. **361**(1809), 1655–1674 (2003). https://doi.org/10.1098/rsta.2003.1227
8. M.A. Nielsen, I.L. Chuang, *Quantum Computation and Quantum Information* (Cambridge University Press, Cambridge, 2010). ISBN: 978-0-511-97666-7. 10.1017/CBO9780511976667
9. G. Kurizki, P. Bertet, Y. Kubo, K. Mølmer, D. Petrosyan, P. Rabl, J. Schmiedmayer, Quantum technologies with hybrid systems. Proc. Natl. Acad. Sci. U. S. A., 2015. 10.1073/pnas.1419326112.
10. Y.-C. Cheng, G.R. Fleming, Dynamics of light harvesting in photosynthesis. Annu. Rev. Phys. Chem. (2009). 10.1146/annurev.physchem.040808.090259
11. G.D. Scholes, G.R. Fleming, A. Olaya-Castro, R. van Grondelle, Lessons from nature about solar light harvesting. Nat. Chem. **3**(10), 763–774 (2011). 10.1038/nchem.1145. PMID: 21941248
12. X.-G. Wen, Topological orders in rigid states. Inter. J. Mod. Phys. B **4**(02), 239–271 (1990). 10.1142/S0217979290000139
13. B. Sutherland, *Beautiful models: 70 years of exactly solved quantum many-body problems* (World Scientific, Singapore, 2004)
14. J.M. Zhang, R.X. Dong, Exact diagonalization: the Bose–Hubbard model as an example. Eur. J. Phys. **31**(3), 591–602 (2010). 10.1088/0143-0807/31/3/016
15. L.P. Kadanoff, More is the same; phase transitions and mean field theories. J. Stat. Phys. (2009). 10.1007/s10955-009-9814-1
16. C. Hätting, Electronic structure: Hartree-Fock and correlation methods. Multiscale Simulation Methods in Molecular Sciences in J. Grotendort, N. Attig, S. Blügel, D. Marx (Eds.), *Institute*

T. Felser, S. Montangero (eds.), *Introduction to Tensor Network Methods*,
Graduate Texts in Physics, https://doi.org/10.1007/978-3-032-17635-6

for Advanced Simulation, Forschungszentrum Jülich, NIC Series, Vol. 42, pp. 77–120 (2009). ISBN: 978-3-9810843-8-2
17. C.F. Fischer, General Hartree-Fock program. Comput. Phys. Commun. (1987). https://doi.org/10.1016/0010-4655(87)90053-1
18. V.L. Ginzburg. Fiz. Tverdogo Tela **2**, 2031 (1960). English Transl. Soviet Physics-Solid State, 1960
19. R.Y. Rubinstein, D.P. Kroese. *Simulation and the Monte Carlo method* (John Wiley & Sons, Hoboken, 2011)
20. E.Y. Loh, J.E. Gubernatis, R.T. Scalettar, S.R. White, D.J. Scalapino, R.L. Sugar, Sign problem in the numerical simulation of many-electron systems. Phys. Rev. B (1990). 10.1103/PhysRevB.41.9301
21. M. Schiró, M. Fabrizio, Real-time diagrammatic Monte Carlo for nonequilibrium quantum transport. Phys. Rev. B (2009). 10.1103/PhysRevB.79.153302
22. M. Troyer, U.-J. Wiese, Computational complexity and fundamental limitations to fermionic quantum Monte Carlo simulations. Phys. Rev. Lett. (2005). 10.1103/PhysRevLett.94.170201
23. K. Wilson, The renormalization group: critical phenomena and the Kondo problem. Rev. Mod. Phys. (1975). 10.1103/RevModPhys.47.773
24. S.R. White, Density matrix formulation for quantum renormalization groups. Phys. Rev. Lett. (1992). 10.1103/PhysRevLett.69.2863
25. U. Schollwöck, The density-matrix renormalization group in the age of matrix product states. Ann. Phys. (N. Y). (2011). 10.1016/j.aop.2010.09.012
26. R. Orús, A practical introduction to tensor networks: Matrix product states and projected entangled pair states. Ann. Phys. **349**, 117–158 (2014). 10.1016/j.aop.2014.06.013
27. I. Peschel, X. Wang, M. Kaulke, K. Hallberg (eds.), *Density-matrix renormalization - A new numerical method in physics*. Lecture Notes in Physics (Springer, Berlin, 1999)
28. K.A. Hallberg, New trends in density matrix renormalization. Adv. Phys. (2006). 10.1080/00018730600766432
29. S. Sachdev, Viewpoint: tensor networks-a new tool for old problems. Physics **2** (2009). 10.1103/physics.2.90
30. I.P. McCulloch, From density-matrix renormalization group to matrix product states. J. Stat. Mech. Theory Exp. (2007). 10.1088/1742-5468/2007/10/P10014
31. R. Orús, Advances on tensor network theory: symmetries, fermions, entanglement, and holography. Eur. Phys. J. B **87**(11) (2014). 10.1140/epjb/e2014-50502-9
32. P. Silvi, F. Tschirsich, M. Gerster, J. Jünemann, D. Jaschke, M. Rizzi, S. Montangero, The Tensor networks anthology: simulation techniques for many-body quantum lattice systems. *SciPost Physics Lecture Notes* (2019). 10.21468/SciPostPhysLectNotes.8
33. M. Srednicki, Entropy and area. Phys. Rev. Lett. (1993). 10.1103/PhysRevLett.71.666
34. T. Xiang, J. Lou, and Z. Su, Two-dimensional algorithm of the density-matrix renormalization group. Phys. Rev. B (2001). 10.1103/PhysRevB.64.104414
35. Ö. Legeza, J. Sólyom, Quantum data compression, quantum information generation, and the density-matrix renormalization-group method. Phys. Rev. B (2004). 10.1103/PhysRevB.70.205118
36. G. Vidal, J.I. Latorre, E. Rico, A. Kitaev, Entanglement in quantum critical phenomena. Phys. Rev. Lett. (2003). 10.1103/PhysRevLett.90.227902
37. M.B. Plenio, J. Eisert, J. Dreißig, M. Cramer, Entropy, entanglement, and area: analytical results for harmonic lattice systems. Phys. Rev. Lett. (2005). 10.1103/PhysRevLett.94.060503
38. J. Eisert, M. Cramer, M.B. Plenio, Area laws for the entanglement entropy. Rev. Mod. Phys. (2010). 10.1103/RevModPhys.82.277
39. M. Lubasch, J.I. Cirac, M.-C. Bañuls, Algorithms for finite projected entangled pair states. Phys. Rev. B (2014). 10.1103/PhysRevB.90.064425
40. F. Verstraete, J.J. García-Ripoll, J.I. Cirac, Matrix product density operators: simulation of finite-temperature and dissipative systems. Phys. Rev. Lett. (2004). 10.1103/PhysRevLett.93.207204

41. M. Kliesch, D. Gross, J. Eisert, Matrix-product operators and states: NP-hardness and undecidability. Phys. Rev. Lett. (2014). 10.1103/PhysRevLett.113.160503
42. L. Tagliacozzo, G. Evenbly, G. Vidal, Simulation of two-dimensional quantum systems using a tree tensor network that exploits the entropic area law. Phys. Rev. B (2009). 10.1103/PhysRevB.80.235127
43. G. Evenbly, G. Vidal, Class of highly entangled many-body states that can be efficiently simulated. Phys. Rev. Lett. (2014). 10.1103/PhysRevLett.112.240502
44. M. Gerster, P. Silvi, M. Rizzi, R. Fazio, T. Calarco, S. Montangero, Unconstrained tree tensor network: an adaptive gauge picture for enhanced performance. Phys. Rev. B (2014). 10.1103/PhysRevB.90.125154
45. G. Vidal, Entanglement renormalization. Phys. Rev. Lett. (2007). 10.1103/PhysRevLett.99.220405
46. J. Eisert, *Entanglement and Tensor Network States* (2013). 10.48550/arXiv.1308.3318
47. S. Östlund, S. Rommer, Thermodynamic limit of density matrix renormalization. Phys. Rev. Lett. (1995). 10.1103/PhysRevLett.75.3537
48. F. Verstraete, D. Porras, J.I. Cirac, Density matrix renormalization group and periodic boundary conditions: a quantum information perspective. Phys. Rev. Lett. (2004). 10.1103/PhysRevLett.93.227205
49. A.J. Daley, C. Kollath, U. Schollwöck, G. Vidal, Time-dependent density-matrix renormalization-group using adaptive effective hilbert spaces. J. Stat. Mech. Theory Exper. (2004). 10.1088/1742-5468/2004/04/P04005
50. D. Perez-Garcia, F. Verstraete, M.M. Wolf, J.I. Cirac, *Matrix Product State Representations* (2007). 10.48550/arXiv.quant-ph/0608197
51. F. Verstraete, J.J. Garcia-Ripoll, J.I. Cirac, Matrix product density operators: simulation of finite-temperature and dissipative systems. Phys. Rev. Lett. (2004). 10.1103/PhysRevLett.93.207204
52. R. Orús, G. Vidal, Infinite time-evolving block decimation algorithm beyond unitary evolution. Phys. Rev. B (2008). 10.1103/PhysRevB.78.155117
53. G. Vidal, Efficient classical simulation of slightly entangled quantum computations. Phys. Rev. Lett. (2003). 10.1103/PhysRevLett.91.147902
54. J. Haegeman, J.I. Cirac, T.J. Osborne, I. Pižorn, H. Verschelde, F. Verstraete, Time-dependent variational principle for quantum lattices. Phys. Rev. Lett. (2011). 10.1103/PhysRevLett.107.070601
55. E. Rico, T. Pichler, M. Dalmonte, P. Zoller, S. Montangero, Tensor networks for lattice gauge theories and atomic quantum simulation. Phys. Rev. Lett. (2014). 10.1103/PhysRevLett.112.201601
56. S. Kühn, J.I. Cirac, M.-C. Bañuls, Quantum simulation of the schwinger model: a study of feasibility. Phys. Rev. A (2014). 10.1103/PhysRevA.90.042305
57. T. Pichler, M. Dalmonte, E. Rico, P. Zoller, S. Montangero, Real-time dynamics in U(1) lattice gauge theories with tensor networks. Phys. Rev. X (2016). 10.1103/PhysRevX.6.011023
58. M.C. Bañuls, K. Cichy, J.I. Cirac, K. Jansen, S. Kühn, Density induced phase transitions in the schwinger model: a study with matrix product states. Phys. Rev. Lett. (2017). 10.1103/PhysRevLett.118.071601
59. P. Silvi, E. Rico, T. Calarco, S. Montangero, Lattice gauge tensor networks. New J. Phys. (2014). 10.1088/1367-2630/16/10/103015
60. M.C. Bañuls, K. Cichy, J.I. Cirac, K. Jansen, S. Kühn, Efficient basis formulation for (1 + 1)-dimensional SU(2) lattice gauge theory: spectral calculations with matrix product states. Phys. Rev. X (2017). 10.1103/PhysRevX.7.041046
61. K. Zapp, R. Orús, Tensor network simulation of QED on infinite lattices: learning from (1+1)d, and prospects for (2+1)d. Phys. Rev. D (2017). 10.1103/PhysRevD.95.114508
62. B. Buyens, S. Montangero, J. Haegeman, F. Verstraete, K. Van Acoleyen, Finite-representation approximation of lattice gauge theories at the continuum limit with tensor networks. Phys. Rev. D (2017). 10.1103/PhysRevD.95.094509

63. M.C. Bañuls, K. Cichy, J.I. Cirac, K. Jansen, S. Kühn, Density induced phase transitions in the schwinger model: a study with matrix product states. Phys. Rev. Lett. (2017). 10.1103/PhysRevLett.118.071601
64. M.C. Bañuls, K. Cichy, J.I. Cirac, K. Jansen, S. Kühn, H. Saito, Towards overcoming the Monte Carlo sign problem with tensor networks. EPJ Web Conf. (2017). 10.1051/epjconf/201713704001
65. B. Buyens, J. Haegeman, H. Verschelde, F. Verstraete, K. Van Acoleyen, Confinement and String Breaking for QED_2 in the Hamiltonian Picture. Phys. Rev. X (2016). 10.1103/PhysRevX.6.041040
66. M.C,. Bañuls, K. Cichy, K. Jansen, H. Saito, Chiral condensate in the Schwinger model with matrix product operators. Phys. Rev. D (2016). 10.1103/PhysRevD.93.094512
67. T.J. Osborne, *Tobiasosborne/Lattice-Gauge-Theory-and-Tensor-Networks* (2024). Github repository: https://github.com/tobiasosborne/lattice-gauge-theory-and-tensor-networks
68. S. Kühn, E. Zohar, J.I. Cirac, M.C. Bañuls, Non-Abelian string breaking phenomena with matrix product states. J. High Energy Phys. (2015). 10.1007/JHEP07(2015)130
69. J. Haegeman, K. Van Acoleyen, N. Schuch, J.I. Cirac, F. Verstraete, Gauging quantum states: from global to local symmetries in many-body systems. Phys. Rev. X (2015). 10.1103/PhysRevX.5.011024
70. L. Tagliacozzo, G. Vidal, Entanglement renormalization and gauge symmetry. Phys. Rev. B (2011). 10.1103/PhysRevB.83.115127
71. L. Tagliacozzo, A. Celi, M. Lewenstein, Tensor networks for lattice gauge theories with continuous groups. Phys. Rev. X (2014). 10.1103/PhysRevX.4.041024
72. T.M.R. Byrnes, P. Sriganesh, R.J. Bursill, C.J. Hamer, Density matrix renormalisation group approach to the massive Schwinger model. Phys. Rev. D (2002)
73. E. Ercolessi, P. Facchi, G. Magnifico, S. Pascazio, F.V. Pepe, Phase transitions in ${Z}_{n}$ Gauge models: towards quantum simulations of the Schwinger-Weyl QED. Phys. Rev. D (2018). 10.1103/PhysRevD.98.074503
74. B. Buyens, J. Haegeman, K. Van Acoleyen, H. Verschelde, F. Verstraete, Matrix product states for Gauge field theories. Phys. Rev. Lett. (2014). 10.1103/PhysRevLett.113.091601
75. M.C. Banuls, K. Cichy, J.I. Cirac, K Jansen, The mass spectrum of the Schwinger model with matrix product states. J. High Energy Phys. (2013). 10.1007/JHEP11(2013)158
76. S. Morita, Y. Ozeki, H. Nishimori, Gauge theory for quantum spin glasses. J. Phys. Soc. Jpn. (2006). 10.1143/JPSJ.75.014001
77. M.C. Bañuls, K. Cichy, J.I. Cirac, K. Jansen, H. Saito, Matrix product states for lattice field theories. *Proceedings of the 31st International Symposium on Lattice Field Theory* (2013)
78. U.-J. Wiese, Towards quantum simulating QCD. Nucl. Phys. A (2014). 10.1016/j.nuclphysa.2014.09.102
79. M. Dalmonte, S. Montangero, Lattice gauge theory simulations in the quantum information era. Contemp. Phys. (2016). 10.1080/00107514.2016.1151199
80. K.G. Wilson, The renormalization group: Critical phenomena and the Kondo problem. Rev. Mod. Phys. (1975) 10.1103/RevModPhys.47.773
81. I. Montvay, G. Muenster. *Quantum Fields on a Lattice* (Cambridge University Press, Cambridge, 1994)
82. M. Creutz. *Quarks, Gluons and Lattices* (Cambridge University Press, Cambridge, 1997)
83. T. DeGrand, C. DeTar. *Lattice Methods for Quantum Chromodynamics* (World Scientific, Singapore, 2006)
84. C. Gattringer, C.B. Lang. *Quantum Chromodynamics on the Lattice* (Springer-Verlag, Berlin, 2010)
85. E. Fradkin, J.E. Moore, Entanglement entropy of 2D conformal quantum critical points: Hearing the shape of a quantum drum. Phys. Rev. Lett. **97**(5), 050404 (2006). 10.1103/PhysRevLett.97.050404
86. J. Kogut, L. Susskind, Hamiltonian formulation of Wilson's lattice guage theories. Phys. Rev. D (1975)

87. J.B. Kogut, An introduction to lattice gauge theory and spin systems. Rev. Mod. Phys. **51**(4), 659–713 (1979). 10.1103/RevModPhys.51.659
88. D. Horn, Finite matrix models with continuous local gauge invariance. Phys. Lett. B **100**(2), 149–151 (1981). https://doi.org/10.1016/0370-2693(81)90763-2 (https://www.sciencedirect.com/science/article/pii/0370269381907632)
89. P. Orland, D. Rohrlich, Lattice gauge magnets: local isospin from spin. Nucl. Phys. B **338**(3), 647–672 (1990). https://doi.org/10.1016/0550-3213(90)90646-U. https://www.sciencedirect.com/science/article/pii/055032139090646U
90. S. Chandrasekharan, U.J. Wiese, Quantum link models: a discrete approach to gauge theories. Nucl. Phys. B (1997). 10.1016/S0550-3213(97)80041-7
91. B. Schlittgen, U.-J. Wiese, Low-energy effective theories of quantum spin and quantum link models. Phys. Rev. D (2001). 10.1103/PhysRevD.63.085007
92. C. Lacroix, P. Mendels, F. Mila (eds.), *Introduction to Frustrated Magnetism*. Springer Series in Solid-State Sciences, vol. 164 (Springer, 2010)
93. R.P. Feynman, Simulating physics with computers. Int. J. Theor. Phys. (1982). 10.1007/BF02650179
94. J.I. Cirac, P. Zoller, Goals and opportunities in quantum simulation. Nat. Phys. (2012). 10.1038/nphys2275
95. I. Bloch, J. Dalibard, S. Nascimbène, Quantum simulations with ultracold quantum gases. Nat. Phys. (2012). 10.1038/nphys2259
96. D. Jaksch, P. Zoller, The cold atom Hubbard toolbox. Ann. Phys. (N.Y). (2005). 10.1016/j.aop.2004.09.010
97. E.H. Lieb, T. Schultz, D Mattis, Two Soluble models of an Antiferromagnetic Chain. Ann. Phys. **16**(3), 407–466 (1961). 10.1016/0003-4916(61)90115-4
98. E. Lieb. *The Hubbard Model: Some Rigorous Results and Open Problemss*. http://arxiv.org/abs/cond-mat/9311033.
99. H.P. Büchler, M. Hermele, S.D. Huber, M.P. Fisher, P. Zoller, Atomic quantum simulator for lattice gauge theories and ring exchange models. Phys. Rev. Lett. (2005). 10.1103/PhysRevLett.95.040402
100. H. Weimer, M. Müller, I. Lesanovsky, P. Zoller, H.P. Büchler, A Rydberg quantum simulator. Nat. Phys. (2010). 10.1038/nphys1614
101. L. Tagliacozzo, A. Celi, P. Orland, M.W. Mitchell, M. Lewenstein, Simulation of non-Abelian gauge theories with optical lattices. Nat. Commun. (2013). 10.1038/ncomms3615
102. A.W. Glaetzle, M. Dalmonte, R. Nath, I. Rousochatzakis, R. Moessner, P. Zoller, Quantum Spin-ice and dimer models with rydberg atoms. Phys. Rev. X (2014). 10.1103/PhysRevX.4.041037
103. P. Schauss, J. Zeiher, T. Fukuhara, S. Hild, M. Cheneau, T. Macri, T. Pohl, I. Bloch, C. Gross, Crystallization in Ising quantum magnets. Science (2015). 10.1126/science.1258351
104. H. Bernien, S. Schwartz, A. Keesling, H. Levine, A. Omran, H. Pichler, S. Choi, A.S. Zibrov, M. Endres, M. Greiner, V. Vuletić, M.D. Lukin, Probing many-body dynamics on a 51-atom quantum simulator. Nature (2017). 10.1038/nature24622
105. J.I. Cirac, P. Maraner, J.K. Pachos, Cold atom simulation of interacting relativistic quantum field theories. Phys. Rev. Lett. (2010). 10.1103/PhysRevLett.105.190403
106. J. Casanova, L. Lamata, I.L. Egusquiza, R. Gerritsma, C.F. Roos, J. J. García-Ripoll, E. Solano, Quantum simulation of quantum field theories in trapped ions. Phys. Rev. Lett. (2011). 10.1103/PhysRevLett.107.260501
107. S.P. Jordan, K.S.M. Lee, J. Preskill, Quantum algorithms for quantum field theories. Science **336**(6085), 1130–1133 (2012). 10.1126/science.1217069
108. D. Banerjee, M. Dalmonte, M. Müller, E. Rico, P. Stebler, U.-J. Wiese, P. Zoller, Atomic quantum simulation of dynamical gauge fields coupled to fermionic matter: from string breaking to evolution after a quench. Phys. Rev. Lett. (2012). 10.1103/PhysRevLett.109.175302
109. D. Banerjee, M. Bügli, M. Dalmonte, E. Rico, P. Stebler, U.-J. Wiese, P. Zoller, Atomic quantum simulation of U(N) and SU(N) non-abelian lattice gauge theories. Phys. Rev. Lett. (2013). 10.1103/PhysRevLett.110.125303

110. S. Barrett, K. Hammerer, S. Harrison, T.E. Northup, T.J. Osborne, Simulating quantum fields with cavity QED. Phys. Rev. Lett. (2013). 10.1103/PhysRevLett.110.090501
111. L. Tagliacozzo, A. Celi, A. Zamora, M. Lewenstein, Optical Abelian lattice gauge theories. Ann. Phys. (N. Y). (2013). 10.1016/j.aop.2012.11.009
112. D. Marcos, P. Rabl, E. Rico, P. Zoller, Superconducting circuits for quantum simulation of dynamical gauge fields. Phys. Rev. Lett. (2013). 10.1103/PhysRevLett.111.110504
113. P. Hauke, D. Marcos, M. Dalmonte, P. Zoller, Quantum simulation of a lattice schwinger model in a chain of trapped ions. Phys. Rev. X (2013). 10.1103/PhysRevX.3.041018
114. E. Zohar, M. Burrello, Formulation of lattice gauge theories for quantum simulations. Phys. Rev. D (2015). 10.1103/PhysRevD.91.054506
115. S. Notarnicola, E. Ercolessi, P. Facchi, G. Marmo, S. Pascazio, F.V. Pepe, Discrete abelian gauge theories for quantum simulations of QED. J. Phys. A Math. Theor. (2015). 10.1088/1751-8113/48/30/30FT01
116. E. Zohar, A. Farace, B. Reznik, J.I. Cirac, "Digital quantum simulation of Z_2 lattice gauge theories with dynamical fermionic matter. Phys. Rev. Lett. (2017). 10.1103/PhysRevLett.118.070501
117. E. Zohar, J.I. Cirac, B. Reznik, Quantum simulations of lattice gauge theories using ultracold atoms in optical lattices. Reports Prog. Phys. (2016). 10.1088/0034-4885/79/1/014401
118. V. Kasper, F. Hebenstreit, F. Jendrzejewski, M.K. Oberthaler, J. Berges, Implementing quantum electrodynamics with ultracold atomic systems. New J. Phys. (2017). https://doi.org/10.1088/1367-2630/aa54e0
119. E.A. Martinez, C.A. Muschik, P. Schindler, D. Nigg, A. Erhard, M. Heyl, P. Hauke, M. Dalmonte, T. Monz, P. Zoller, R. Blatt, Real-time dynamics of lattice gauge theories with a few-qubit quantum computer. Nature (2016). 10.1038/nature18318
120. S. Succi, *An Introduction to Computational Physics* (Scuola Normale Superiore, Pisa, 2002)
121. S. Oliveira, D. Stewart, *Writing Scientific Software – A guide to good style* (Cambridge University Press, Cambridge, 2006)
122. W.H. Press, S.A. Teukolsky, W.T. Vetterling, B.P. Flannery, *Numerical Recipes in FORTRAN; The Art of Scientific Computing*, 2nd edn. (Cambridge University Press, New York, 1993). ISBN: 0-52143-716-4
123. W.R. Gibbs, *Computation in Modern Physics*, 3rd edn. (World Scientific Publishing Company, Singapore, 2006). ISBN: 978-9-813-10670-3
124. R.A. Van de Geijn, E.S. Quintana-Ortí, *The Science of programming matrix computations* (2008). www.lulu.com. https://www.cs.utexas.edu/~rvdg/tmp/TSoPMC.pdf
125. D.P. Bertsekas, *Constrained Optimization and Lagrange Multiplier Methods* (Academic Press, Cambridge, 1982)
126. K. Dowd, C.R. Severance, *High Performance Computing*. A Nutshell Handbook (O'Reilly, Sebastopol, 1998). ISBN: 978-1-565-92312-6
127. K. Mehlhorn, P. Sanders, *Algorithms and Data Structures: The Basic Toolbox*. SpringerLink: Springer e-Books (Springer, Berlin, 2008). ISBN: 978-3-540-77977-3
128. C. Johnson, *Numerical Solution of Partial Differential Equations by the Finite Element Method*. Dover Books on Mathematics Series (Dover Publications, Garden City, 2012). Incorporated. ISBN: 978-0-486-13159-7
129. S.C. Brenner, C. Carstensen, Finite element methods, in *Encyclopedia of Computational Mechanics* (John Wiley & Sons, Ltd, Cambridge, 2004). ISBN: 978-0-470-09135-7 10.1002/0470091355.ecm003
130. E. Anderson, Z. Bai, C. Bischof, L.S. Blackford, J. Demmel, J. Dongarra, J. Du Croz, A. Greenbaum, S. Hammarling, A. McKenney, D. Sorensen. *LAPACK Users' Guide*, 3rd edn. Software, Environments, and Tools (Society for Industrial and Applied Mathematics (SIAM), Philadelphia, 1999). ISBN: 978-0-898-71960-4 10.1137/1.9780898719604
131. C. Lanczos, An iteration method for the solution of the eigenvalue problem of linear differential and integral operators. J. Res. Natl. Bureau of Standards **45**(4), 255–282 (1950). 10.6028/jres.045.026

132. J. Liesen, Z. Strakos, *Krylov Subspace Methods: Principles and Analysis*. Numerical Mathematics and Scientific Computation (Oxford University Press, Oxford, 2012). ISBN: 978-0-191-63032-3
133. G.W. Anderson, A. Guionnet, O. Zeitouni, *An Introduction to Random Matrices* (Cambridge University Press, Cambridge, 2009). ISBN: 978-0-511-80133-4 10.1017/CBO9780511801334
134. C.F. Fischer, General Hartree-Fock program. Comput. Phys. Commun. **43**, 355–365 (1987)
135. C. Cohen-Tannoudji, B. Diu, F. Laloë, *Quantum Mechanics* (Wiley, Cambridge, 1977). ISBN: 978-2-705-65834-2
136. R. Bulla, T. Costi, T. Pruschke, Numerical renormalization group method for quantum impurity systems. Rev. Mod. Phys. (2008) 10.1103/RevModPhys.80.395
137. L. Carr, *Understanding Quantum Phase Transitions* (CRC Press, Boca Raton, 2010). ISBN: 978-1-439-80261-8
138. E. Barouch, B.M. McCoy, Statistical mechanics of the X y model. II. Spin-correlation functions. Phys. Rev. A (1971). 10.1103/PhysRevA.3.786
139. P. Pfeuty, The one-dimensional Ising model with a transverse field. Ann. Phys. (1970). 10.1016/0003-4916(70)90270-8
140. L. Amico, R. Fazio, A. Osterloh, V. Vedral, Entanglement in many-body systems. Rev. Mod. Phys. (2008). 10.1103/RevModPhys.80.517
141. G. Mussardo, *Statistical Field Theory* (Oxford University Press, Oxford, 2009)
142. D.J. Amit, V. Martin-Mayor, *Field Theory, the Renormalization Group, and Critical Phenomena: Graphs to Computers*, 3rd edn. (World Scientific, Singapore, 2005). ISBN: 978-9-813-10207-1
143. I. Peschel, X. Wang, M. Kaulke, K. Hallberg, *Density-Matrix Renormalization - A New Numerical Method in Physics: Lectures of a Seminar and Workshop held at the Max-Planck-Institut für Physik komplexer Systeme, Dresden, Germany, August 24th to September 18th, 1998*. Lecture Notes in Physics (Springer Berlin Heidelberg, 2014). ISBN: 978-3-662-14232-5
144. S.R. White, Density-matrix algorithms for quantum renormalization groups. Phys. Rev. B (1993). 10.1103/PhysRevB.48.10345
145. U. Schollwöck, The density-matrix renormalization group. Rev. Mod. Phys. (2005). 10.1103/RevModPhys.77.259
146. A.L. Malvezzi, An introduction to numerical methods in low-dimensional quantum systems. Br. J. Phys. (2003). 10.1590/S0103-97332003000100004
147. G. Evenbly, G. Vidal, Algorithms for entanglement renormalization. Phys. Rev. B (2009). 10.1103/PhysRevB.79.144108
148. G. Evenbly, G. Vidal, Entanglement renormalization in two spatial dimensions. Phys. Rev. Lett. (2009). 10.1103/PhysRevLett.102.180406
149. T. Felser, S. Notarnicola, S. Montangero, *Efficient Tensor Network Ansatz for High-Dimensional Quantum Many-Body Problems* (2021). 10.1103/PhysRevLett.126.170603
150. T. Felser, Tree Tensor Networks for High-Dimensional Quantum Systems and Beyond. Doctoral Thesis. Saarländische Universitäts- und Landesbibliothek, 2021. 10.22028/D291-35211
151. V. Strassen, Gaussian elimination is not optimal. Numer. Math. (1969). 10.1007/BF02165411
152. A. Royer, Wigner function in Liouville space: a canonical formalism. Phys. Rev. A (1991). 10.1103/PhysRevA.43.44
153. I. Bengtsson, K. Zyczkowski, *Geometry of Quantum States: An Introduction to Quantum Entanglement* (Cambridge University Press, Cambridge, 2007). ISBN: 978-1-139-45346-2
154. E.C.G. Sudarshan, P.M. Mathews, J. Rau, Stochastic dynamics of quantum-mechanical systems. Phys. Rev. **121**(3), 920–294 (1961). 10.1103/PhysRev.121.920
155. A. Jamiolkowski, Linear transformations which preserve trace and positive semidefiniteness of operators. Rep. Math. Phys. **3**(4), 275–278 (1972). 10.1016/0034-4877(72)90011-0
156. M.-D. Choi, Completely positive linear maps on complex matrices. Lin. Alg. Appl. **10**(3), 285–290 (1975). https://doi.org/10.1016/0024-3795(75)90075-0. https://www.sciencedirect.com/science/article/pii/0024379575900750

157. P. Arrighi, C. Patricot, On quantum operations as quantum states. Ann. Phys. (N. Y). (2004). 10.1016/j.aop.2003.11.005
158. S. Lang, *Complex Analysis*. Graduate Texts in Mathematics (Springer, New York, 2013). ISBN: 978-1-475-73083-8
159. K. Van Acoleyen, M. Mariën, F. Verstraete, Entanglement rates and area laws. Phys. Rev. Lett. **111**(17), 170501 (2013). 10.1103/physrevlett.111.170501. https://link.aps.org/doi/10.1103/PhysRevLett.111.170501
160. S. Michalakis, *Stability of the Area Law for the Entropy of Entanglement* (2012). https://arxiv.org/abs/1206.6900
161. M.B. Plenio, J. Eisert, J. Dreißig, M. Cramer, Entropy, entanglement, and area: analytical results for harmonic lattice systems. Phys. Rev. Lett. (2005). 10.1103/PhysRevLett.94.060503
162. M. Cramer, J. Eisert, Correlations, spectral gap and entanglement in harmonic quantum systems on generic lattices. New J. Phys. (2006). 10.1088/1367-2630/8/5/071
163. M. Cramer, J. Eisert, M.B. Plenio, Statistics dependence of the entanglement entropy. Phys. Rev. Lett. (2007). 10.1103/PhysRevLett.98.220603
164. C.E. Shannon, A mathematical theory of communication. Bell Syst. Technol. J. (1948). 10.1002/j.1538-7305.1948.tb01338.x
165. C.E. Shannon, A mathematical theory of communication. Bell Syst. Technol. J. (1948). 10.1002/j.1538-7305.1948.tb00917.x
166. M. Nielsen, I. Chuang (eds.), *Quantum Computation and Quantum Information* (Cambridge University Press, Cambridge, 2000). ISBN: 978-0-521-63503-5
167. G. Vidal, Entanglement renormalization. Phys. Rev. Lett. (2007). 10.1103/PhysRevLett.99.220405
168. B. Swingle, Entanglement renormalization and holography. Phys. Rev. D (2012). 10.1103/PhysRevD.86.065007
169. G. Evenbly, G. Vidal, Tensor network states and geometry. J. Stat. Phys. **145**(4), 891–918 (2011). http://dx.doi.org/10.1007/s10955-011-0237-4
170. P. Silvi, F. Tschirsich, M. Gerster, J. Jünemann, D. Jaschke, M. Rizzi, S. Montangero, The tensor networks anthology: simulation techniques for many-body quantum lattice systems. SciPost Phys. Lect. Notes (2019). 10.21468/SciPostPhysLectNotes.8
171. S.-J. Ran, E. Tirrito, C. Peng, X. Chen, L. Tagliacozzo, G. Su, M. Lewenstein, Two-dimensional tensor networks and contraction algorithms, in *Tensor Network Contractions: Methods and Applications to Quantum Many-Body Systems* (Springer, International Publishing Cham, 2020), pp. 63–86. ISBN: 978-3-030-34489-4 10.1007/978-3-030-34489-4_3
172. F. Verstraete, J.I. Cirac, Valence-bond states for quantum computation. Phys. Rev. A (2004). 10.1103/PhysRevA.70.060302
173. F. Verstraete, M.M. Wolf, D. Perez-Garcia, J.I. Cirac, Criticality, the area law, and the computational power of projected entangled pair states. Phys. Rev. Lett. (2006). 10.1103/PhysRevLett.96.220601
174. R. Orús, A practical introduction to tensor networks: matrix product states and projected entangled pair states. Ann. Phys. (2014). 10.1016/j.aop.2014.06.013
175. F. Verstraete, J.I. Cirac, *Renormalization Algorithms for Quantum-Many Body Systems in Two and Higher Dimensions* (2004). https://arxiv.org/abs/cond-mat/0407066
176. D. Perez-Garcia, F. Verstraete, J.I. Cirac, M.M. Wolf, PEPS as Unique Ground States of Local Hamiltonians. Quantum Information & Computation **8**(6-7), 650–663 (2007). arXiv:0707.2260
177. Y.-Y. Shi, L.-M. Duan, G. Vidal, Classical simulation of quantum many-body systems with a tree tensor network. Phys. Rev. A (2006). 10.1103/PhysRevA.74.022320
178. P. Silvi, V. Giovannetti, S. Montangero, M. Rizzi, J.I. Cirac, R. Fazio, Homogeneous binary trees as ground states of quantum critical Hamiltonians. Phys. Rev. A (2010). 10.1103/PhysRevA.81.062335
179. M. Gerster, P. Silvi, M. Rizzi, R. Fazio, T. Calarco, S. Montangero, Unconstrained tree tensor network: an adaptive gauge picture for enhanced performance. Phys. Rev. B (2014). 10.1103/PhysRevB.90.125154

180. M. Gerster, M. Rizzi, P. Silvi, M. Dalmonte, S. Montangero, Fractional quantum Hall effect in the interacting Hofstadter model via tensor networks. Phys. Rev. B (2017). 10.1103/PhysRevB.96.195123
181. T. Felser, P. Silvi, M. Collura, S. Montangero, Two-dimensional quantum-link lattice quantum electrodynamics at finite density. Phys. Rev. X (2020). 10.1103/PhysRevX.10.041040
182. S. Östlund, S. Rommer, Thermodynamic limit of density matrix renormalization. Phys. Rev. Lett. (1995). 10.1103/PhysRevLett.75.3537
183. I.V. Oseledets, Tensor-train decomposition. SIAM J. Sci. Comput. **33**(5), 2295–2317 (2011). https://doi.org/10.1137/090752286
184. S.R. White, Density matrix formulation for quantum renormalization groups. Phys. Rev. Lett. (1992). 10.1103/PhysRevLett.69.2863
185. U. Schollwöck, The density-matrix renormalization group in the age of matrix product states. Ann. Phys. (2011). 10.1016/j.aop.2010.09.012
186. S.R. White, A.E. Feiguin, Real-time evolution using the density matrix renormalization group. Phys. Rev. Lett. (2004). 10.1103/PhysRevLett.93.076401
187. G. Vidal, Efficient classical simulation of slightly entangled quantum computations. Phys. Rev. Lett. (2003). 10.1103/PhysRevLett.91.147902
188. E.M. Stoudenmire, S.R. White, Studying two-dimensional systems with the density matrix renormalization group. Annu. Rev. Condens. Matter Phys. (2012). 10.1146/annurev-conmatphys-020911-125018
189. G. Cataldi, A. Abedi, G. Magnifico, S. Notarnicola, N.D. Pozza, V. Giovannetti, S. Montangero, Hilbert curve vs hilbert space: exploiting fractal 2D covering to increase tensor network efficiency. Quantum (2021). 10.22331/q-2021-09-29-556
190. Y.-Y. Shi, L.-M. Duan, G. Vidal, Classical simulation of quantum many-body systems with a tree tensor network. Phys. Rev. A (2006). 10.1103/PhysRevA.74.022320
191. A.J. Ferris, Area law and real-space renormalization. Phys. Rev. B (2013). 10.1103/PhysRevB.87.125139
192. M. Gerster, M. Rizzi, F. Tschirsich, P. Silvi, R. Fazio, S. Montangero, Superfluid density and quasi-long-range order in the one-dimensional disordered Bose–Hubbard model. New J. Phys. (2016). 10.1088/1367-2630/18/1/015015
193. V. Murg, F. Verstraete, Ö Legeza, R.M. Noack, Simulating strongly correlated quantum systems with tree tensor networks. Phys. Rev. B (2010). 10.1103/PhysRevB.82.205105
194. G. Magnifico, T. Felser, P. Silvi, S. Montangero,. Lattice quantum electrodynamics in (3+1)-dimensions at finite density with tensor networks. Nature Commun. (2021). 10.1038/s41467-021-23646-3
195. N. Schuch, M.M. Wolf, F. Verstraete, J.I. Cirac, Computational complexity of projected entangled pair states. Phys. Rev. Lett. (2007). 10.1103/PhysRevLett.98.140506
196. J. Haferkamp, D. Hangleiter, J. Eisert, M. Gluza, Contracting projected entangled pair states is average-case hard. Phys. Rev. Res. (2020). 10.1103/PhysRevResearch.2.013010
197. B. Swingle, *Constructing Holographic Spacetimes Using Entanglement Renormalization* (2012). https://arxiv.org/abs/1209.3304
198. M. Nozaki, S. Ryu, T. Takayanagi, Holographic geometry of entanglement renormalization in quantum field theories. J. High Energy Phys. (2012). 10.1007/JHEP10(2012)193
199. A. Mollabashi, M. Naozaki, S. Ryu, T. Takayanagi, Holographic geometry of cMERA for quantum quenches and finite temperature. J. High Energy Phys. (2014). 10.1007/JHEP03(2014)098
200. P. Hayden, S. Nezami, X.-L. Qi, N. Thomas, M. Walter, Z. Yang, holographic duality from random tensor networks. J. High Energy Phys. (2016). 10.1007/JHEP11(2016)009
201. N. Nakatani, G.K.-L. Chan, Efficient tree tensor network states (TTNS) for quantum chemistry: generalizations of the density matrix renormalization group algorithm. J. Chem. Phys. (2013). 10.1063/1.4798639
202. G.M. Crosswhite, A.C. Doherty, G. Vidal, Applying matrix product operators to model systems with long-range interactions. Phys. Rev. B (2008). 10.1103/PhysRevB.78.035116

203. I.P. McCulloch, From density-matrix renormalization group to matrix product states. J. Stat. Mech. Theory Exper. (2007). 10.1088/1742-5468/2007/10/p10014
204. P. Czarnik, J. Dziarmaga, Variational approach to projected entangled pair states at finite temperature. Phys. Rev. B (2015). 10.1103/PhysRevB.92.035152
205. P. Czarnik, M.M. Rams, J. Dziarmaga, Variational tensor network renormalization in imaginary time: benchmark results in the Hubbard model at finite temperature. Phys. Rev. B (2016). 10.1103/PhysRevB.94.235142
206. A. Kshetrimayum, H. Weimer, R. Orós, A simple tensor network algorithm for two-dimensional steady states. Nature Commun. (2017). 10.1038/s41467-017-01511-6
207. L. Kohn, P. Silvi, M. Gerster, M. Keck, R. Fazio, G. E. Santoro, S. Montangero, Superfluid-to-mott transition in a Bose-Hubbard ring: persistent currents and defect formation. Phys. Rev. A (2020). 10.1103/PhysRevA.101.023617
208. P. Barmettler, M. Punk, V. Gritsev, E. Demler, E. Altman, Relaxation of antiferromagnetic order in spin-$1/2$ chains following a quantum quench. Phys. Rev. Lett. (2009). 10.1103/PhysRevLett.102.130603
209. C. Kollath, A.M. Läuchli, E. Altman, Quench dynamics and nonequilibrium phase diagram of the bose-hubbard model. Phys. Rev. Lett. (2007). 10.1103/PhysRevLett.98.180601
210. F. Pellegrini, S. Montangero, G.E. Santoro, R. Fazio, Adiabatic quenches through an extended quantum critical region. Phys. Rev. B (2008). 10.1103/PhysRevB.77.140404
211. B. Gardas, J. Dziarmaga, W.H. Zurek, Dynamics of the quantum phase transition in the one-dimensional Bose-Hubbard model: excitations and correlations induced by a quench". Phys. Rev. B (2017). 10.1103/PhysRevB.95.104306
212. P. Doria, T. Calarco, S. Montangero, Optimal control technique for many-body quantum dynamics. Phys. Rev. Lett. (2011). 10.1103/PhysRevLett.106.190501
213. L. Arceci, P. Silvi, S. Montangero, *Entanglement of Formation of Mixed Many-body Quantum States via Tree Tensor Operators* (2020)
214. F. Verstraete, J.J. García-Ripoll, J.I. Cirac, Matrix product density operators: simulation of finite-temperature and dissipative systems. Phys. Rev. Lett. (2004). 10.1103/PhysRevLett.93.207204
215. M. Zwolak, G. Vidal, Mixed-state dynamics in one-dimensional quantum lattice systems: a time-dependent superoperator renormalization algorithm. Phys. Rev. Lett. (2004). 10.1103/PhysRevLett.93.207205
216. A.H. Werner, D. Jaschke, P. Silvi, M. Kliesch, T. Calarco, J. Eisert, S. Montangero, Positive tensor network approach for simulating open quantum many-body systems. Phys. Rev. Lett. (2016). 10.1103/PhysRevLett.116.237201
217. A.J. Daley, Quantum trajectories and open many-body quantum systems. Adv. Phys. (2014). 10.1080/00018732.2014.933502
218. M. Collura, L. Dell'Anna, T. Felser, S. Montangero, On the descriptive power of neural-networks as constrained tensor networks with exponentially large bond dimension. SciPost Phys. Core (2021). 10.21468/scipostphyscore.4.1.001
219. J. Chen, S. Cheng, H. Xie, L. Wang, T. Xiang, Equivalence of restricted Boltzmann machines and tensor network states. Phys. Rev. B (2018). 10.1103/physrevb.97.085104
220. Y. Levine, D. Yakira, N. Cohen, A. Shashua,. *Deep Learning and Quantum Entanglement: Fundamental Connections with Implications to Network Design*, in *International Conference on Learning Representations* (2018). https://openreview.net/forum?id=SywXXwJAb
221. E. Stoudenmire, D.J. Schwab, Supervised learning with tensor networks, in *Advances in Neural Information Processing Systems 29*, ed. by D.D. Lee, M. Sugiyama, U.V. Luxburg, I. Guyon, R. Garnett (Curran Associates, Inc., Red Hook, 2016), pp. 4799–4807
222. E.M. Stoudenmire, Learning relevant features of data with multi-scale tensor networks. Quantum Sci. Technol. (2018). 10.1088/2058-9565/aaba1a
223. A. Novikov, M. Trofimov, I. Oseledets, Exponential Machines (2017). https://arxiv.org/abs/1605.03795
224. A. Cichocki, N. Lee, I. Oseledets, A.-H. Phan, Q. Zhao, M. Sugiyama, D.P. Mandic, Tensor networks for dimensionality reduction and large-scale optimization: part 2 applications and future perspectives. Found. Trends Mach. Learn. (2017). 10.1561/2200000067

225. D. Liu, S.-J. Ran, P. Wittek, C. Peng, R.B. García, G. Su, M. Lewenstein, Machine learning by unitary tensor network of hierarchical tree structure. New J. Phys. (2019). 10.1088/1367-2630/ab31ef
226. C. Roberts, A. Milsted, M. Ganahl, A. Zalcman, B. Fontaine, Y. Zou, J. Hidary, G. Vidal, S. Leichenauer, *TensorNetwork: A Library for Physics and Machine Learning* (2019). https://arxiv.org/abs/1905.01330
227. T. Felser, M. Trenti, L. Sestini, A. Gianelle, D. Zuliani, D. Lucchesi, S. Montangero, Quantum-inspired machine learning on high-energy physics data. npj. Quantum Inf. **7**, 111 (2021). https://doi.org/10.1038/s41534-021-00443-w
228. M. Trenti, B. Petersen, T. Felser, *A Quantum-inspired Camera Perception Method in Autonomous Driving* (2023)
229. P. Calabrese, J. Cardy, Entanglement entropy and quantum field theory. J. Stat. Mech. Theory Exp. (2004). 10.1088/1742-5468/2004/06/P06002
230. R. Orús, T.-C. Wei, O. Buerschaper, A. García-Saez, Topological transitions from multipartite entanglement with tensor networks: a procedure for sharper and faster characterization. Phys. Rev. Lett. (2014). 10.1103/PhysRevLett.113.257202
231. A.J. Ferris, Fourier transform for fermionic systems and the spectral tensor network. Phys. Rev. Lett. (2014). 10.1103/PhysRevLett.113.010401
232. G. Ehlers, J. Sólyom, Ö. Legeza, R.M. Noack, Entanglement structure of the Hubbard model in momentum space. Phys. Rev. B (2015). 10.1103/PhysRevB.92.235116
233. G. Ehlers, S.R. White, R.M. Noack, Hybrid-space density matrix renormalization group study of the doped two-dimensional Hubbard model. Phys. Rev. B (2017). 10.1103/PhysRevB.95.125125
234. T.J. Osborne, *A Renormalisation-Group Algorithm for Eigenvalue Density Functions of Interacting Quantum Systems*. https://arxiv.org/abs/cond-mat/0605194
235. F. Schrodi, P. Silvi, F. Tschirsich, R. Fazio, S. Montangero, Density of states of many-body quantum systems from tensor networks. Phys. Rev. B (2017). 10.1103/PhysRevB.96.094303
236. A.J. Daley, Quantum trajectories and open many-body quantum systems. Adv. Phys. (2014). 10.1080/00018732.2014.933502
237. A. Molnar, N. Schuch, F. Verstraete, J.I. Cirac, Approximating Gibbs states of local Hamiltonians efficiently with projected entangled pair states. Phys. Rev. B - Condens. Matter Mater. Phys. (2015). 10.1103/PhysRevB.91.045138
238. S.R. White, Minimally entangled typical quantum states at finite temperature. Phys. Rev. Lett. (2009). 10.1103/PhysRevLett.102.190601
239. M.C. Bañuls, K. Cichy, J.I. Cirac, K. Jansen, H. Saito, Thermal evolution of the Schwinger model with matrix product operators. Phys. Rev. D (2015). 10.1103/PhysRevD.92.034519
240. B.-B. Chen, Y.-J. Liu, Z. Chen, W. Li, Series-expansion thermal tensor network approach for quantum lattice models. Phys. Rev. B (2017). 10.1103/PhysRevB.95.161104
241. A.H. Werner, D. Jaschke, P. Silvi, M. Kliesch, T. Calarco, J. Eisert, S. Montangero, Sup. mat. positive tensor network approach for simulating open quantum many-body systems. Phys. Rev. Lett. (2016). 10.1103/PhysRevLett.116.237201
242. S. Iblisdir, R. Orús, J. Latorre, Matrix product states algorithms and continuous systems. Phys. Rev. B (2007). 10.1103/PhysRevB.75.104305
243. M. Ganahl, J. Rincón, G. Vidal, Continuous matrix product states for quantum fields: an energy minimization algorithm. Phys. Rev. Lett. (2017). 10.1103/PhysRevLett.118.220402
244. P. Silvi, T. Calarco, G. Morigi, S. Montangero, Ab initio characterization of the quantum linear-zigzag transition using density matrix renormalization group calculations. Phys. Rev. B (2014). 10.1103/PhysRevB.89.094103
245. G.K.-L. Chan, S. Sharma, The density matrix renormalization group in quantum chemistry. Annu. Rev. Phys. Chem. (2011). 10.1146/annurev-physchem-032210-103338
246. N. Nakatani, G.K.-L. Chan, Efficient tree tensor network states (TTNS) for quantum chemistry: generalizations of the density matrix renormalization group algorithm. J. Chem. Phys. (2013). 10.1063/1.4798639

247. S. Wouters, D. Van Neck, The density matrix renormalization group for ab initio quantum chemistry. Eur. Phys. J. D (2014). 10.1140/epjd/e2014-50500-1
248. M. Motta, D.M. Ceperley, G.K.-L. Chan, J.A. Gomez, E. Gull, S. Guo, C.A. Jiménez-Hoyos, T.N. Lan, J. Li, F. Ma, A.J. Millis, N.V. Prokof'ev, U. Ray, G.E. Scuseria, S. Sorella, E.M. Stoudenmire, Q. Sun, I.S. Tupitsyn, S.R. White, D. Zgid, S. Zhang, Towards the solution of the many-electron problem in real materials: equation of state of the hydrogen chain with state-of-the-art many-body methods. Phys. Rev. X (2017). 10.1103/PhysRevX.7.031059
249. B. Swingle, Entanglement renormalization and holography. Phys. Rev. D (2012). 10.1103/PhysRevD.86.065007
250. J. Molina-Vilaplana, P. Sodano, Holographic view on quantum correlations and mutual information between disjoint blocks of a quantum critical system. J. High Energy Physics (2011). 10.1007/JHEP10(2011)011
251. C. Beny, Causal structure of the entanglement renormalization ansatz. New J. Phys. (2013)
252. F. Pastawski, B. Yoshida, D. Harlow, J. Preskill, Holographic quantum error-correcting codes: toy models for the bulk/boundary correspondence. J. High Energy Phys. (2015). 10.1007/JHEP06(2015)149
253. S. Singh, Tensor network state correspondence and holography. Phys. Rev. D (2018). 10.1103/PhysRevD.97.026012
254. S. Singh, N.A. McMahon, G.K. Brennen, Holographic spin networks from tensor network states. Phys. Rev. D (2018). 10.1103/PhysRevD.97.026013
255. M. Fishman, S. White, E.M. Stoudenmire, The ITensor software library for tensor network calculations. SciPost Phys. Codebases (2022). 10.21468/SciPostPhysCodeb.4
256. M. Fishman, S. White, E.M. Stoudenmire, Codebase Release 0.3 for ITensor. SciPost Phys. Codebases (2022). 10.21468/SciPostPhysCodeb.4-r0.3
257. S. Al-Assam, S.R. Clark, D. Jaksch, The tensor network theory library. J. Stat. Mech. Theory Exp. (2017). 10.1088/1742-5468/aa7df3
258. D. Jaschke, M.L. Wall, L.D. Carr, Open source matrix product states: opening ways to simulate entangled many-body quantum systems in one dimension. Comput. Phys. Commun. (2018). 10.1016/j.cpc.2017.12.015
259. M. Dolfi, B. Bauer, S. Keller, A. Kosenkov, T. Ewart, A. Kantian, T. Giamarchi, M. Troyer, Matrix product state applications for the ALPS project. Comput. Phys. Commun. (2014). https://doi.org/10.1016/j.cpc.2014.08.019
260. S.F. Keller, M. Reiher, Determining factors for the accuracy of DMRG in chemistry. Chim. Int. J. Chem. (2014). 10.2533/chimia.2014.200
261. S. Keller, M. Dolfi, M. Troyer, M. Reiher, An efficient matrix product operator representation of the quantum chemical Hamiltonian. J. Chem. Phys. (2015). 10.1063/1.4939000
262. G. De Chiara, M. Rizzi, D. Rossini, S. Montangero, Density Matrix Renormalization Group for Dummies. J. Comput. Theor. Nanosci. (2008). 10.1166/jctn.2008.011
263. E. Noether, Invariant variation problems. Transp. Theory Stat. Phys. (1971). 10.1080/00411457108231446
264. T. Lancaster, S.J. Blundell, *Quantum Field Theory for the Gifted Amateur* (Oxford University Press, Oxford, 2014). ISBN: 978-0-191-51093-9
265. M. Tinkham, *Group Theory and Quantum Mechanics*. Dover Books on Chemistry (Dover Publications, Garden City, 2012). ISBN: 978-0-486-13166-5
266. S. Singh, H.-Q. Zhou, G. Vidal, Simulation of one-dimensional quantum systems with a global SU(2) symmetry. New J. Phys. (2010). 10.1088/1367-2630/12/3/033029
267. A. Weichselbaum, Non-abelian symmetries in tensor networks: a quantum symmetry space approach. Ann. Phys. (2012). 10.1016/j.aop.2012.07.009
268. S. Singh, G. Vidal, Tensor network states and algorithms in the presence of a global SU(2) symmetry. Phys. Rev. B (2012). 10.1103/PhysRevB.86.195114
269. P. Corboz, A.M. Läuchli, K. Penc, M. Troyer, F. Mila, Simultaneous dimerization and SU(4) symmetry breaking of 4-color fermions on the square lattice. Phys. Rev. Lett. (2011). 10.1103/PhysRevLett.107.215301

270. S. Singh, G. Vidal, Tensor network states and algorithms in the presence of a global SU(2) symmetry. Phys. Rev. B (2012). 10.1103/PhysRevB.86.195114
271. A. Weichselbaum, Non-abelian symmetries in tensor networks: A quantum symmetry space approach. Ann. Phys. (N. Y). (2012). 10.1016/j.aop.2012.07.009
272. S. Chandrasekharan, Confinement, chiral symmetry breaking and continuum limits in quantum link models. Nucl. Phys. B - Proc. Suppl. (1999). 10.1016/S0920-5632(99)85189-5
273. U.-J. Wiese, Ultracold quantum gases and lattice systems: quantum simulation of lattice gauge theories. Ann. Phys. (2013). 10.1002/andp.201300104
274. K.G. Wilson, Confinement of quarks. Phys. Rev. D (1974). 10.1103/PhysRevD.10.2445
275. M. Creutz, L. Jacobs, C. Rebbi, Monte Carlo computations in lattice gauge theories. Phys. Rep. (1983). 10.1016/0370-1573(83)90016-9
276. O. Philipsen, The QCD equation of state from the lattice. Prog. Part. Nucl. Phys. (2013). 10.1016/j.ppnp.2012.09.003
277. K.A. Ross, L. Savary, B.D. Gaulin, L. Balents, Quantum Excitations in Quantum Spin Ice. Phys. Rev. X (2011). 10.1103/PhysRevX.1.021002
278. S.B. Lee, S. Onoda, L. Balents, Generic quantum spin ice. Phys. Rev. B (2012). 10.1103/PhysRevB.86.104412
279. N. Goldman, J.C. Budich, P. Zoller, Topological quantum matter with ultracold gases in optical lattices. Nat. Phys. (2016). 10.1038/nphys3803
280. P. Silvi, E. Rico, M. Dalmonte, F. Tschirsich, S. Montangero, Finite-density phase diagram of a non-abelian lattice gauge theory with tensor networks. Quantum (2017). 10.22331/q-2017-04-25-9
281. T. Pichler, M. Dalmonte, E. Rico, P. Zoller, S. Montangero, Real-time dynamics in U(1) lattice gauge theories with tensor networks. Phys. Rev. X (2016). 10.1103/PhysRevX.6.011023
282. T. Felser, P. Silvi, M. Collura, S. Montangero, Two-dimensional quantum-link lattice quantum electrodynamics at finite density. Phys. Rev. X (2020). 10.1103/PhysRevX.10.041040
283. B. Pirvu, V. Murg, J.I. Cirac, F. Verstraete, Matrix product operator representations. New J. Phys. (2010). 10.1088/1367-2630/12/2/025012
284. F. Fröwis, V. Nebendahl, W. Dür, Tensor operators: constructions and applications for long-range interaction systems. Phys. Rev. A (2010). 10.1103/PhysRevA.81.062337
285. J. Cui, J.I. Cirac, M.C. Bañuls, Variational matrix product operators for the steady state of dissipative quantum systems. Phys. Rev. Lett. (2015). 10.1103/PhysRevLett.114.220601
286. E. Mascarenhas, H. Flayac, V. Savona, Matrix-product-operator approach to the nonequilibrium steady state of driven-dissipative quantum arrays. Phys. Rev. A (2015). 10.1103/PhysRevA.92.022116
287. P. Pippan, S.R. White, H.G. Evertz, Efficient matrix product state method for periodic boundary conditions. Phys. Rev. B (2008). 10.1103/PhysRevB.81.081103
288. G. De las Cuevas, N. Schuch, D. Pérez-García, J.I. Cirac, Purifications of multipartite states: limitations and constructive methods. New J. Phys. (2013). 10.1088/1367-2630/15/12/123021
289. F. Verstraete, M.M. Wolf, D. Perez-Garcia, J.I. Cirac, Criticality, the area law, and the computational power of projected entangled pair states. Phys. Rev. Lett. (2006). 10.1103/PhysRevLett.96.220601
290. N. Schuch, M.M. Wolf, F. Verstraete, J.I. Cirac, Computational complexity of projected entangled pair states. Phys. Rev. Lett. (2007). 10.1103/PhysRevLett.98.140506
291. G. Vidal, Entanglement renormalization. Phys. Rev. Lett. (2007). 10.1103/PhysRevLett.99.220405
292. G. Vidal, Class of quantum many-body states that can be efficiently simulated. Phys. Rev. Lett. (2008). 10.1103/PhysRevLett.101.110501
293. M. Rizzi, S. Montangero, G. Vidal, Simulation of Time evolution with multiscale entanglement renormalization ansatz. Phys. Rev. A (2008). 10.1103/PhysRevA.77.052328
294. G. Evenbly, G. Vidal, Class of highly entangled many-body states that can be efficiently simulated. Phys. Rev. Lett. (2014). 10.1103/PhysRevLett.112.240502
295. S. Anders, H.J. Briegel, W. Dür, A variational method based on weighted graph states. New J. Phys. (2007). 10.1088/1367-2630/9/10/361

296. F. Mezzacapo, N. Schuch, M. Boninsegni, J.I. Cirac, Ground-state properties of quantum many-body systems: entangled-plaquette states and variational Monte Carlo. New J. Phys. (2009). 10.1088/1367-2630/11/8/083026
297. N. Schuch, M.M. Wolf, F. Verstraete, J.I. Cirac, Simulation of quantum many-body systems with strings of operators and Monte Carlo tensor contractions. Phys. Rev. Lett. (2008). 10.1103/PhysRevLett.100.040501
298. G. Evenbly, Hyperinvariant tensor networks and holography. Phys. Rev. Lett. (2017). 10.1103/PhysRevLett.119.141602
299. M. Suzuki, Fractal decomposition of exponential operators with applications to many-body theories and Monte Carlo simulations. Phys. Lett. A (1990). https://doi.org/10.1016/0375-9601(90)90962-N
300. J. Huyghebaeert, H. De Raedt, Product formula methods for time-dependent Schrödinger problems. J. Phys. A. Math. Gen. **23**(24), 5777–5793 (1990). https://iopscience.iop.org/article/10.1088/0305-4470/23/24/019
301. D. Poulin, A. Qarry, R. Somma, F. Verstraete, Quantum simulation of time-dependent hamiltonians and the convenient illusion of hilbert space. Phys. Rev. Lett. (2011). 10.1103/PhysRevLett.106.170501
302. S.R. White, A.E. Feiguin, Real-time evolution using the density matrix renormalization group. Phys. Rev. Lett. (2004). 10.1103/PhysRevLett.93.076401
303. G. Vidal, Efficient simulation of one-dimensional quantum many-body systems. Phys. Rev. Lett. (2004). 10.1103/PhysRevLett.93.040502
304. A.H. Werner, D. Jaschke, P. Silvi, M. Kliesch, T. Calarco, J. Eisert, S. Montangero, Positive tensor network approach for simulating open quantum many-body systems. Phys. Rev. Lett. (2016). 10.1103/PhysRevLett.116.237201
305. F. Fröwis, V. Nebendahl, W. Dür, Tensor operators: constructions and applications for long-range interaction systems. Phys. Rev. A (2010). 10.1103/PhysRevA.81.062337
306. M.P. Zaletel, R.S.K. Mong, C. Karrasch, J.E. Moore, F. Pollmann, *Time-Evolving a Matrix Product State with Long-ranged Interactions* (2015). 10.1103/PhysRevB.91.165112
307. M. Rizzi, S. Montangero, G. Vidal, Simulation of time evolution with multiscale entanglement renormalization ansatz. Phys. Rev. A (2008). 10.1103/PhysRevA.77.052328
308. J. Haegeman, C. Lubich, I. Oseledets, B. Vandereycken, F. Verstraete, Unifying time evolution and optimization with matrix product states. Phys. Rev. B - Condens. Matter Mater. Phys. (2016). 10.1103/PhysRevB.94.165116
309. J. Haegeman, J.I. Cirac, T.J. Osborne, I. Pizorn, H. Verschelde, F. Verstraete, Time-dependent variational principle for quantum lattices. Phys. Rev. Lett. (2011). 10.1103/PhysRevLett.107.070601
310. E.R. Davidson, The iterative calculation of a few of the lowest eigenvalues and corresponding eigenvectors of large real-symmetric matrices. J. Comput. Phys. (1975). https://doi.org/10.1016/0021-9991(75)90065-0
311. W.E. Arnoldi, The principle of minimized iterations in the solution of the matrix eigenvalue problem. Q. Appl. Math. (1951). https://doi.org/10.1090/qam/42792
312. D. Sorensen, R. Lehoucq, C. Yang, K. Maschhoff, S. Ledru, A. Cornet. *arpack-ng* (2017). Software Repository: https://github.com/opencollab/arpack-ng
313. J. Cervený, *gilbert* (2018). Software Repository: https://github.com/jakubcerveny/gilbert
314. J. Zhang, S.-I. Kamata, Y. Ueshige, A pseudo-hilbert scan algorithm for arbitrarily-sized rectangle region, in *Advances in Machine Vision, Image Processing, and Pattern Analysis*, ed. by N. Zheng, X. Jiang, X. Lan (Springer Berlin Heidelberg, Berlin/Heidelberg, 2006), pp. 290–299. ISBN: 978-3-540-37598-2
315. T. Felser, Tensor Network analysis on non-Abelien quantum many-body systems. Master Thesis at the University of Ulm (Institute for Complex Quantum Systems), 2017.
316. S. Singh, *Tensor Network States and Algorithms in the presence of Abelian and non-Abelian Symmetries* (2012). https://arxiv.org/abs/1203.2222
317. A. Weichselbaum, Non-abelian symmetries in tensor networks: a quantum symmetry space approach. Ann. Phys. **327**(12), 2972–3047 (2012). http://dx.doi.org/10.1016/j.aop.2012.07.009

318. E.M. Stoudenmire, S.R. White, Real-space parallel density matrix renormalization group. Phys. Rev. B (2013). 10.1103/PhysRevB.87.155137
319. A. Browaeys, T. Lahaye, Many-body physics with individually controlled Rydberg atoms. Nature Phys. (2020). 10.1038/s41567-019-0733-z
320. P. Schauß, M. Cheneau, M. Endres, T. Fukuhara, S. Hild, A. Omran, T. Pohl, C. Gross, S. Kuhr, I. Bloch, Observation of spatially ordered structures in a two-dimensional Rydberg gas. Nature (2012). 10.1038/nature11596
321. T. Olsacher, L. Postler, P. Schindler, T. Monz, P. Zoller, L.M. Sieberer, Scalable and parallel tweezer gates for quantum computing with long ion strings. PRX Quantum (2020). 10.1103/prxquantum.1.020316
322. F. Arute, K. Arya, R. Babbush, D. Bacon, J.C. Bardin, R. Barends, R. Biswas, S. Boixo, F.G.S.L. Brandao, D.A. Buell, B. Burkett, Y. Chen, Z. Chen, B. Chiaro, R. Collins, W. Courtney, A. Dunsworth, E. Farhi, B. Foxen, A. Fowler, C. Gidney, M. Giustina, R. Graff, K. Guerin, S. Habegger, M.P. Harrigan, M.J. Hartmann, A. Ho, M. Hoffmann, T. Huang, T.S. Humble, S.V. Isakov, E. Jeffrey, Z. Jiang, D. Kafri, K. Kechedzhi, J. Kelly, P.V. Klimov, S. Knysh, A. Korotkov, F. Kostritsa, D. Landhuis, M. Lindmark, E. Lucero, D. Lyakh, S. Mandrá, J.R. McClean, M. McEwen, A. Megrant, X. Mi, K. Michielsen, M. Mohseni, J. Mutus, O. Naaman, M. Neeley, C. Neill, M.Y. Niu, E. Ostby, A. Petukhov, J.C. Platt, C. Quintana, E.G. Rieffel, P. Roushan, N.C. Rubin, D. Sank, K.J. Satzinger, V. Smelyanskiy, K.J. Sung, M.D. Trevithick, A. Vainsencher, B. Villalonga, T. White, Z.J. Yao, P. Yeh, A. Zalcman, H. Neven, J.M. Martinis, Quantum supremacy using a programmable superconducting processor. Nature (2019). 10.1038/s41586-019-1666-5
323. T. Barthel, M. Kliesch, J. Eisert, Real-space renormalization yields finite correlations. Phys. Rev. Lett. (2010). 10.1103/PhysRevLett.105.010502
324. A.W. Sandvik, Finite-size scaling of the ground-state parameters of the two-dimensional Heisenberg model. Phys. Rev. B (1997). 10.1103/physrevb.56.11678
325. G. Carleo, M. Troyer, Solving the quantum many-body problem with artificial neural networks. Science (2017). 10.1126/science.aag2302
326. F. Mezzacapo, N. Schuch, M. Boninsegni, J.I. Cirac, Ground-state properties of quantum many-body systems: entangled-plaquette states and variational Monte Carlo. New J. Phys. (2009). 10.1088/1367-2630/11/8/083026
327. J. Jordan, R. Orús, G. Vidal, F. Verstraete, J.I. Cirac, Classical simulation of infinite-size quantum lattice systems in two spatial dimensions. Phys. Rev. Lett. (2008). 10.1103/PhysRevLett.101.250602
328. R. Orús, G. Vidal, Simulation of two-dimensional quantum systems on an infinite lattice revisited: corner transfer matrix for tensor contraction. Phys. Rev. B (2009). 10.1103/PhysRevB.80.094403
329. O. Sharir, Y. Levine, N. Wies, G. Carleo, A. Shashua, Deep autoregressive models for the efficient variational simulation of many-body quantum systems. Phys. Rev. Lett. (2020). 10.1103/PhysRevLett.124.020503
330. L. He, H. An, C. Yang, F. Wang, J. Chen, C. Wang, W. Liang, S. Dong, Q. Sun, W. Han, W. Liu, Y. Han, W. Yao, PEPS++: towards extreme-scale simulations of strongly correlated quantum many-particle models on sunway TaihuLight. IEEE Trans. Parallel Distrib. Syst. (2018). 10.1109/TPDS.2018.2848618
331. W.-Y. Liu, Y.-Z. Huang, S.-S. Gong, Z.-C. Gu, Accurate simulation for finite projected entangled pair states in two dimensions. Phys. Rev. B **103**(23) (2021). http://dx.doi.org/10.1103/PhysRevB.103.235155
332. S. Singh, G. Vidal, Global symmetries in tensor network states: symmetric tensors versus minimal bond dimension. Phys. Rev. B (2013). 10.1103/PhysRevB.88.115147
333. S. Sachdev, *Quantum Phase Transitions* (Cambridge University Press, Cambridge, 2011). ISBN: 978-1-139-50021-0
334. E. Fradkin, *Field Theories of Condensed Matter Physics*. Field Theories of Condensed Matter Physics (Cambridge University Press, Cambridge, 2013). ISBN: 978-0-521-76444-5
335. L.D. Landau, On the theory of phase transitions. Zh. Eksp. Teor. Fiz. **7**, 19–32 (1937)

336. S. Blundell. *Magnetism in Condensed Matter*. Oxford Master Series in Condensed Matter Physics (Oxford University Press, Oxford, 2001). ISBN: 978-0-198-50591-4
337. S.L. Sondhi, S.M. Girvin, J.P. Carini, D. Shahar, Continuous quantum phase transitions. Rev. Mod. Phys. **69**(1), 315–333 (1997). https://link.aps.org/doi/10.1103/RevModPhys.69.315
338. M. Suzuki, Relationship between d-dimensional quantal spin systems and (d+1)-dimensional ising systems: equivalence, critical exponents and systematic approximants of the partition function and spin correlations. Prog. Theor. Phys. (1976). 10.1143/PTP.56.1454
339. A. Osterloh, L. Amico, G. Falci, R. Fazio, Scaling of entanglement close to a quantum phase transition. Nature (2002). 10.1038/416608a
340. M. Fisher, M. Barber, Scaling theory for finite-size effects in the critical region. Phys. Rev. Lett. (1972). https://doi.org/10.1103/PhysRevLett.28.1516
341. T.J. Osborne, M.A. Nielsen, Entanglement in a simple quantum phase transition. Phys. Rev. A (2002). 10.1103/PhysRevA.66.032110
342. G. Refael, J.E. Moore, Entanglement entropy of random quantum critical points in one dimension. Phys. Rev. Lett. (2004). 10.1103/PhysRevLett.93.260602
343. D. Binosi, G. De Chiara, S. Montangero, A. Recati, Increasing entanglement through engineered disorder in the random Ising chain. Phys. Rev. B (2007). 10.1103/PhysRevB.76. 140405
344. J.C. Getelina, F.C. Alcaraz, J.A. Hoyos, Entanglement properties of correlated random spin chains and similarities with conformally invariant systems. Phys. Rev. B (2016). 10.1103/ PhysRevB.93.045136
345. J.A. Hoyos, N. Laflorencie, A.P. Vieira, T. Vojta, Protecting clean critical points by local disorder correlations. EPL (Europhysics Lett.) (2011). 10.1209/0295-5075/93/30004
346. N. Laflorencie, Quantum entanglement in condensed matter systems. Phys. Rep. (2015). 10. 1016/j.physrep.2016.06.008
347. A. Kitaev, J. Preskill, Topological entanglement entropy. Phys. Rev. Lett. (2006). 10.1103/ PhysRevLett.96.110404
348. M. Levin, X.-G. Wen, Detecting topological order in a ground state wave function. Phys. Rev. Lett. (2006). 10.1103/PhysRevLett.96.110405
349. H.-C. Jiang, Z. Wang, L. Balents, Identifying topological order by entanglement entropy. Nat. Phys. (2012). 10.1038/nphys2465
350. H.-C. Jiang, R.R.P. Singh, L. Balents, Accuracy of topological entanglement entropy on finite cylinders. Phys. Rev. Lett. (2013). 10.1103/PhysRevLett.111.107205
351. S. Depenbrock, I.P. McCulloch, U. Schollwoeck, Nature of the spin-liquid ground state of the S=1/2 Heisenberg model on the kagome lattice. Phys. Rev. Lett. (2012). 10.1103/ PhysRevLett.109.067201
352. R. Orus, T.C. Wei, O. Buerschaper, A. Garcia-Saez, Topological transitions from multipartite entanglement with tensor networks: a procedure for sharper and faster characterization. Phys. Rev. Lett. (2014). 10.1103/PhysRevLett.113.257202
353. M. Gerster, M. Rizzi, P. Silvi, M. Dalmonte, S. Montangero, Fractional quantum Hall effect in the interacting Hofstadter model via tensor networks. Phys. Rev. B (2017). 10.1103/ PhysRevB.96.195123
354. J.B. Kogut, An introduction to lattice gauge theory and spin systems. Rev. Modern Phys. (1979). 10.1103/RevModPhys.51.659
355. H.J. Rothe. *Lattice Gauge Theories: An Introduction*, 4th edn. (World Scientific Publishing Company, Singapore, 2012). ISBN: 978-9-814-36587-1, 978-9-814-36585-7
356. J.B. Kogut, The lattice gauge theory approach to quantum chromodynamics. Rev. Modern Phys. (1983). 10.1103/RevModPhys.55.775
357. M.D. Schwartz. *Quantum Field Theory and the Standard Model* (Cambridge University Press, Cambridge, 2014). ISBN: 978-1-107-03473-0, 978-1-107-03473-0
358. T.P. Cheng, L.F. Li. *Gauge Theory of Elementary Particle Physics* (Oxford University Press, Oxford, 2006)
359. I.J.R. Aitchison, A.J.G. Hey, *Gauge Theories in Particle Physics: A Practical Introduction, vol. 1: From Relativistic Quantum Mechanics to QED* (CRC Press, Bristol, 2003). ISBN: 978-1-466-51299-3

360. A.S. Kronfeld, C. Quigg, Resource letter QCD-1: quantum chromodynamics. Am. J. Phys. (2010). 10.1119/1.3454865
361. A.S. Kronfeld, Twenty-first century lattice gauge theory: results from the quantum chromodynamics Lagrangian. Ann. Rev. Nucl. Particle Sci. (2012). 10.1146/annurev-nucl-102711-094942
362. X.Y. Xu, Y. Qi, L. Zhang, F.F. Assaad, C. Xu, Z.Y. Meng, Monte Carlo study of lattice compact quantum electrodynamics with fermionic matter: the parent state of quantum phases. Phys. Rev. X (2019). 10.1103/PhysRevX.9.021022
363. R. Gupta, *Introduction to Lattice QCD* (1998). https://arxiv.org/abs/hep-lat/9807028
364. M.P. HernÁndez, 1 lattice field theory fundamentals, in *Modern Perspectives in Lattice QCD: Quantum Field Theory and High Performance Computing: Lecture Notes of the Les Houches Summer School, August 2009*, vol. 93, ed. by L. Lellouch, R. Sommer, B. Svetitsky, A. Vladikas, L.F. Cugliandolo (Oxford University Press, Oxford, 2011), p. 0. ISBN: 978-0-199-69160-9. https://doi.org/10.1093/acprof:oso/9780199691609.003.0001
365. K. Nagata, Finite-density lattice QCD and sign problem: current status and open problems. Progress Particle Nucl. Phys. (2022). 10.1016/j.ppnp.2022.103991
366. K.G. Wilson, Confinement of quarks. Phys. Rev. D (1974). 10.1103/PhysRevD.10.2445
367. M. Biswal, M. Deka, S. Digal, P.S. Saumia, Confinement - deconfinement transition in an $SU(2)$ Higgs theory. Phys. Rev. D (2017). 10.1103/PhysRevD.96.014503
368. V.G. Bornyakov, V.V. Braguta, E.M. Ilgenfritz, A.Y. Kotov, I.E. Kudrov, A.V. Molochkov, A.A. Nikolaev, R.N. Rogalyov, Confinement-deconfinement transition in dense SU(2) QCD. EPJ Web Conf. (2018). 10.1051/epjconf/201817507009
369. M. Engelhardt, K. Langfeld, H. Reinhardt, O. Tennert, Deconfinement in SU(2) Yang-Mills theory as a center vortex percolation transition. Phys. Rev. D (2000). 10.1103/PhysRevD.61.054504
370. J. Ambjøorn, P. Olesen, C. Peterson, Stochastic confinement and dimensional reduction (I). Four-dimensional SU(2) Lattice Gauge theory. Nucl. Phys. B (1984). 10.1016/0550-3213(84)90475-9
371. J. Ambjørn, P. Olesen, C. Peterson, Stochastic confinement and dimensional reduction (II). Three-dimensional SU(2) Lattice Gauge theory. Nucl. Phys. B (1984). 10.1016/0550-3213(84)90242-6
372. M. Mitter, J.M. Pawlowski, N. Strodthoff, Chiral symmetry breaking in continuum QCD. Phys. Rev. D (2015). 10.1103/PhysRevD.91.054035
373. S. Gottlieb, W. Liu, D. Toussaint, R.L. Renken, R.L. Sugar, Chiral-symmetry breaking in lattice QCD with two and four fermion flavors. Phys. Rev. D (1987). 10.1103/PhysRevD.35.3972
374. M. Alford, K. Rajagopal, F. Wilczek, Color-flavor locking and chiral symmetry breaking in high density QCD. Nucl. Phys. B (1999). 10.1016/S0550-3213(98)00668-3
375. J. Kogut, M. Stone, H.W. Wyld, J. Shigemitsu, S.H. Shenker, D.K. Sinclair, Scales of chiral symmetry breaking in quantum chromodynamics. Phys. Rev. Lett. (1982). 10.1103/PhysRevLett.48.1140
376. Z. Davoudi, E.T. Neil, C.W. Bauer, T. Bhattacharya, T. Blum, P. Boyle, R.C. Brower, S. Catterall, N.H. Christ, V. Cirigliano, G. Colangelo, C. DeTar, W. Detmold, R.G. Edwards, A.X. El-Khadra, S. Gottlieb, R. Gupta, D.C. Hackett, A. Hasenfratz, T. Izubuchi, W.I. Jay, L. Jin, C. Kelly, A.S. Kronfeld, C. Lehner, H.-W. Lin, M. Lin, A.T. Lytle, S. Meinel, Y. Meurice, S. Mukherjee, A. Nicholson, S. Prelovsek, M.J. Savage, P.E. Shanahan, R.S. Van De Water, M.L. Wagman, O. Witzel, *Report of the Snowmass 2021 Topical Group on Lattice Gauge Theory* (2022). 10.48550/arXiv.2209.10758
377. G. Baskaran, P.W. Anderson, Gauge theory of high-temperature superconductors and strongly correlated Fermi systems. Phys. Rev. B (1988). 10.1103/PhysRevB.37.580
378. A. Mezzacapo, E. Rico, C. Sabín, I.L. Egusquiza, L. Lamata, E. Solano, Non-Abelian SU(2) Lattice Gauge theories in superconducting circuits. Phys. Rev. Lett. (2015). 10.1103/PhysRevLett.115.240502

379. D. Marcos, P. Widmer, E. Rico, M. Hafezi, P. Rabl, U.-J. Wiese, P. Zoller, Two-dimensional lattice gauge theories with superconducting quantum circuits. Ann. Phys. (2014). https://doi.org/10.1016/j.aop.2014.09.011
380. C. Chamon, D. Green, Z.-C. Yang, Constructing quantum spin liquids using combinatorial Gauge symmetry. Phys. Rev. Lett. (2020). 10.1103/PhysRevLett.125.067203
381. Y. Zhou, K. Kanoda, T.-K. Ng, Quantum spin liquid states. Rev. Mod. Phys. (2017). 10.1103/RevModPhys.89.025003
382. M. Creutz, L. Jacobs, C. Rebbi, Monte Carlo study of abelian lattice gauge theories. Phys. Rev. D (1979). 10.1103/PhysRevD.20.1915
383. M. Creutz, Monte Carlo study of quantized SU(2) Gauge theory. Phys. Rev. D (1980). 10.1103/PhysRevD.21.2308
384. B. Berg, J. Stehr, SU(2) Lattice Gauge Theory and Monte Carlo calculations. Zeitschrift für Physik C Particles and Fields (1981). 10.1007/BF01548769
385. M. Creutz, L. Jacobs, C. Rebbi, Monte Carlo computations in Lattice Gauge theories. Phys. Rep. (1983). 10.1016/0370-1573(83)90016-9
386. M. Creutz. *Lattice Gauge Theory and Monte Carlo Methods*. Technical Report, BNL-42086. Brookhaven National Lab. (BNL), Upton, NY (United States), Nov. 1988. 10.2172/6530895
387. M. Creutz, Lattice Gauge theories and Monte Carlo algorithms. Nucl. Phys. B - Proc. Suppl. (1989). 10.1016/0920-5632(89)90061-3
388. T.D. Kieu, C.J. Griffin, Monte Carlo simulations with indefinite and complex-valued measures. Phys. Rev. E (1994). 10.1103/PhysRevE.49.3855
389. E. Ghobadpour, M. Kolb, M.R. Ejtehadi, R. Everaers, Monte Carlo simulation of a lattice model for the dynamics of randomly branching double-folded ring polymers. Phys. Rev. E (2021). 10.1103/PhysRevE.104.014501
390. H.J.M. van Bemmel, D.F.B. ten Haaf, W. van Saarloos, J.M.J. van Leeuwen, G. An, Fixed-node quantum Monte Carlo method for lattice fermions. Phys. Rev. Lett. (1994). 10.1103/PhysRevLett.72.2442
391. X.Y. Xu, Y. Qi, L. Zhang, F.F. Assaad, C. Xu, Z.Y. Meng, Monte Carlo study of lattice compact quantum electrodynamics with fermionic matter: the parent state of quantum phases. Phys. Rev. X (2019). 10.1103/PhysRevX.9.021022
392. M. Loan, M. Brunner, C. Sloggett, C. Hamer, Path integral Monte Carlo approach to the U(1) Lattice Gauge theory in 2+1 dimensions. Phys. Rev. D (2003). 10.1103/PhysRevD.68.034504
393. J.E. Lynn, I. Tews, S. Gandolfi, A. Lovato, Quantum Monte Carlo methods in nuclear physics: recent advances. Ann. Rev. Nucl. Particle Sci. (2019). 10.1146/annurev-nucl-101918-023600
394. E.Y. Loh, J.E. Gubernatis, R.T. Scalettar, S.R. White, D.J. Scalapino, R.L. Sugar, Sign problem in the numerical simulation of many-electron systems. Phys. Rev. B (1990). 10.1103/PhysRevB.41.9301
395. C. Gattringer, K. Langfeld, Approaches to the sign problem in lattice field theory. Int. J. Modern Phys. A (2016). 10.1142/S0217751X16430077
396. G. Magnifico, G. Cataldi, M. Rigobello, P. Majcen, D. Jaschke, P. Silvi, S. Montangero,. *Tensor Networks for Lattice Gauge Theories beyond One Dimension: A Roadmap* (2024). 10.48550/arXiv.2407.03058
397. G. Magnifico, T. Felser, P. Silvi, S. Montangero, Lattice quantum electrodynamics in (3+1)-dimensions at finite density with tensor networks. Nature Commun. (2021). 10.1038/s41467-021-23646-3
398. M. Rigobello, S. Notarnicola, G. Magnifico, S. Montangero, Entanglement generation in $(1+1)\mathrm{D}$ QED scattering processes. Phys. Rev. D (2021). 10.1103/PhysRevD.104.114501
399. G. Cataldi, G. Magnifico, P. Silvi, S. Montangero, Simulating (2+1)D SU(2) Yang-Mills Lattice Gauge theory at finite density with tensor networks. Phys. Rev. Res. (2024). 10.1103/PhysRevResearch.6.033057
400. M. Rigobello, G. Magnifico, P. Silvi, S. Montangero, *Hadrons in (1+1)D Hamiltonian Hardcore Lattice QCD* (2023). 10.48550/arXiv.2308.04488

401. G. Calajò, G. Cataldi, M. Rigobello, D. Wanisch, G. Magnifico, P. Silvi, S. Montangero, J.C. Halimeh, *Quantum Many-Body Scarring in a Non-Abelian Lattice Gauge Theory* (2024). 10.48550/arXiv.2405.13112
402. G. Calajò, G. Magnifico, C. Edmunds, M. Ringbauer, S. Montangero, P. Silvi, Digital quantum simulation of a (1+1)D SU(2) Lattice Gauge theory with ion qudits. PRX Quantum (2024). 10.1103/PRXQuantum.5.040309
403. M. Ballarin, G. Cataldi, G. Magnifico, D. Jaschke, M. Di Liberto, I. Siloi, S. Montangero, P. Silvi, Digital quantum simulation of lattice fermion theories with local encoding. Quantum (2024). 10.22331/q-2024-09-04-1460
404. G. Cataldi, *Hamiltonian Lattice Gauge Theories: Emergent Properties from Tensor Network Methods* (2025). 10.48550/arXiv.2501.11115
405. J. Kogut, L. Susskind, Hamiltonian formulation of Wilson's Lattice Gauge theories. Phys. Rev. D (1975). 10.1103/PhysRevD.11.395
406. M. Creutz, *Quantum Fields on the Computer* (World Scientific, Singapore, 1992). ISBN: 978-981-0-20940-7
407. G. Clemente, A. Crippa, K. Jansen, Strategies for the determination of the running coupling of ($2+1$)-dimensional QED with quantum computing. Phys. Rev. D (2022). 10.1103/PhysRevD.106.114511
408. A. Crippa, S. Romiti, L. Funcke, K. Jansen, S. Kühn, P. Stornati, C. Urbach,. *Towards Determining the (2+1)-Dimensional Quantum Electrodynamics Running Coupling with Monte Carlo and Quantum Computing Methods* (2024). 10.48550/arXiv.2404.17545
409. L. Susskind, Lattice fermions. Phys. Rev. D (1977). 10.1103/PhysRevD.16.3031
410. H.B. Nielsen, M. Ninomiya, Absence of neutrinos on a lattice: (I). Proof by homotopy theory. Nucl. Phys. B (1981). 10.1016/0550-3213(81)90361-8
411. H.B. Nielsen, M. Ninomiya, Absence of neutrinos on a lattice: (II). Intuitive topological proof. Nucl. Phys. B (1981). 10.1016/0550-3213(81)90524-1
412. A. Bermudez, L. Mazza, M. Rizzi, N. Goldman, M. Lewenstein, M.A. Martin-Delgado, W. Fermions, Axion Electrodynamics in optical lattices. Phys. Rev. Lett. (2010). 10.1103/PhysRevLett.105.190404
413. L. Mazza, A. Bermudez, N. Goldman, M. Rizzi, M.A. Martin-Delgado, M. Lewenstein, An optical-lattice-based quantum simulator for relativistic field theories and topological insulators. New J. Phys. (2012). 10.1088/1367-2630/14/1/015007
414. Y. Kuno, I. Ichinose, Y. Takahashi, Generalized Lattice Wilson–Dirac fermions in (1 + 1) dimensions for atomic quantum simulation and topological phases. Sci. Rep. (2018). 10.1038/s41598-018-29143-w
415. T.V. Zache, F. Hebenstreit, F. Jendrzejewski, M.K. Oberthaler, J. Berges, P. Hauke, Quantum simulation of lattice gauge theories using wilson fermions. Quantum Sci. Technol. (2018). 10.1088/2058-9565/aac33b
416. P. Silvi, Y. Sauer, F. Tschirsich, S. Montangero, Tensor network simulation of an SU(3) Lattice Gauge theory in 1D. Phys. Rev. D (2019). 10.1103/PhysRevD.100.074512
417. R. Brauer, H. Weyl, Spinors in n dimensions. Am. J. Math. (1935). 10.2307/2371218
418. E. Zohar, Quantum simulation of Lattice Gauge theories in more than one space dimension—requirements, challenges and methods. Philos. Trans. R. Soc. A Math. Phys. Eng. Sci. (2021). 10.1098/rsta.2021.0069
419. J. Bender, E. Zohar, Gauge redundancy-free formulation of compact QED with dynamical matter for quantum and classical computations. Phys. Rev. D (2020). 10.1103/PhysRevD.102.114517
420. D. Horn, Finite matrix models with continuous local gauge invariance. Phys. Lett. B (1981). 10.1016/0370-2693(81)90763-2
421. P. Orland, D. Rohrlich, Lattice Gauge magnets: local isospin from spin. Nucl. Phys. B (1990). 10.1016/0550-3213(90)90646-U
422. S. Chandrasekharan, U.J. Wiese, Quantum link models: a discrete approach to gauge theories. Nucl. Phys. B (1997). 10.1016/S0550-3213(97)80041-7

423. R. Brower, S. Chandrasekharan, U.-J. Wiese, QCD as a quantum link model. Phys. Rev. D (1999). 10.1103/PhysRevD.60.094502
424. L. Tagliacozzo, A. Celi, M. Lewenstein, Tensor networks for lattice gauge theories with continuous groups. Phys. Rev. X (2014). 10.1103/PhysRevX.4.041024
425. G. Magnifico, M. Dalmonte, P. Facchi, S. Pascazio, F.V. Pepe, E. Ercolessi, Real time dynamics and confinement in the $\mathbb{Z}_{n}$ Schwinger-Weyl lattice model for 1+1 QED. Quantum (2020). 10.22331/q-2020-06-15-281
426. J.F. Haase, L. Dellantonio, A. Celi, D. Paulson, A. Kan, K. Jansen, C.A. Muschik, A resource efficient approach for quantum and classical simulations of Gauge theories in particle physics. Quantum (2021). 10.22331/q-2021-02-04-393
427. D.C. Hackett, K. Howe, C. Hughes, W. Jay, E.T. Neil, J.N. Simone, Digitizing Gauge fields: lattice Monte Carlo results for future quantum computers. Phys. Rev. A (2019). 10.1103/PhysRevA.99.062341
428. T.V. Zache, D. González-Cuadra, P. Zoller, *Quantum and Classical Spin Network Algorithms for q-Deformed Kogut-Susskind Gauge Theories* (2023). 10.48550/arXiv.2304.02527
429. F. Arute, K. Arya, R. Babbush, D. Bacon, J.C. Bardin, R. Barends, A. Bengtsson, S. Boixo, M. Broughton, B.B. Buckley, D.A. Buell, B. Burkett, N. Bushnell, Y. Chen, Z. Chen, Y.-A. Chen, B. Chiaro, R. Collins, S.J. Cotton, W. Courtney, S. Demura, A. Derk, A. Dunsworth, D. Eppens, T. Eckl, C. Erickson, E. Farhi, A. Fowler, B. Foxen, C. Gidney, M. Giustina, R. Graff, J.A. Gross, S. Habegger, M.P. Harrigan, A. Ho, S. Hong, T. Huang, W. Huggins, L.B. Ioffe, S.V. Isakov, E. Jeffrey, Z. Jiang, C. Jones, D. Kafri, K. Kechedzhi, J. Kelly, S. Kim, P.V. Klimov, A.N. Korotkov, F. Kostritsa, D. Landhuis, P. Laptev, M. Lindmark, E. Lucero, M. Marthaler, O. Martin, J.M. Martinis, A. Marusczyk, S. McArdle, J.R. McClean, T. McCourt, M. McEwen, A. Megrant, C. Mejuto-Zaera, X. Mi, M. Mohseni, W. Mruczkiewicz, J. Mutus, O. Naaman, M. Neeley, C. Neill, H. Neven, M. Newman, M.Y. Niu, T.E. O'Brien, E. Ostby, B. Pató, A. Petukhov, H. Putterman, C. Quintana, J.-M. Reiner, P. Roushan, N.C. Rubin, D. Sank, K.J. Satzinger, V. Smelyanskiy, D. Strain, K.J. Sung, P. Schmitteckert, M. Szalay, N.M. Tubman, A. Vainsencher, T. White, N. Vogt, Z.J. Yao, P. Yeh, A. Zalcman, S. Zanker,. *Observation of Separated Dynamics of Charge and Spin in the Fermi-Hubbard Model* (2020). 10.48550/arXiv.2010.07965
430. R. Barends, L. Lamata, J. Kelly, L. García-Álvarez, A.G. Fowler, A. Megrant, E. Jeffrey, T.C. White, D. Sank, J.Y. Mutus, B. Campbell, Y. Chen, Z. Chen, B. Chiaro, A. Dunsworth, I.-C. Hoi, C. Neill, P. J.J. O'Malley, C. Quintana, P. Roushan, A. Vainsencher, J. Wenner, E. Solano, J.M. Martinis, Digital quantum simulation of fermionic models with a superconducting circuit. Nature Commun. (2015). 10.1038/ncomms8654
431. Y. Salathé, M. Mondal, M. Oppliger, J. Heinsoo, P. Kurpiers, A. Potočnik, A. Mezzacapo, U. Las Heras, L. Lamata, E. Solano, S. Filipp, A. Wallraff, Digital quantum simulation of spin models with circuit quantum electrodynamics. Phys. Rev. X (2015). 10.1103/PhysRevX.5.021027
432. P.J.J. O'Malley, R. Babbush, I.D. Kivlichan, J. Romero, J.R. McClean, R. Barends, J. Kelly, P. Roushan, A. Tranter, N. Ding, B. Campbell, Y. Chen, Z. Chen, B. Chiaro, A. Dunsworth, A.G. Fowler, E. Jeffrey, E. Lucero, A. Megrant, J.Y. Mutus, M. Neeley, C. Neill, C. Quintana, D. Sank, A. Vainsencher, J. Wenner, T.C. White, P.V. Coveney, P.J. Love, H. Neven, A. Aspuru-Guzik, J.M. Martinis, Scalable quantum simulation of molecular energies. Phys. Rev. X (2016). 10.1103/PhysRevX.6.031007
433. S. Stanisic, J.L. Bosse, F.M. Gambetta, R.A. Santos, W. Mruczkiewicz, T.E. O'Brien, E. Ostby, A. Montanaro, Observing ground-state properties of the fermi-hubbard model using a scalable algorithm on a quantum computer. Nature Commun. (2022). 10.1038/s41467-022-33335-4
434. P. Jordan, E. Wigner, Über das Paulische Äquivalenzverbot". Zeitschrift für Physik (1928). 10.1007/BF01331938
435. M.A. Nielsen, M.R. Dowling, M. Gu, A.C. Doherty, Quantum computation as geometry. Science (2006). 10.1126/science.1121541

436. S.B. Bravyi, A. Yu, Kitaev, fermionic quantum computation. Ann. Phys. (2002). 10.1006/aphy.2002.6254
437. Z. Jiang, J. McClean, R. Babbush, H. Neven, Majorana loop stabilizer codes for error mitigation in fermionic quantum simulations. Phys. Rev. Appl. (2019). 10.1103/PhysRevApplied.12.064041
438. F. Verstraete, J.I. Cirac, Mapping local hamiltonians of fermions to local hamiltonians of spins. J. Stat. Mech. Theory Exper. (2005). 10.1088/1742-5468/2005/09/P09012
439. P. Corboz, R. Orús, B. Bauer, G. Vidal, Simulation of strongly correlated fermions in two spatial dimensions with fermionic projected entangled-pair states. Phys. Rev. B (2010). 10.1103/PhysRevB.81.165104
440. C.V. Kraus, N. Schuch, F. Verstraete, J.I. Cirac, Fermionic projected entangled pair states. Phys. Rev. A (2010). 10.1103/PhysRevA.81.052338
441. L. Tagliacozzo, A. Celi, P. Orland, M.W. Mitchell, M. Lewenstein, Simulations of non-abelian gauge theories with optical lattices. Nature Commun. (2013). 10.1038/ncomms3615
442. P. Silvi, E. Rico, T. Calarco, S. Montangero, Lattice Gauge tensor networks. New J. Phys. (2014). 10.1088/1367-2630/16/10/103015
443. E. Zohar, J.I. Cirac, Eliminating fermionic matter fields in lattice Gauge theories. Phys. Rev. B (2018). 10.1103/PhysRevB.98.075119
444. I. Raychowdhury, J.R. Stryker, Loop, string, and hadron dynamics in SU(2) hamiltonian lattice Gauge theories. Phys. Rev. D (2020). 10.1103/PhysRevD.101.114502
445. E. Zohar, J.I. Cirac, Removing staggered fermionic matter in U(N) and SU(N) Lattice Gauge theories. Phys. Rev. D (2019). 10.1103/PhysRevD.99.114511
446. E. Zohar, M. Burrello, Formulation of Lattice Gauge theories for quantum simulations. Phys. Rev. D (2015). 10.1103/PhysRevD.91.054506
447. P. Sala, T. Shi, S. Kühn, M.C. Bañuls, E. Demler, J.I. Cirac, Variational study of U(1) and SU(2) Lattice Gauge theories with Gaussian states in \$1+1\$ dimensions. Phys. Rev. D (2018). 10.1103/PhysRevD.98.034505
448. T. Byrnes, Y. Yamamoto, Simulating Lattice Gauge theories on a quantum computer. Phys. Rev. A (2006). 10.1103/PhysRevA.73.022328
449. S.V. Mathis, G. Mazzola, I. Tavernelli, Toward scalable simulations of Lattice Gauge theories on quantum computers". Phys. Rev. D (2020). 10.1103/PhysRevD.102.094501
450. Z. Davoudi, M. Hafezi, C. Monroe, G. Pagano, A. Seif, A. Shaw, Towards analog quantum simulations of Lattice Gauge theories with trapped ions. Phys. Rev. Res. (2020). 10.1103/PhysRevResearch.2.023015
451. G. Mazzola, S.V. Mathis, G. Mazzola, I. Tavernelli, Gauge-invariant quantum circuits for \$U\$(1) and Yang-Mills Lattice Gauge theories. Phys. Rev. Res. (2021). 10.1103/PhysRevResearch.3.043209
452. A. Kan, L. Funcke, S. Kühn, L. Dellantonio, J. Zhang, J.F. Haase, C.A. Muschik, K. Jansen, Investigating a \$(3+1)\mathrm{D}\$ topological \$\ensuremath{\theta}\$-term in the hamiltonian formulation of Lattice Gauge theories for quantum and classical simulations. Phys. Rev. D (2021). 10.1103/PhysRevD.104.034504
453. A. Mariani, S. Pradhan, E. Ercolessi, Hamiltonians and Gauge-invariant Hilbert space for Lattice Yang-Mills-like theories with finite Gauge group. Phys. Rev. D (2023). 10.1103/PhysRevD.107.114513
454. D. Pomarico, L. Cosmai, P. Facchi, C. Lupo, S. Pascazio, F.V. Pepe, Dynamical quantum phase transitions of the schwinger model: real-time dynamics on IBM quantum. Entropy (2023). 10.3390/e25040608
455. C.W. Bauer, Z. Davoudi, N. Klco, M.J. Savage, Quantum simulation of fundamental particles and forces. Nature Rev. Phys. (2023). 10.1038/s42254-023-00599-8
456. C.W. Bauer, Z. Davoudi, A.B. Balantekin, T. Bhattacharya, M. Carena, W.A. de Jong, P. Draper, A. El-Khadra, N. Gemelke, M. Hanada, D. Kharzeev, H. Lamm, Y.-Y. Li, J. Liu, M. Lukin, Y. Meurice, C. Monroe, B. Nachman, G. Pagano, J. Preskill, E. Rinaldi, A. Roggero, D.I. Santiago, M.J. Savage, I. Siddiqi, G. Siopsis, D. Van Zanten, N. Wiebe, Y. Yamauchi, K. Yeter-Aydeniz, S. Zorzetti, Quantum simulation for high-energy physics. PRX Quantum (2023). 10.1103/PRXQuantum.4.027001

457. P. Fontana, J.C.P. Barros, A. Trombettoni, Quantum simulator of link models using spinor dipolar ultracold atoms. Phys. Rev. A (2023). 10.1103/PhysRevA.107.043312
458. F. Strocchi, Non-perturbative foundations of quantum field theory, in *An Introduction to Non-Perturbative Foundations of Quantum Field Theory*, ed. by F. Strocchi (Oxford University Press, Oxford, 2013), p. 0. ISBN: 978-0-199-67157-1. https://doi.org/10.1093/acprof:oso/9780199671571.003.0003
459. G. Cataldi, *Ed-Lgt. Exact Diagonalization Code for Lattice Gauge Theories and Quantum Many Body Hamiltonians* (2024). 10.5281/ZENODO.11145318
460. F. Verstraete, V. Murg, J.I. Cirac, Matrix product states, projected entangled pair states, and variational renormalization group methods for quantum spin systems. Adv. Phys. (2008). 10.1080/14789940801912366
461. D. Paulson, L. Dellantonio, J.F. Haase, A. Celi, A. Kan, A. Jena, C. Kokail, R. van Bijnen, K. Jansen, P. Zoller, C.A. Muschik, Simulating 2D effects in Lattice Gauge theories on a quantum computer. PRX Quantum (2021). 10.1103/PRXQuantum.2.030334
462. C. Senko, P. Richerme, J. Smith, A. Lee, I. Cohen, A. Retzker, C. Monroe, Realization of a quantum integer-spin chain with controllable interactions. Phys. Rev. X (2015). 10.1103/PhysRevX.5.021026
463. Y. Kuno, K. Kasamatsu, Y. Takahashi, I. Ichinose, T. Matsui, Real-time dynamics and proposal for feasible experiments of Lattice Gauge–Higgs model simulated by cold atoms. New J. Phys. (2015). 10.1088/1367-2630/17/6/063005
464. Y. Kuno, S. Sakane, K. Kasamatsu, I. Ichinose, T. Matsui, Atomic quantum simulation of a three-dimensional U(1) Gauge-Higgs model. Phys. Rev. A (2016). 10.1103/PhysRevA.94.063641
465. J. Zhang, J. Unmuth-Yockey, J. Zeiher, A. Bazavov, S.-W. Tsai, Y. Meurice, Quantum simulation of the universal features of the polyakov loop. Phys. Rev. Lett. (2018). 10.1103/PhysRevLett.121.223201
466. S. Notarnicola, M. Collura, S. Montangero,. Real-time-dynamics quantum simulation of $(1+1)\text{-dimensional}$ Lattice QED with Rydberg atoms. Phys. Rev. Res. (2020). 10.1103/PhysRevResearch.2.013288
467. J. Hauschild, F. Pollmann, Efficient numerical simulations with tensor networks: tensor network python (TeNPy). *SciPost Physics Lecture Notes* (2018). 10.21468/SciPostPhysLectNotes.5
468. T. Felser, S. Notarnicola, S. Montangero, Efficient tensor network ansatz for high-dimensional quantum many-body problems. Phys. Rev. Lett. (2021). 10.1103/PhysRevLett.126.170603
469. E. Zohar, J.I. Cirac, B. Reznik, Quantum simulations of Lattice Gauge theories using ultracold atoms in optical lattices. Rep. Progress Phys. (2015). 10.1088/0034-4885/79/1/014401
470. G. Burgio, R. De Pietri, H.A. Morales-Técotl, L.F. Urrutia, J. D. Vergara, The basis of the physical hilbert space of Lattice Gauge theories. Nucl. Phys. B (2000). 10.1016/S0550-3213(99)00533-7
471. X. Yao, *SU(2) Non-Abelian Gauge Theory on a Plaquette Chain Obeys Eigenstate Thermalization Hypothesis* (2023). 10.48550/arXiv.2303.14264
472. B. Müller, X. Yao, *Simple Hamiltonian for Quantum Simulation of Strongly Coupled 2+1D SU(2) Lattice Gauge Theory on a Honeycomb Lattice* (2023). 10.48550/arXiv.2307.00045
473. M. Ringbauer, M. Meth, L. Postler, R. Stricker, R. Blatt, P. Schindler, T. Monz, A universal qudit quantum processor with trapped ions. Nature Phys. (2022). 10.1038/s41567-022-01658-0
474. B. Buyens, S. Montangero, J. Haegeman, F. Verstraete, K. Van Acoleyen, Finite-representation approximation of Lattice Gauge theories at the continuum limit with tensor networks. Phys. Rev. D (2017). 10.1103/PhysRevD.95.094509
475. A. Ciavarella, N. Klco, M.J. Savage, Trailhead for quantum simulation of SU(3) Yang-Mills Lattice Gauge theory in the local multiplet basis. Phys. Rev. D (2021). 10.1103/PhysRevD.103.094501
476. Z. Davoudi, I. Raychowdhury, A. Shaw, Search for efficient formulations for hamiltonian simulation of non-abelian Lattice Gauge theories. Phys. Rev. D (2021). 10.1103/PhysRevD.104.074505

477. Y. Tong, V.V. Albert, J.R. McClean, J. Preskill, Y. Su, Provably accurate simulation of Gauge theories and bosonic systems. Quantum (2022). 10.22331/q-2022-09-22-816
478. R. Fiore, P. Giudice, D. Giuliano, D. Marmottini, A. Papa, P. Sodano, QED_3 on a space-time Lattice: a comparison between compact and noncompact formulation, in *Proceedings of XXIIIrd International Symposium on Lattice Field Theory — PoS(LAT2005)*, vol. 20 (SISSA Medialab, 2005), p. 243. 10.22323/1.020.0243
479. O. Raviv, Y. Shamir, B. Svetitsky, Nonperturbative beta function in three-dimensional electrodynamics. Phys. Rev. D (2014). 10.1103/PhysRevD.90.014512
480. B. Svetitsky, O. Raviv, Y. Shamir, Beta function of three-dimensional QED, in *Proceedings of The 32nd International Symposium on Lattice Field Theory — PoS(LATTICE2014)*, vol. 214 (SISSA Medialab, 2015), p. 051. 10.22323/1.214.0051
481. L. Funcke, C.F. Groß, K. Jansen, S. Kühn, S. Romiti, C. Urbach, Hamiltonian limit of Lattice QED in 2+1 dimensions, in *Proceedings of The 39th International Symposium on Lattice Field Theory — PoS(LATTICE2022)*, vol. 430 (SISSA Medialab, 2023), p. 292. 10.22323/1.430.0292
482. C. Strouthos, J.B. Kogut, *The Phases of Non-Compact QED(3)* (2008). 10.48550/arXiv.0804.0300
483. J. Bender, *Quantum and Classical Methods for Lattice Gauge Theories in Higher Dimensions*. Ph.D. Thesis. Technische Universität München, 2023
484. C. Kang, M.B. Soley, E. Crane, S.M. Girvin, N. Wiebe, *Leveraging Hamiltonian Simulation Techniques to Compile Operations on Bosonic Devices* (2023). 10.48550/arXiv.2303.15542
485. A. Lucas, Ising formulations of many NP problems. Front. Phys. (2014). 10.3389/fphy.2014.00005
486. D. Aharonov, W. van Dam, J. Kempe, Z. Landau, S. Lloyd, O. Regev, Adiabatic quantum computation is equivalent to standard quantum computation. Siam Rev. (2008). 10.1137/080734479
487. D. Kienzler, H.Y. Lo, B. Keitch, L. de Clercq, F. Leupold, F. Lindenfelser, M. Marinelli, V. Negnevitsky, J.P. Home, Quantum harmonic oscillator state synthesis by reservoir engineering. Science (2015). 10.1126/science.1261033
488. H. Krauter, C.A. Muschik, K. Jensen, W. Wasilewski, J.M. Petersen, J.I. Cirac, E.S. Polzik, Entanglement generated by dissipation and steady state entanglement of two macroscopic objects. Phys. Rev. Lett. (2011). 10.1103/PhysRevLett.107.080503
489. J.F. Poyatos, J.I. Cirac, P. Zoller, Quantum reservoir engineering with laser cooled trapped ions. Phys. Rev. Lett. (1996). 10.1103/PhysRevLett.77.4728
490. D. Tannor, *Introduction to Quantum Mechanics: A Time-Dependent Perspective* (University Science Books, Melville, 2016). ISBN: 978-1-891-38999-3
491. J. Roland, *Adiabatic Quantum Compputation*. Ph.D. Theiss, Université libre de Brussels, 2004.
492. J. Roland, N.J. Cerf, Quantum search by local adiabatic evolution. Phys. Rev. A (2002). 10.1103/PhysRevA.65.042308
493. T. Caneva, T. Calarco, R. Fazio, G.E. Santoro, S. Montangero, Speeding up critical system dynamics through optimized evolution. Phys. Rev. A (2011). 10.1103/PhysRevA.84.012312
494. C. Zener, Non-Adiabatic crossing of energy levels. Proc. R. Soc. Lond. A Math. Phys. Eng. Sci. (1932). 10.1098/rspa.1932.0165
495. L.D. Landau, Zur Theorie der Energieubertragung. II. Phys. Soviet Union **2**, 46–51 (1932)
496. E. Farhi, J. Goldstone, S. Gutmann, J. Lapan, A. Lundgren, D. Preda, A quantum adiabatic evolution algorithm applied to random instances of an NP-complete problem. Science (2001). 10.1126/science.1057726
497. G.E. Santoro, E. Tosatti, Optimization using quantum mechanics: quantum annealing through adiabatic evolution. J. Phys. A. Math. Gen. (2006). 10.1088/0305-4470/39/36/R01
498. G.E. Santoro, Theory of quantum annealing of an ising spin glass. Science (2002). 10.1126/science.1068774
499. W.H. Zurek, U. Dorner, P. Zoller, Dynamics of a quantum phase transition. Phys. Rev. Lett. (2005). 10.1103/PhysRevLett.95.105701

500. W.H. Zurek, Cosmological experiments in superfluid helium? Nature (1985). 10.1038/317505a0
501. T.W.B. Kibble, Topology of cosmic domains and strings. J. Phys. A. Math. Gen. (1976). 10.1088/0305-4470/9/8/029
502. B. Damski, The simplest quantum model supporting the kibble-zurek mechanism of topological defect production: Landau-zener transitions from a new perspective. Phys. Rev. Lett. (2005). 10.1103/PhysRevLett.95.035701
503. E. Canovi, E. Ercolessi, P. Naldesi, L. Taddia, D. Vodola, Dynamics of entanglement entropy and entanglement spectrum crossing a quantum phase transition. Phys. Rev. B (2014). 10.1103/PhysRevB.89.104303
504. B. Gardas, J. Dziarmaga, W.H. Zurek, Dynamics of the quantum phase transition in the one-dimensional Bose-Hubbard model: Excitations and correlations induced by a quench. Phys. Rev. B (2017). 10.1103/PhysRevB.95.104306
505. J. Dziarmaga, J. Meisner, W.H. Zurek, Winding up of the wave-function phase by an insulator-to-superfluid transition in a ring of coupled bose-einstein condensates. Phys. Rev. Lett. (2008). 10.1103/PhysRevLett.101.115701
506. J.-M. Cui, Y.-F. Huang, Z. Wang, D.-Y. Cao, J. Wang, W.-M. Lv, L. Luo, A. del Campo, Y.-J. Han, C.-F. Li, G.-C. Guo, Experimental trapped-ion quantum simulation of the Kibble-Zurek dynamics in momentum space. Sci. Rep. (2016). 10.1038/srep33381
507. A. Dutta, A. Rahmani, A. del Campo. Anti-Kibble-Zurek behavior in crossing the quantum critical point of a thermally isolated system driven by a noisy control field. Phys. Rev. Lett. (2016). 10.1103/PhysRevLett.117.080402
508. P. Silvi, G. Morigi, T. Calarco, S. Montangero, Crossover from classical to quantum Kibble-Zurek scaling. Phys. Rev. Lett. (2016). 10.1103/PhysRevLett.116.225701
509. P.M. Chesler, A.M. García-García, H. Liu, Defect formation beyond Kibble-Zurek mechanism and holography. Phys. Rev. X (2015). 10.1103/PhysRevX.5.021015
510. S. Ulm, J. Roßnagel, G. Jacob, C. Degünther, S.T. Dawkins, U.G. Poschinger, R. Nigmatullin, A. Retzker, M.B. Plenio, F. Schmidt-Kaler, K. Singer, Observation of the Kibble-Zurek scaling law for defect formation in ion crystals. Nat. Commun. (2013). 10.1038/ncomms3290
511. K. Pyka, J. Keller, H.L. Partner, R. Nigmatullin, T. Burgermeister, D.M. Meier, K. Kuhlmann, A. Retzker, M.B. Plenio, W.H. Zurek, A. del Campo, T.E. Mehlstäubler, Topological defect formation and spontaneous symmetry breaking in ion Coulomb crystals. Nat. Commun. (2013). 10.1038/ncomms3291
512. J. Dziarmaga, W.H. Zurek, Quench in the 1D Bose-Hubbard model: topological defects and excitations from the Kosterlitz-Thouless phase transition dynamics. Sci. Rep. (2015). 10.1038/srep05950
513. J. Dziarmaga, Density of Kinks just after a quench in an underdamped system. Phys. Rev. Lett. (1998). 10.1103/PhysRevLett.81.1551
514. N. Navon, A.L. Gaunt, R.P. Smith, Z. Hadzibabic, Critical dynamics of spontaneous symmetry breaking in a homogeneous Bose gas. Science (2015). 10.1126/science.1258676
515. S. Braun, M. Friesdorf, S.S. Hodgman, M. Schreiber, J.P. Ronzheimer, A. Riera, M. del Rey, I. Bloch, J. Eisert, U. Schneider, Emergence of coherence and the dynamics of quantum phase transitions. Proc. Natl. Acad. Sci. (2015). 10.1073/pnas.1408861112.
516. F. Pellegrini, S. Montangero, G.E. Santoro, R. Fazio, Adiabatic quenches through an extended quantum critical region. Phys. Rev. B (2008). 10.1103/PhysRevB.77.140404
517. J. Dziarmaga, Dynamics of a quantum phase transition and relaxation to a steady state. Adv. Phys. (2010). 10.1080/00018732.2010.514702
518. J. Dziarmaga, M. Tylutki, W.H. Zurek, Quench from Mott insulator to superfluid. Phys. Rev. B (2012). 10.1103/PhysRevB.86.144521
519. P. Silvi, G. De Chiara, T. Calarco, G. Morigi, S. Montangero, Full characterization of the quantum linear-zigzag transition in atomic chains. Ann. Phys. (2013). 10.1002/andp.201300090
520. C. Schneider, D. Porras, T. Schaetz, Experimental quantum simulations of many-body physics with trapped ions. Rep. Prog. Phys. **75**(2), 024401 (2012). 10.1088/0034-4885/75/2/024401

521. M. Mielenz, J. Brox, S. Kahra, G. Leschhorn, M. Albert, T. Schaetz, H. Landa, B. Reznik, Trapping of topological-structural defects in coulomb crystals. Phys. Rev. Lett. (2013). 10.1103/PhysRevLett.110.133004
522. L. Corman, L. Chomaz, T. Bienaimé, R. Desbuquois, C. Weitenberg, S. Nascimbène, J. Dalibard, J. Beugnon, Quench-Induced Supercurrents in an Annular Bose Gas. Phys. Rev. Lett. (2014). 10.1103/PhysRevLett.113.135302
523. E. Shimshoni, G. Morigi, S. Fishman, Quantum zigzag transition in ion chains. Phys. Rev. Lett. (2011). 10.1103/PhysRevLett.106.010401
524. E. Shimshoni, G. Morigi, S. Fishman, Quantum structural phase transition in chains of interacting atoms. Phys. Rev. A (2011). 10.1103/PhysRevA.83.032308
525. J. Als-Nielsen, R.J. Birgeneau, Mean field theory, the Ginzburg criterion, and marginal dimensionality of phase transitions. Am. J. Phys. (1977). https://doi.org/10.1119/1.11019
526. D.J. Amit, The Ginzburg criterion-rationalized. J. Phys. C Solid State Phys. (1974)
527. S. Campbell, S. Deffner, Trade-off between speed and cost in shortcuts to adiabaticity. Phys. Rev. Lett. (2017). 10.1103/PhysRevLett.118.100601
528. J.G. Muga, X. Chen, E. Torrontegui, S. Ibáñez, I. Lizuain, A. Ruschhaupt, Shortcuts to quantum adiabatic processes. J. Phys. Conf. Ser. (2011). 10.1088/1742-6596/306/1/012022
529. A. del Campo, Shortcuts to adiabaticity by counterdiabatic driving. Phys. Rev. Lett. (2013). 10.1103/PhysRevLett.111.100502
530. E. Torrontegui, S. Martínez-Garaot, A. Ruschhaupt, J.G. Muga, Shortcuts to adiabaticity: fast-forward approach. Phys. Rev. A (2012). 10.1103/PhysRevA.86.013601
531. E. Torrontegui, S. Ibáñez, S. Martínez-Garaot, M. Modugno, A. del Campo, D. Guéry-Odelin, A. Ruschhaupt, X. Chen, J. Gonzalo, Muga, shortcuts to adiabaticity. Adv. At. Mol. Opt. Phys. (2013). 10.1016/B978-0-12-408090-4.00002-5
532. C. Jarzynski, Generating shortcuts to adiabaticity in quantum and classical dynamics. Phys. Rev. A (2013). 10.1103/PhysRevA.88.040101
533. E. Torrontegui, I. Lizuain, S. González-Resines, A. Tobalina, A. Ruschhaupt, R. Kosloff, J.G. Muga, Energy consumption for shortcuts to adiabaticity. Phys. Rev. A (2017). 10.1103/PhysRevA.96.022133
534. A. del Campo, M.G. Boshier, Shortcuts to adiabaticity in a time-dependent box. Sci. Rep. (2012). 10.1038/srep00648
535. W. Rohringer, D. Fischer, F. Steiner, I.E. Mazets, J. Schmiedmayer, M. Trupke, Non-equilibrium scale invariance and shortcuts to adiabaticity in a one-dimensional Bose gas. Sci. Rep. (2015). 10.1038/srep09820
536. H. Saberi, T. Opatrný, K. Mølmer, A. Del Campo, Adiabatic tracking of quantum many-body dynamics. Phys. Rev. A - At. Mol. Opt. Phys. (2014). 10.1103/PhysRevA.90.060301
537. D. D'Alessandro, *Introduction to Quantum Control and Dynamics*. Chapman & Hall/CRC Applied Mathematics & Nonlinear Science (CRC Press, Boca Raton, 2007). ISBN: 978-1-584-88883-3
538. C. Altafini, F. Ticozzi, Modeling and control of quantum systems: an introduction. IEEE Trans. Automat. Contr. (2012). 10.1109/TAC.2012.2195830
539. C. Brif, R. Chakrabarti, H. Rabitz, Control of quantum phenomena: past, present and future. New J. Phys. (2010). 10.1088/1367-2630/12/7/075008
540. S. van Frank, M. Bonneau, J. Schmiedmayer, S. Hild, C. Gross, M. Cheneau, I. Bloch, T. Pichler, A. Negretti, T. Calarco, S. Montangero, Optimal control of complex atomic quantum systems. Sci. Rep. (2016). 10.1038/srep34187
541. M.G. Bason, M. Viteau, N. Malossi, P. Huillery, E. Arimondo, D. Ciampini, R. Fazio, V. Giovannetti, R. Mannella, O. Morsch, High-fidelity quantum driving. Nat. Phys. (2011). 10.1038/nphys2170
542. Z. Sun, J. Liu, J. Ma, X. Wang, Quantum speed limits in open systems: non-Markovian dynamics without rotating-wave approximation. Sci. Rep. (2015). 10.1038/srep08444
543. T. Caneva, T. Calarco, S. Montangero, Chopped random-basis quantum optimization. Phys. Rev. A (2011). 10.1103/PhysRevA.84.022326

544. N. Wu, A. Nanduri, H. Rabitz, Optimal suppression of defect generation during a passage across a quantum critical point. Phys. Rev. B - Condens. Matter Mater. Phys. (2015). 10.1103/PhysRevB.91.041115
545. C.P. Koch, Controlling open quantum systems: tools, achievements, and limitations. J. Phys. Condens. Matter (2016). 10.1088/0953-8984/28/21/213001
546. J. Cui, R. van Bijnen, T. Pohl, S. Montangero, T. Calarco, Optimal control of Rydberg lattice gases. Quantum Sci. Technol. (2017). 10.1088/2058-9565/aa7daf
547. S. Deffner, E. Lutz, Energy–time uncertainty relation for driven quantum systems. J. Phys. A Math. Theor. (2013). 10.1088/1751-8113/46/33/335302
548. V. Giovannetti, S. Lloyd, L. Maccone, Quantum limits to dynamical evolution. Phys. Rev. A (2003). 10.1103/PhysRevA.67.052109
549. P.M. Poggi, D.A. Wisniacki, Optimal control of many-body quantum dynamics: chaos and complexity. Phys. Rev. A (2016). 10.1103/PhysRevA.94.033406
550. M. Hirose, P. Cappellaro, Time-optimal control with finite bandwidth. Quantum Inf. Process. **17**(88), (2018)
551. R. Uzdin, R. Kosloff, Speed limits in Liouville space for open quantum systems. EPL (Europhysics Lett. (2016). 10.1209/0295-5075/115/40003
552. S. Lloyd, S. Montangero, Information theoretical analysis of quantum optimal control. Phys. Rev. Lett. (2014). 10.1103/PhysRevLett.113.010502
553. M.R. Frey, Quantum speed limits—primer, perspectives, and potential future directions. Quantum Inf. Process. (2016). 10.1007/s11128-016-1405-x
554. D.P. Pires, M. Cianciaruso, L.C. Céleri, G. Adesso, D.O. Soares-Pinto, Generalized geometric quantum speed limits. Phys. Rev. X (2016). 10.1103/PhysRevX.6.021031
555. M.M. Taddei, B.M. Escher, L. Davidovich, R.L. de Matos Filho, Quantum speed limit for physical processes. Phys. Rev. Lett. (2013). 10.1103/PhysRevLett.110.050402
556. S. Deffner, E. Lutz, Quantum speed limit for non-markovian dynamics. Phys. Rev. Lett. (2013). 10.1103/PhysRevLett.111.010402
557. J.J.W.H. Sørensen, M.K. Pedersen, M. Munch, P. Haikka, J.H. Jensen, T. Planke, M.G. Andreasen, M. Gajdacz, K. Mølmer, A. Lieberoth, J.F. Sherson, Exploring the quantum speed limit with computer games. Nature (2016). 10.1038/nature17620
558. I. Brouzos, A.I. Streltsov, A. Negretti, R.S. Said, T. Caneva, S. Montangero, T. Calarco, Quantum speed limit and optimal control of many-boson dynamics. Phys. Rev. A (2015). 10.1103/PhysRevA.92.062110
559. S. van Frank, A. Negretti, T. Berrada, R. Bücker, S. Montangero, J.-F. Schaff, T. Schumm, T. Calarco, J. Schmiedmayer, Interferometry with non-classical motional states of a Bose-Einstein condensate. Nat. Commun. (2014). 10.1038/ncomms5009
560. A. Rahmani, C. Chamon, Optimal control for unitary preparation of many-body states: application to luttinger liquids. Phys. Rev. Lett. (2011). 10.1103/PhysRevLett.107.016402
561. P. Doria, T. Calarco, S. Montangero, Optimal control technique for many-body quantum dynamics. Phys. Rev. Lett. (2011). 10.1103/PhysRevLett.106.190501
562. T. Caneva, A. Silva, R. Fazio, S. Lloyd, T. Calarco, S. Montangero, Complexity of controlling quantum many-body dynamics. Phys. Rev. A (2014). 10.1103/PhysRevA.89.042322
563. S. Lloyd, S. Montangero, Information theoretical analysis of quantum optimal control. Phys. Rev. Lett. (2014). 10.1103/PhysRevLett.113.010502
564. T. Caneva, T. Calarco, S. Montangero, Entanglement-storage units. New J. Phys. (2012). 10.1088/1367-2630/14/9/093041
565. M.H. Goerz, F. Motzoi, K.B. Whaley, C.P. Koch, Charting the circuit QED design landscape using optimal control theory. npj Quantum Inf. (2017). 10.1038/s41534-017-0036-0
566. A. del Campo, K. Sengupta, Controlling quantum critical dynamics of isolated systems. Eur. Phys. J. Spec. Top. (2015). 10.1140/epjst/e2015-02350-4
567. S. Rosi, A. Bernard, N. Fabbri, L. Fallani, C. Fort, M. Inguscio, T. Calarco, S. Montangero, Fast closed-loop optimal control of ultracold atoms in an optical lattice. Phys. Rev. A (2013). 10.1103/PhysRevA.88.021601

568. A. Rahmani, T. Kitagawa, E. Demler, C. Chamon, Cooling through optimal control of quantum evolution. Phys. Rev. A (2013). 10.1103/PhysRevA.87.043607
569. M.C. Tichy, M.K. Pedersen, K. Mølmer, J.F. Sherson, Nonadiabatic quantum state control of many bosons in few wells. Phys. Rev. A (2013). 10.1103/PhysRevA.87.063422
570. A. del Campo, M.M. Rams, W.H. Zurek, Assisted finite-rate adiabatic passage across a quantum critical point: exact solution for the quantum ising model. Phys. Rev. Lett. (2012). 10.1103/PhysRevLett.109.115703
571. A. Castro, J. Werschnik, E.K.U. Gross, Controlling the dynamics of many-electron systems from first principles: a combination of optimal control and time-dependent density-functional theory. Phys. Rev. Lett. (2012). 10.1103/PhysRevLett.109.153603
572. T. Caneva, M. Murphy, T. Calarco, R. Fazio, S. Montangero, V. Giovannetti, G.E. Santoro, Optimal control at the quantum speed limit. Phys. Rev. Lett. (2009). 10.1103/PhysRevLett.103.240501
573. M.M. Müller, A. Külle, R. Löw, T. Pfau, T. Calarco, S. Montangero, Room-temperature Rydberg single-photon source. Phys. Rev. A (2013). 10.1103/PhysRevA.87.053412
574. N. Rach, M.M. Müller, T. Calarco, S. Montangero, Dressing the chopped-random-basis optimization: a bandwidth-limited access to the trap-free landscape. Phys. Rev. A (2015). 10.1103/PhysRevA.92.062343
575. K.W. Moore, H. Rabitz, Exploring constrained quantum control landscapes. J. Chem. Phys. (2012). 10.1063/1.4757133
576. C. Kittel, H. Shore, Development of a phase transition for a rigorously solvable many-body system. Phys. Rev. (1965). 10.1103/PhysRev.138.A1165
577. S. Deffner, S. Campbell, Quantum speed limits: from Heisenberg's uncertainty principle to optimal quantum control. J. Phys. A Math. Theor. (2017). 10.1088/1751-8121/aa86c6
578. N. Margolus, L.B. Levitin, The maximum speed of dynamical evolution. Phys. D Nonlinear Phenomena (1998). https://doi.org/10.1016/S0167-2789(98)00054-2
579. G.C. Hegerfeldt, Driving at the quantum speed limit: optimal control of a two-level system. Phys. Rev. Lett. (2013). 10.1103/PhysRevLett.111.260501
580. M. Murphy, S. Montangero, V. Giovannetti, T. Calarco, Communication at the quantum speed limit along a spin chain. Phys. Rev. A (2010). 10.1103/PhysRevA.82.022318
581. S. Ashhab, P.C. de Groot, F. Nori, Speed limits for quantum gates in multiqubit systems. Phys. Rev. A (2012). 10.1103/PhysRevA.85.052327
582. X. Wang, M. Allegra, K. Jacobs, S. Lloyd, C. Lupo, M. Mohseni, Quantum brachistochrone curves as geodesics: obtaining accurate minimum-time protocols for the control of quantum systems. Phys. Rev. Lett. **114**(17) (2015). http://dx.doi.org/10.1103/PhysRevLett.114.170501
583. L. Levitin, T. Toffoli, Fundamental limit on the rate of quantum dynamics: the unified bound is tight. Phys. Rev. Lett. (2009). 10.1103/PhysRevLett.103.160502
584. T. Cohen, D. McGady, Schwinger mechanism revisited. Phys. Rev. D (2008). 10.1103/PhysRevD.78.036008
585. F. Hebenstreit, J. Berges, D. Gelfand, Real-time dynamics of string breaking. Phys. Rev. Lett. (2013). 10.1103/PhysRevLett.111.201601
586. M. Heyl, A. Polkovnikov, S. Kehrein, Dynamical quantum phase transitions in the transverse-field ising model. Phys. Rev. Lett. (2013). 10.1103/PhysRevLett.110.135704
587. V. Gurarie, Viewpoint: quantum phase transitions go dynamical. Physics **10**, 95 (2017). https://aps.org/10.1103/Physics.10.95
588. A.D. Greentree, C. Tahan, J.H. Cole, L.C.L. Hollenberg, Quantum phase transitions of light. Nat. Phys. (2006). 10.1038/nphys466
589. M.J. Hartmann, F.G.S.L. Brandão, M.B. Plenio, Strongly interacting polaritons in coupled arrays of cavities. Nat. Phys. (2006). 10.1038/nphys462
590. D. Rossini, R. Fazio, Mott-insulating and glassy phases of polaritons in 1D arrays of coupled cavities. Phys. Rev. Lett. (2007). 10.1103/PhysRevLett.99.186401
591. P. Jurcevic, H. Shen, P. Hauke, C. Maier, T. Brydges, C. Hempel, B.P. Lanyon, M. Heyl, R. Blatt, C.F. Roos, Direct observation of dynamical quantum phase transitions in an interacting many-body system. Phys. Rev. Lett. (2017). 10.1103/PhysRevLett.119.080501

592. C. Gogolin, *Equilibration and Thermalization in Quantum Systems*. Ph.D. Thesis. Freien Universität Berlin, 2014. 10.13140/2.1.1636.4485
593. A. Polkovnikov, K. Sengupta, A. Silva, M. Vengalattore, Colloquium: nonequilibrium dynamics of closed interacting quantum systems. Rev. Mod. Phys. (2011). 10.1103/RevModPhys.83.863
594. G. Benenti, G. Casati, S. Montangero, D. Shepelyansky, Efficient quantum computing of complex dynamics. Phys. Rev. Lett. (2001). 10.1103/PhysRevLett.87.227901
595. E. Canovi, D. Rossini, R. Fazio, G.E. Santoro, A. Silva, Quantum quenches, thermalization, and many-body localization. Phys. Rev. B (2011). 10.1103/PhysRevB.83.094431
596. L. Santos, A. Polkovnikov, M. Rigol, Entropy of isolated quantum systems after a quench. Phys. Rev. Lett. (2011). 10.1103/PhysRevLett.107.040601
597. R. Dorner, J. Goold, C. Cormick, M. Paternostro, V. Vedral, Emergent thermodynamics in a quenched quantum many-body system. Phys. Rev. Lett. (2012). 10.1103/PhysRevLett.109.160601
598. J.A. Kjäll, J.H. Bardarson, F. Pollmann, Many-body localization in a disordered quantum ising chain. Phys. Rev. Lett. (2014). 10.1103/PhysRevLett.113.107204
599. J. Eisert, M. Friesdorf, C. Gogolin, Quantum many-body systems out of equilibrium. Nat. Phys. (2015). 10.1038/nphys3215
600. A.M. Läuchli, C. Kollath, Spreading of correlations and entanglement after a quench in the one-dimensional Bose–Hubbard model. J. Stat. Mech. Theory Exp. (2008). 10.1088/1742-5468/2008/05/P05018
601. S. Wolff, A. Sheikhan, C. Kollath, Dissipative time evolution of a chiral state after a quantum quench. Phys. Rev. A (2016). 10.1103/PhysRevA.94.043609
602. A.W. Chin, Á. Rivas, S.F. Huelga, M.B. Plenio, Exact mapping between system-reservoir quantum models and semi-infinite discrete chains using orthogonal polynomials. J. Math. Phys. (2010). 10.1063/1.3490188
603. J. Prior, A.W. Chin, S.F. Huelga, M.B. Plenio, Efficient simulation of strong system-environment interactions. Phys. Rev. Lett. (2010). 10.1103/PhysRevLett.105.050404
604. P. Lodahl, S. Mahmoodian, S. Stobbe, A. Rauschenbeutel, P. Schneeweiss, J. Volz, H. Pichler, P. Zoller, Chiral quantum optics. Nature (2017). 10.1038/nature21037
605. H. Pichler, P. Zoller, Photonic circuits with time delays and quantum feedback. Phys. Rev. Lett. (2016). 10.1103/PhysRevLett.116.093601
606. T. Ramos, H. Pichler, A.J. Daley, P. Zoller, Quantum spin dimers from chiral dissipation in cold-atom chains. Phys. Rev. Lett. (2014). 10.1103/PhysRevLett.113.237203
607. P. Arrighi, J. Grattage, The quantum game of life. Phys. World (2012). 10.1209/0295-5075/97/20012
608. D. Bleh, T. Calarco, S. Montangero, Quantum game of life. EPL (Europhysics Lett. (2012). 10.1209/0295-5075/97/20012
609. U. Alvarez-Rodriguez, M. Sanz, L. Lamata, E. Solano, Artificial life in quantum technologies. Sci. Rep. (2016). 10.1038/srep20956
610. M.A. Valdez, D. Jaschke, D.L. Vargas, L.D. Carr, Quantifying complexity in quantum phase transitions via mutual information complex networks. Phys. Rev. Lett. (2017). 10.1103/PhysRevLett.119.225301
611. E. Stoudenmire, D.J. Schwab, Supervised learning with tensor networks, in *Advances in Neural Information Processing Systems*, vol. 29, ed. by D.D. Lee, M. Sugiyama, U.V. Luxburg, I. Guyon, R. Garnett (Curran Associates, Inc., Red Hook, 2016), pp. 4799–4807
612. J. Steppan, *MNIST Examples* (2017). https://commons.wikimedia.org/wiki/File:MnistExamples.png
613. Y. LeCun, C. Cortes, *MNIST dataset*
614. S. Ruder, *An Overview of Gradient Descent Optimization Algorithms* (2017). https://arxiv.org/abs/1609.04747
615. T. Hastie, R. Tibshirani, J. Friedman, *The Elements of Statistical Learning* (Springer, New York, 2009). ISBN: 978-0-387-84857-0. https://doi.org/10.1007/978-0-387-84858-7

616. P. Kharyuk, D. Nazarenko, I. Oseledets, *Comparative Study of Discrete Wavelet Transforms and Wavelet Tensor Train Decomposition to Feature Extraction of FTIR Data of Medicinal Plants* (2018). https://arxiv.org/abs/1807.07099
617. W. Wang, V. Aggarwal, S. Aeron, *Principal Component Analysis with Tensor Train Subspace* (2018). https://arxiv.org/abs/1803.05026
618. Q. Zhang, L.T. Yang, Z. Chen, P. Li, A tensor-train deep computation model for industry informatics big data feature learning. IEEE Trans. Ind. Inf. (2018). 10.1109/TII.2018.2791423
619. W. Hackbusch, S. Kühn, A new scheme for the tensor representation. J. Fourier Anal. Appl. (2009). 10.1007/s00041-009-9094-9
620. Ivan Glasser, Nicola Pancotti, and J. Ignacio Cirac. *From Probabilistic Graphical Models to Generalized Tensor Networks for Supervised Learning* (2019). https://arxiv.org/abs/1806.05964
621. K.-R. Muller, S. Mika, G. Ratsch, K. Tsuda, B. Scholkopf, An introduction to kernel-based learning algorithms. IEEE Trans. Neural Netw. (2001). 10.1109/72.914517
622. V.N. Vapnik. *The Nature of Statistical Learning Theory* (Springer, New York, 2000). ISBN: 978-1-441-93160-3, 978-1-475-73264-1 10.1007/978-1-4757-3264-1
623. W. Waegeman, T. Pahikkala, A. Airola, T. Salakoski, M. Stock, B. De Baets, A kernel-based framework for learning graded relations from data. IEEE Trans. Fuzzy Syst. (2012). 10.1109/TFUZZ.2012.2194151
624. J.T. Rolfe, M. Cook, Multifactor expectation maximization for factor graphs, in *Artificial Neural Networks – ICANN 2010*, ed. by K. Diamantaras, W. Duch, L.S. Iliadis (Springer Berlin Heidelberg, Berlin/Heidelberg, 2010), pp. 267–276. ISBN: 978-3-642-15825-4
625. M.A. Nielsen, I.L. Chuang. *Quantum Computation and Quantum Information* (Cambridge University Press, Cambridge, 2010)
626. IEEE Standard for Binary Floating-Point Arithmetic, *ANSI/IEEE Std 754-1985* (1985). 10.1109/IEEESTD.1985.82928
627. B.J. Bloom, T.L. Nicholson, J.R. Williams, S.L. Campbell, M. Bishof, X. Zhang, W. Zhang, S.L. Bromley, J. Ye, An optical lattice clock with accuracy and stability at the 10(-18) level. Nature (2014). 10.1038/nature12941
628. P.J. Denning, The locality principle. Commun. ACM (2005). 10.1145/1070838.1070856
629. E. Strohmaier, J. Dongarra, H. Simon, M. Meuer, *The top 500 list*. www.top500.org
630. R. van der Pas, E. Stotzer, C. Terboven. *Using OpenMP—The Next Step: Affinity, Accelerators, Tasking, and SIMD*. Scientific and Engineering Computation (MIT Press, Cambridge, 2017). ISBN: 978-0-262-53478-9
631. L.S. Blackford, J. Demmel, J. Dongarra, I. Duff, S. Hammarling, G. Henry, M. Heroux, L. Kaufman, A. Lumsdaine, A. Petitet, R. Pozo, K. Remington, R.C. Whaley, An updated set of basic linear algebra subprograms (BLAS). ACM Trans. Math. Softw. (2002). 10.1145/567806.567807
632. W. Gropp, E. Lusk, A. Skjellum. *Using MPI: Portable Parallel Programming with the Message-Passing Interface*. Scientific and Engineering Computation (MIT Press, Cambridge, 2014). ISBN: 978-0-262-52739-2
633. G.M. Amdahl, *Proceedings AFIPS* (1967)
634. T.L. Sterling. *Beowulf Cluster Computing with Linux*. Scientific and Computational Engineering Series (2002). https://www.books24x7.com/books24x7.asp. ISBN: 978-0-262-69274-8
635. M. Scarpino, *OpenCL in Action: How to Accelerate Graphics and Computation* (Manning, Shelter Island, 2012). ISBN: 978-1-617-29017-6
636. A. Sheppard, *Programming GPUs* (O'Reilly Media, Sebastopol, 2012). Incorporated. ISBN: 978-1-449-30235-1
637. A. Robbins, *GDB Pocket Reference*. Pocket Reference (O'Reilly Media, Sebastopol, 2005). ISBN: 978-0-596-55335-7
638. R. Stallman, Cygnus Support. *Debugging with GDB: The GNU Source-level Debugger* (Free Software Foundation, Boston, 1996). ISBN: 978-1-882-11409-2
639. F. Haake, *Quantum Signatures of Chaos*. Springer Series in Synergetics (Springer, Berlin/Heidelberg, 2010). ISBN: 978-3-642-05428-0

640. E.M. Burke, B.M. Coyner, *Java Extreme Programming Cookbook*. Cookbook Series (O'Reilly, Sebastopol, 2003). ISBN: 978-0-596-00387-6
641. S. Aaronson. *Quantum Computing Since Democritus* (Cambridge University Press, Cambridge, 2013). ISBN: 978-0-521-19956-8
642. D.A. Spielman, S.-H. Teng, Smoothed analysis: an attempt to explain the behavior of algorithms in practice. Commun. ACM (2009). https://doi.org/10.1145/1562764.1562785
643. M. Bläser, B. Manthey, Smoothed complexity theory. Proceedings of the 37th International Conference on Mathematical Foundations of Computer Science (MFCS'12) (Springer-Verlag, Berlin/Heidelberg, 2012), pp. 198–209. https://doi.org/10.1007/978-3-642-32589-2_20
644. D. Spielman, S.-H. Teng, Smoothed analysis of algorithms: why the simplex algorithm usually takes polynomial time. Proceedings of the Thirty-Third Annual ACM Symposium on Theory of Computing (STOC '01) (Association for Computing Machinery, New York, 2001), pp. 296–305. https://doi.org/10.1145/380752.380813

Zeitfracht Medien GmbH
Ferdinand-Jühlke-Straße 7
99095 Erfurt, Deutschland
produktsicherheit@kolibri360.de